T0137796

NATO Science Series

A Series presenting the results of scientific meetings supported under the NATO Science Programme.

The Series is published by IOS Press, Amsterdam, and Kluwer Academic Publishers in conjunction with the NATO Scientific Affairs Division

Sub-Series

I. **Life and Behavioural Sciences**	IOS Press
II. **Mathematics, Physics and Chemistry**	Kluwer Academic Publishers
III. **Computer and Systems Science**	IOS Press
IV. **Earth and Environmental Sciences**	Kluwer Academic Publishers
V. **Science and Technology Policy**	IOS Press

The NATO Science Series continues the series of books published formerly as the NATO ASI Series.

The NATO Science Programme offers support for collaboration in civil science between scientists of countries of the Euro-Atlantic Partnership Council. The types of scientific meeting generally supported are "Advanced Study Institutes" and "Advanced Research Workshops", although other types of meeting are supported from time to time. The NATO Science Series collects together the results of these meetings. The meetings are co-organized bij scientists from NATO countries and scientists from NATO's Partner countries – countries of the CIS and Central and Eastern Europe.

Advanced Study Institutes are high-level tutorial courses offering in-depth study of latest advances in a field.
Advanced Research Workshops are expert meetings aimed at critical assessment of a field, and identification of directions for future action.

As a consequence of the restructuring of the NATO Science Programme in 1999, the NATO Science Series has been re-organised and there are currently Five Sub-series as noted above. Please consult the following web sites for information on previous volumes published in the Series, as well as details of earlier Sub-series.

http://www.nato.int/science
http://www.wkap.nl
http://www.iospress.nl
http://www.wtv-books.de/nato-pco.htm

Series II: Mathematics, Physics and Chemistry – Vol. 123

Techniques and Concepts of High-Energy Physics XII

edited by

Harrison B. Prosper

Florida State University,
Tallahassee, Florida

and

Michael Danilov

ITEP,
Moscow, Russia

Kluwer Academic Publishers

Dordrecht / Boston / London

Published in cooperation with NATO Scientific Affairs Division

Proceedings of the NATO Advanced Study Institute on
Techniques and Concepts of High-Energy Physics XII
St. Croix, Virgin Islands, U.S.A.
13–24 June 2002

A C.I.P. Catalogue record for this book is available from the Library of Congress.

ISBN 1-4020-1590-9 (HB)
ISBN 1-4020-1591-7 (PB)

Published by Kluwer Academic Publishers,
P.O. Box 17, 3300 AA Dordrecht, The Netherlands.

Sold and distributed in North, Central and South America
by Kluwer Academic Publishers,
101 Philip Drive, Norwell, MA 02061, U.S.A.

In all other countries, sold and distributed
by Kluwer Academic Publishers,
P.O. Box 322, 3300 AH Dordrecht, The Netherlands.

Printed on acid-free paper

Printed in the Netherlands.

Contents

Preface

The twelfth Advanced Study Institute (ASI) on Techniques and Concepts of High Energy Physics was held at the Hotel on the Cay in St. Croix, U.S. Virgin Islands in June 2002. The Institute attracted 11 lecturers and 42 advanced PhD students and recent PhD recipients in experimental particle physics from 14 different countries. The scientific program covered a broad sweep of topics that are expected to remain of interest for many years to come. The topics in this volume complement those in earlier volumes (published by Kluwer) and should be of interest to many physicists.

The main financial support for the Institute was provided by the Scientific Affairs Division of the North Atlantic Treaty Organization (NATO). The Institute was co-sponsored by the U.S. Department of Energy (DOE), the Fermi National Accelerator Laboratory (Fermilab), the U.S. National Science Foundation (NSF), the Florida State University (FSU) — Offices of the Provost and the Dean of Arts and Sciences, the Department of Physics and the FSU High Energy Physics Group — and the Institute for Theoretical and Experimental Physics (ITEP, Moscow).

We thank the students for their part in making this ASI a success and the lecturers for their hard work in preparing lectures, enthusiastically delivered, and for effectively engaging the students in discussion. We thank those who contributed the splendid manuscripts. We acknowledge, and are thankful for, the sound advice we received from the members of our Advisory Committee, whose names are listed in the back of the volume. We thank Gavin Nesom and Diane Weikal for organizing the pleasant, and civilized, wine tasting evening, Yildirim Mutaf for organizing the excellent student presentations and Zandy-Marie Hillis of the United States National Park Service for her fascinating description of the marine life and geology of St. Croix.

We thank Pat Rapp for support from the United States Department of Energy. We thank Kathy Mork, the Institute Secretary, for her tireless effort in keeping track of the numerous details we forgot and Marion

Hazlewood for her hospitality, and that of her staff, at the Hotel on the Cay, as well as Earl Powell at the King Christian Hotel. Hurchell Greenaway and his staff worked hard to keep us fed and entertained. We thank him heartily.

Finally, we thank Dr. Fausto Pedrazzini for his patience and strong support and the NATO Division of Scientific Affairs for their cooperation and support.

HARRISON B. PROSPER

MICHAEL DANILOV

THE STANDARD MODEL AND THE TOP QUARK

Scott Willenbrock

Physics Department
University of Illinois at Urbana-Champaign
1110 W. Green St., Urbana, IL 61801
willen@uiuc.edu

Abstract

The top quark is one of the least well-studied components of the standard model. In these lectures I discuss the expected properties of the top quark, which will be tested at the Fermilab Tevatron and the CERN Large Hadron Collider. I begin with a modern review of the standard model, emphasizing the underlying concepts. I then discuss the role the top quark plays in precision electroweak analyses. The last two lectures address the strong and weak interactions of the top quark, with an emphasis on the top-quark spin.

1.   Introduction

The top quark is the least well-studied of the quarks. Why is the top quark an interesting and worthwhile object to study? Here are four of the most compelling reasons:

- A more accurate measurement of the top-quark mass is valuable as an input to precision electroweak analyses.

- We would like to know if the top quark is just an ordinary quark, or if it is exotic in some way.

- The top quark may be useful to discover new particles. For example, of all the fermions, the Higgs boson couples most strongly to the top quark. It might be possible to observe the Higgs boson produced in association with a $t\bar{t}$ pair.

- Events containing top quarks are backgrounds to new physics that we hope to discover. This may sound mundane, but it is extremely

H.B. Prosper and M. Danilov (eds.), Techniques and Concepts of High-Energy Physics XII, 1–41.

important. For example, the discovery of the top quark itself was only possible once we understood the background from W+jets.

Although these lectures are principally about the top quark, I have chosen to broaden them to include a review of the standard model. The top quark is very much a part of the standard model, and it is useful to discuss the physics of the top quark from that perspective. The physics of the top quark is a vast subject, and cannot be covered in a few lectures. Instead, I have chosen several subjects of broad interest related to the top quark, and discuss them in some depth, with an emphasis on the underlying concepts. I have also included several exercises with each lecture, which I strongly urge you to perform. They will engage you with the material in a way that will help solidify your understanding. The exercises can be performed using only material contained in these lectures. The exercises are of various levels of difficulty, indicated by $*$ (easy), $**$ (moderate), and $***$ (hard). Solutions are provided in an appendix.

The first lecture is a review of the standard model from a modern point of view. It assumes the reader already has some familiarity with the standard model, and concentrates on the concepts that underlie the theory. The second lecture discusses the role the top quark plays in precision electroweak analyses via one-loop processes. The third and fourth lectures discuss the strong and weak interactions of the top quark, respectively, with an emphasis on the top-quark spin.

2. The Standard Model

In Table 1 I list the fermion fields that make up the standard model, along with their $SU(3) \times SU(2) \times U(1)_Y$ quantum numbers. The index $i = 1, 2, 3$ on each field refers to the generation, and the subscript L, R refers to the chirality of the field ($\psi_{L,R} \equiv \frac{1}{2}(1 \mp \gamma_5)\psi$). The left-chiral and right-chiral fields corresponding to a given particle have different $SU(2) \times U(1)$ quantum numbers, which leads to parity violation in the weak interaction.

Let's break the Lagrangian of the standard model into pieces. First consider the pure gauge interactions, given by

$$\mathcal{L}_{Gauge} = \frac{1}{2g_S^2} \text{Tr}\, G^{\mu\nu}G_{\mu\nu} + \frac{1}{2g^2}\text{Tr}\, W^{\mu\nu}W_{\mu\nu} - \frac{1}{4}B^{\mu\nu}B_{\mu\nu}\,, \qquad (1)$$

where $G^{\mu\nu}$ is the field-strength tensor of the gluon field, $W^{\mu\nu}$ is that of the weak-boson field, and $B^{\mu\nu}$ is that of the hypercharge-boson field. These terms contain the kinetic energy of the gauge bosons and their self interactions. Next comes the gauge interactions of the fermion ("matter") fields,

Table 1. The fermion fields of the standard model and their gauge quantum numbers.

				$SU(3)$	$SU(2)$	$U(1)_Y$
$Q_L^i =$	$\begin{pmatrix} u_L \\ d_L \end{pmatrix}$	$\begin{pmatrix} c_L \\ s_L \end{pmatrix}$	$\begin{pmatrix} t_L \\ b_L \end{pmatrix}$	3	2	$\frac{1}{6}$
$u_R^i =$	u_R	c_R	t_R	3	1	$\frac{2}{3}$
$d_R^i =$	d_R	s_R	b_R	3	1	$-\frac{1}{3}$
$L_L^i =$	$\begin{pmatrix} \nu_{eL} \\ e_L \end{pmatrix}$	$\begin{pmatrix} \nu_{\mu L} \\ \mu_L \end{pmatrix}$	$\begin{pmatrix} \nu_{\tau L} \\ \tau_L \end{pmatrix}$	1	2	$-\frac{1}{2}$
$e_R^i =$	e_R	μ_R	τ_R	1	1	-1

$$\mathcal{L}_{Matter} = i\bar{Q}_L^i \not{D} Q_L^i + i\bar{u}_R^i \not{D} u_R^i + i\bar{d}_R^i \not{D} d_R^i + i\bar{L}_L^i \not{D} L_L^i + i\bar{e}_R^i \not{D} e_R^i \ , \quad (2)$$

These terms contain the kinetic energy and gauge interactions of the fermions, which depend on the fermion quantum numbers. For example,

$$\not{D} Q_L = \gamma^\mu (\partial_\mu + i g_S G_\mu + i g W_\mu + i\frac{1}{6} g' B_\mu) Q_L \quad (3)$$

since the field Q_L participates in all three gauge interactions. A sum on the index i, which represents the generation, is implied in the Lagrangian.

We have constructed the simplest and most general Lagrangian, given the fermion fields and gauge symmetries.[1] The gauge symmetries forbid masses for any of the particles. In the case of the fermions, masses are forbidden by the fact that the left-chiral and right-chiral components of a given fermion field have different $SU(2) \times U(1)_Y$ quantum numbers. For example, a mass term for the up quark,

$$\mathcal{L} = -m\bar{u}_L u_R + h.c. \ , \quad (4)$$

is forbidden by the fact that u_L is part of the $SU(2)$ doublet Q_L, so such a term violates the $SU(2)$ gauge symmetry (it also violates $U(1)_Y$).

---

[1] I will give a precise definition to "simplest" later in this lecture. For now, it means the minimum number of fields and derivatives are used in each term in the Lagrangian.

Although we only imposed the gauge symmetry on the Lagrangian, it turns out that it has a good deal of global symmetry as well, associated with the three generations. Because all fermions are massless thus far in our analysis, there is no difference between the three generations - they are physically indistinguishable. This manifests itself as a global flavor symmetry of the matter Lagrangian, Eq. (2), which is invariant under the transformations

$$Q_L^i \rightarrow U_{Q_L}^{ij} Q_L^j$$
$$u_R^i \rightarrow U_{u_R}^{ij} u_R^j$$
$$d_R^i \rightarrow U_{d_R}^{ij} d_R^j$$
$$L_L^i \rightarrow U_{L_L}^{ij} L_L^j$$
$$e_R^i \rightarrow U_{e_R}^{ij} e_R^j, \tag{5}$$

where each U is an arbitrary 3×3 unitary matrix.

Exercise 1.1 (*) Show this.

Since there are five independent $U(3)$ symmetries, the global flavor symmetry of the Lagrangian is $[U(3)]^5$.

The Lagrangian thus far contains only three parameters, the couplings of the three gauge interactions. Their approximate values (evaluated at M_Z) are

$$g_S \approx 1$$
$$g \approx \frac{2}{3}$$
$$g' \approx \frac{2}{3\sqrt{3}}.$$

These couplings are all of order unity.

Electroweak symmetry breaking – The theory thus far is very simple and elegant, but it is incomplete - all particles are massless. We now turn to electroweak symmetry breaking, which is responsible for generating the masses of the gauge bosons and fermions.

In the standard model, electroweak symmetry breaking is achieved by introducing another field into the model, the Higgs field ϕ, with the quantum numbers shown in Table 2. The simplest and most general Lagrangian for the Higgs field, consistent with the gauge symmetry, is

$$\mathcal{L}_{Higgs} = (D^\mu \phi)^\dagger D_\mu \phi + \mu^2 \phi^\dagger \phi - \lambda (\phi^\dagger \phi)^2. \tag{6}$$

The first term contains the Higgs-field kinetic energy and gauge interactions. The remaining terms are (the negative of) the Higgs potential,

Figure 1. The Higgs potential. The neutral component of the Higgs field acquires a vacuum-expectation value $\langle\phi^0\rangle = v/\sqrt{2}$ on the circle of minima in Higgs-field space.

shown in Fig. 1. The quadratic term in the potential has been chosen such that the minimum of the potential lies not at zero, but on a circle of minima

$$\langle\phi^0\rangle = \mu/\sqrt{2\lambda} \equiv \frac{v}{\sqrt{2}} \tag{7}$$

where ϕ^0 is the lower (neutral) component of the Higgs doublet field. This equation defines the parameter $v \approx 246$ GeV, the Higgs-field vacuum-expectation value. Making the substitution $\phi = (0, v/\sqrt{2})$ in the Higgs Lagrangian, Eq. (6), one finds that the W and Z bosons have acquired masses

$$M_W = \frac{1}{2}gv \qquad\qquad M_Z = \frac{1}{2}\sqrt{g^2 + g'^2}\,v \tag{8}$$

from the interaction of the gauge bosons with the Higgs field. Since we know g and g', these equations determine the numerical value of v.

The Higgs sector of the theory, Eq. (6), introduces just two new parameters, μ and λ. Rather than μ, we will use the parameter v intro-

Table 2. The Higgs field and its gauge quantum numbers.

	$SU(3)$	$SU(2)$	$U(1)_Y$
$\phi = \begin{pmatrix} \phi^+ \\ \phi^0 \end{pmatrix}$	1	2	$\frac{1}{2}$

duced in Eq. (7). The parameter λ is the Higgs-field self interaction, and will not figure into our discussion.

Fermion masses and mixing – In quantum field theory, anything that is not forbidden is mandatory. With that in mind, there is one more set of interactions, involving the Higgs field and the fermions. The simplest and most general Lagrangian, consistent with the gauge symmetry, is

$$\mathcal{L}_{Yukawa} = -\Gamma_u^{ij} \bar{Q}_L^i \epsilon \phi^* u_R^j - \Gamma_d^{ij} \bar{Q}_L^i \phi d_R^j - \Gamma_e^{ij} \bar{L}_L^i \phi e_R^j + h.c. \qquad (9)$$

where $\Gamma_u, \Gamma_d, \Gamma_e$ are 3×3 complex matrices in generation space.[2] We have therefore apparently introduced $3 \times 3 \times 3 \times 2 = 54$ new parameters into the theory, but as we shall see, only a subset of these parameters are physically relevant. These so-called Yukawa interactions of the Higgs field with fermions violate almost all of the $[U(3)]^5$ global symmetry of the fermion gauge interactions, Eq. (2). The only remaining global symmetries are the subset corresponding to baryon number

$$Q_L^i \to e^{i\theta/3} Q_L^i$$
$$u_R^i \to e^{i\theta/3} u_R^i$$
$$d_R^i \to e^{i\theta/3} d_R^i \qquad (10)$$

and lepton number

$$L_L^i \to e^{i\phi} L_L^i$$
$$e_R^i \to e^{i\phi} e_R^i . \qquad (11)$$

Exercise 1.2 (*) Show this.

The conservation of baryon number and lepton number follow from these symmetries. These symmetries are accidental; they are not put in by hand, but rather follow automatically from the field content and gauge symmetries of the theory. Thus we can say that we understand why baryon number and lepton number are conserved in the standard model.

Replacing the Higgs field with its vacuum-expectation value, $\phi = (0, v/\sqrt{2})$, in Eq. (9) yields

$$\mathcal{L}_M = -M_u^{ij} \bar{u}_L^i u_R^j - M_d^{ij} \bar{d}_L^i d_R^j - M_e^{ij} \bar{e}_L^i e_R^j + h.c. , \qquad (12)$$

where

$$M^{ij} = \Gamma^{ij} \frac{v}{\sqrt{2}} \qquad (13)$$

---

[2]The matrix $\epsilon = \begin{pmatrix} 0 & 1 \\ -1 & 0 \end{pmatrix}$ in $SU(2)$ space is needed in order for the first term in Eq. (9) to respect $SU(2)$ gauge invariance.

are fermion mass matrices. The Yukawa interactions are therefore responsible for providing the charged fermions with mass; the neutrinos, however, remain massless (we will discuss neutrino masses shortly).

The complete Lagrangian of the standard model is the sum of the gauge, matter, Higgs, and Yukawa interactions,

$$\mathcal{L}_{SM} = \mathcal{L}_{Gauge} + \mathcal{L}_{Matter} + \mathcal{L}_{Higgs} + \mathcal{L}_{Yukawa} . \qquad (14)$$

This is the simplest and most general Lagrangian, given the field content and gauge symmetries of the standard model.

Given this Lagrangian, one can proceed to calculate any physical process of interest. However, it is convenient to first perform field redefinitions to make the physical content of the theory manifest. These field redefinitions do not change the predictions of the theory; they are analogous to a change of variables when performing an integration. To make the masses of the fermions manifest, we perform unitary field redefinitions on the fields in order to diagonalize the mass matrices in Eq. (12):

$$u_L^i = A_{u_L}^{ij} u_L'^j \qquad u_R^i = A_{u_R}^{ij} u_R'^j$$
$$d_L^i = A_{d_L}^{ij} d_L'^j \qquad d_R^i = A_{d_R}^{ij} d_R'^j$$
$$e_L^i = A_{e_L}^{ij} e_L'^j \qquad e_R^i = A_{e_R}^{ij} e_R'^j$$
$$\nu_L^i = A_{\nu_L}^{ij} \nu_L'^j \qquad (15)$$

Exercise 1.3 (∗) Show that each matrix A must be unitary in order to preserve the form of the kinetic-energy terms in the matter Lagrangian, Eq. (2), e.g.

$$\mathcal{L}_{KE} = i\bar{u}_L^i \, \slashed{\partial} u_L^i . \qquad (16)$$

Once the mass matrices are diagonalized, the masses of the fermions are manifest. These transformations also diagonalize the Yukawa matrices Γ, since they are proportional to the mass matrices [see Eq. (13)]. However, we must consider what impact these field redefinitions have on the rest of the Lagrangian. They have no effect on the pure gauge or Higgs parts of the Lagrangian, Eqs. (1) and (6), which are independent of the fermion fields. They do impact the matter part of the Lagrangian, Eq. (2). However, a subset of these field redefinitions is the global $[U(3)]^5$ symmetry of the matter Lagrangian; this subset therefore has no impact.

One can count how many physically-relevant parameters remain after the field redefinitions are performed [1]. Let's concentrate on the quark sector. The number of parameters contained in the complex matrices Γ_u, Γ_d is $2 \times 3 \times 3 \times 2 = 36$. The unitary symmetries $U_{Q_L}, U_{u_R}, U_{d_R}$ are a subset of the quark field redefinitions; this subset will not affect the matter part of the Lagrangian. There are $3 \times 3 \times 3$ degrees of freedom in these symmetries (a unitary $N \times N$ matrix has N^2 free parameters), so the total number of parameters that remain in the full Lagrangian after field redefinitions is

$$2 \times 3 \times 3 \times 2 - (3 \times 3 \times 3 - 1) = 10 \qquad (17)$$

where I have subtracted baryon number from the subset of field redefinitions that are symmetries of the matter Lagrangian. Baryon number is a symmetry of the Yukawa Lagrangian, Eq. (9), and hence cannot be used to diagonalize the mass matrices.

Exercise 1.4 (*) Show that the quark field redefinitions are the symmetries $U_{Q_L}, U_{u_R}, U_{d_R}$ if $A_{u_L} = A_{d_L}$.

The ten remaining parameters correspond to the six quark masses and the four parameters of the Cabibbo-Kobayashi-Maskawa (CKM) matrix (three mixing angles and one CP-violating phase). The CKM matrix is $V \equiv A_{d_L}^\dagger A_{u_L}$; we see that this matrix is unity if $A_{u_L} = A_{d_L}$, as expected from Exercise 1.4.

Exercise 1.5 (*) Show that V is unitary.

The mass matrices are related to the Yukawa matrices by Eq. (13). If we make the natural assumption that the Yukawa matrices contain elements of order unity (like the gauge couplings), we expect the fermion masses to be of $\mathcal{O}(v)$, just like M_W and M_Z [see Eq. (8)]. This is not the case; only the top quark has such a large mass. We see that, from the point of view of the standard model, the question is not why the top quark is so heavy, but rather why the other fermions are so light.

Similarly, for a generic Yukawa matrix, one expects the field redefinitions that diagonalize the mass matrices to yield a CKM matrix with large mixing angles. Again, this is not the case; the measured angles are [2]

$$\theta_{12} \approx 13°$$
$$\theta_{23} \approx 2.3°$$
$$\theta_{13} \approx 0.23°$$
$$\delta \approx 60°$$

which, with the exception of the CP-violating phase δ, are small.[3] The question is not why these angles are nonzero, but rather why they are so small.

---

[3]The phase δ is the same as the angle γ of the so-called unitarity triangle.

The fermion masses and mixing angles strongly suggest that there is a deeper structure underlying the Yukawa sector of the standard model. Surely there is some explanation of the peculiar pattern of fermion masses and mixing angles. Since the standard model can accommodate any masses and mixing angles, we must seek an explanation from physics beyond the standard model.

Beyond the Standard Model – Let us back up and ask: why did we stick to the simplest terms in the Lagrangian? The obsolete answer is that these are the renormalizable terms. Renormalizability is a stronger constraint than is really necessary. The modern answer, which is much simpler, is dimensional analysis [3].

We'll work with units such that $\hbar = c = 1$.

Exercise 1.6 (∗) - Show that length has units of mass^{-1}, and hence $\partial_\mu = \partial/\partial x^\mu$ has units of mass.

Since the action has units of $\hbar = 1$, the Lagrangian must have units of mass^4, since

$$S = \int d^4x\, \mathcal{L}\,. \tag{18}$$

From the kinetic energy terms in the Lagrangian for a generic scalar (ϕ), fermion (ψ), and gauge boson (A^μ),

$$\mathcal{L}_{KE} = \partial^\mu \phi^* \partial_\mu \phi + i\bar{\psi}\,\slashed{\partial}\psi - \frac{1}{2}(\partial^\mu A^\nu \partial_\mu A_\nu + \partial^\mu A^\nu \partial_\nu A_\mu) \tag{19}$$

we can deduce the dimensionality of the various fields:

$$\dim \phi = \text{mass}$$
$$\dim \psi = \text{mass}^{3/2}$$
$$\dim A^\mu = \text{mass}\,.$$

All operators (products of fields) in the Lagrangian of the Standard Model are of dimension four, except the operator $\phi^\dagger \phi$ in the Higgs potential, which is of dimension two. The coefficient of this term, μ^2, is the only dimensionful parameter in the standard model; it (or, equivalently, $v \equiv \mu/\sqrt{\lambda}$) sets the scale of all particle masses.

Imagine that the Lagrangian at the weak scale is an expansion in some large mass scale M,

$$\mathcal{L} = \mathcal{L}_{SM} + \frac{1}{M}\dim 5 + \frac{1}{M^2}\dim 6 + \cdots, \tag{20}$$

where dim n represents all operators of dimension n. By dimensional analysis, the coefficient of an operator of dimension n has dimension mass^{4-n}, since the Lagrangian has dimension mass^4. At energies much less than M, the dominant terms in this Lagrangian will be those of $\mathcal{L}_{SM}$; the other terms are suppressed by an inverse power of M. This is the modern reason why we believe the "simplest" terms in the Lagrangian are the dominant ones.

The least suppressed terms in the Lagrangian beyond the standard model are of dimension five. We should therefore expect our first observation of physics beyond the standard model to come from these terms. Given the field content and gauge symmetries of the standard model, there is only one such term:

$$\mathcal{L}_5 = \frac{c^{ij}}{M} L_L^{iT} \epsilon \phi C \phi^T \epsilon L_L^j + h.c. \,, \tag{21}$$

where c^{ij} is a dimensionless matrix in generation space.[4]

Exercise 1.7 (**) - Show that a similar term, with L_L replaced by Q_L, is forbidden by $SU(3) \times U(1)_Y$ gauge symmetry.

This dimension-five operator contains the Higgs-doublet field twice and the lepton-doublet field twice.

Exercise 1.8 (*) - Show that $\mathcal{L}_5$ violates lepton number.

Replacing the Higgs-doublet field with its vacuum-expectation value, $\phi = (0, v/\sqrt{2})$, yields

$$\mathcal{L}_5 = -\frac{c^{ij}}{2} \frac{v^2}{M} \nu_L^{iT} C \nu_L^j + h.c. \,. \tag{22}$$

This is a Majorana mass term for the neutrinos. The recent observation of neutrino oscillations, which requires nonzero neutrino mass, is indeed our first observation of physics beyond the standard model.

Exercise 1.9 (***) Show that the Maki-Nakagawa-Sakata (MNS) matrix (the analogue of the CKM matrix in the lepton sector) has 6 physically-relevant parameters. (Note: c^{ij} is a complex, symmetric matrix.)

The moral is that when we are searching for deviations from the standard model, what we are really doing is looking for the effects of higher-dimension operators. Although there is only one operator of dimension five, there are dozens of operators of dimension six, some of which are listed below [4]:

$$\bar{L}^i \gamma^\mu L^j \bar{L}^k \gamma_\mu L^m$$

---

[4]The 2×2 matrix ϵ in $SU(2)$ space was introduced in an earlier footnote. The 4×4 matrix $C = \begin{pmatrix} -\epsilon & 0 \\ 0 & \epsilon \end{pmatrix}$ in Dirac space is needed for Lorentz invariance.

$$\bar{L}^i \gamma^\mu L^j \bar{Q}^k \gamma_\mu Q^m$$
$$i\bar{Q}^i \gamma_\mu D_\nu G^{\mu\nu} Q^j$$
$$\text{Tr } G^{\mu\nu} G_{\nu\rho} G^\rho_\mu$$
$$\phi^\dagger \phi \text{Tr } W^{\mu\nu} W_{\mu\nu}$$
$$\phi^\dagger D_\mu \phi \bar{e}^i_R \gamma^\mu e^j_R .$$

Thus far, none of the effects of any of these operators have been observed. The best we can do is set lower bounds on M (assuming some dimensionless coefficient). These lower bounds range from 1 TeV to 10^{16} GeV, depending on the operator. As we explore nature at higher energy and with higher accuracy, we hope to begin to see the effects of some of these dimension-six operators.

The mass scale M corresponds to the mass of a particle that is too heavy to observe directly. At energies greater than M, the expansion of Eq. (20) is no longer useful, as each successive term is larger than the previous. Instead, one must explicitly add the new field of mass M to the model. For example, if nature is supersymmetric at the weak scale, one must add the superpartners of the standard-model fields to the theory and include their interactions in the Lagrangian. If we raise the mass scale of the superpartners to be much greater than the weak scale, then we can no longer directly observe the superpartners, and we return to a description in terms of standard-model fields, with an expansion of the Lagrangian in inverse powers of the mass scale of the superpartners, M.

3.   Virtual Top Quark

The top quark plays an important role in precision electroweak analyses. In this lecture I hope to clarify this sometimes confusing subject.

Recall from the previous lecture that the gauge, matter, and Higgs sectors of the standard model depend on only five parameters: the three gauge couplings, g_S, g, g', and the Higgs-field vacuum-expectation value and self interaction, v and λ. At tree level, all electroweak quantities depend on just three of these parameters, g, g', and v. We use the three best-measured electroweak quantities to determine these three parameters at tree level:

$$\alpha = \frac{1}{4\pi} \frac{g^2 g'^2}{g^2 + g'^2} = \frac{1}{137.03599976(50)}$$
$$G_F = \frac{1}{\sqrt{2} v^2} = 1.16637(1) \times 10^{-5} \text{ GeV}^{-2}$$

$$M_Z = \frac{1}{2}\sqrt{g^2 + g'^2}\,v = 91.1876(21)\ \text{GeV} ,$$

where the uncertainty is given in parentheses. The value of α is extracted from low-energy experiments, G_F is extracted from the muon lifetime, and M_Z is measured from e^+e^- annihilation near the Z mass. From these three quantities, we can predict all other electroweak quantities at tree level. For example, the W mass is

$$M_W^2 = \frac{1}{4}g^2 v^2 = \frac{1}{2}M_Z^2 \left(1 + \sqrt{1 - \frac{4\pi\alpha}{\sqrt{2}G_F M_Z^2}}\right) . \tag{23}$$

Exercise 2.1 (*) Verify the expression for M_W in terms of α, G_F, and M_Z.

A more civilized expression for M_W is obtained by *defining*

$$s_W^2 \equiv 1 - \frac{M_W^2}{M_Z^2} . \tag{24}$$

This is the so-called "on-shell" definition[5] of $\sin^2\theta_W$; it has a numerical value of $s_W^2 = 0.2228(4)$. Using this parameter, we can write a simpler expression than Eq. (23) for M_W at tree level:

$$M_W^2 = \frac{\frac{\pi\alpha}{\sqrt{2}G_F}}{s_W^2} . \tag{25}$$

Exercise 2.2 (*) - Verify this equation.

At one loop this expression is modified:

$$M_W^2 = \frac{\frac{\pi\alpha}{\sqrt{2}G_F}}{s_W^2(1 - \Delta r)} , \tag{26}$$

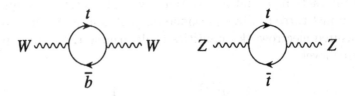

Figure 2.   Virtual top-quark loops contribute to the W and Z masses.

---

[5] So called because it is defined in terms of physical, or "on shell," quantities.

Figure 3.   Virtual Higgs-boson loops contribute to the W and Z masses.

where Δr contains the one-loop corrections. The top quark makes a contribution to Δr via the one-loop diagrams shown in Fig. 2, which contribute to the W and Z masses:

$$(\Delta r)_{\text{top}} \approx -\frac{3G_F m_t^2}{8\sqrt{2}\pi^2}\frac{1}{t_W^2} , \qquad (27)$$

where $t_W^2 \equiv \tan^2\theta_W$. This one-loop correction depends quadratically on the top-quark mass.

The Higgs boson also contributes to Δr via the one-loop diagrams in Fig. 3:

$$(\Delta r)_{\text{Higgs}} \approx \frac{11G_F M_Z^2 c_W^2}{24\sqrt{2}\pi^2}\ln\frac{m_h^2}{M_Z^2} , \qquad (28)$$

where $c_W^2 \equiv \cos^2\theta_W$. This one-loop correction depends only logarithmically on the Higgs-boson mass, so Δr is not nearly as sensitive to m_h as it is to m_t.

Due to the contributions of the top quark and the Higgs boson to Δr, in order to predict M_W at one loop via Eq. (26) we need not just α, G_F, M_Z, but also m_t and m_h. Turning this around, in order to predict m_h, we need α, G_F, M_Z, and m_t, M_W. Thus a precision measurement of m_t and M_W can be used to predict the Higgs mass.

I show in Fig. 4 a plot of M_W vs. m_t, indicating lines of constant Higgs mass.[6] The dashed ellipse indicates the 68% CL measurements of M_W and m_t,

$$M_W = 80.451(33) \text{ GeV}$$
$$m_t = 174.3(5.1) \text{ GeV}$$

(I will return to the solid ellipse momentarily). As you can see, the direct measurements of M_W and m_t favor a light Higgs boson.

Exercise 2.3 (* *) - Derive the slope of the lines of constant Higgs*

---

[6]The small arrow labeled $\Delta\alpha$ in that plot indicates the uncertainty in the lines of constant Higgs mass due to the uncertainty in $\alpha(M_Z)$.

14

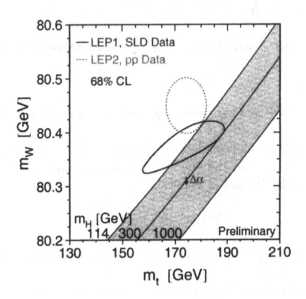

Figure 4. Lines of constant Higgs mass on a plot of M_W vs. m_t. The dashed ellipse is the 68% CL direct measurement of M_W and m_t. The solid ellipse is the 68% CL indirect measurement from precision electroweak data. From http://lepewwg.web.cern.ch/LEPEWWG/.

mass on the plot of M_W vs. m_t. Evaluate it numerically and compare with the plot. [Hint: Derive dM_W^2/dm_t^2 and then evaluate $dM_W/dm_t = (m_t/M_W)dM_W^2/dm_t^2$ numerically. Be careful to use Eq. (24) for s_W^2 in Eq. (26) (you can neglect the dependence of t_W^2 on M_W in Eq. (27) since $(\Delta r)_{top}$ is a small correction).]

Neutral current – Rather than using the direct measurements of M_W and m_t to infer the Higgs-boson mass, one can use other electroweak quantities. The Fermi constant, G_F, is extracted from muon decay, which is a charged-current weak interaction. That leaves the neutral-current weak interaction as another quantity of interest. There is an enormous wealth of data on neutral-current weak interactions, such as e^+e^- annihilation near the Z mass, νN and eN deep-inelastic scattering, νe elastic scattering, atomic parity violation, and so on [2].

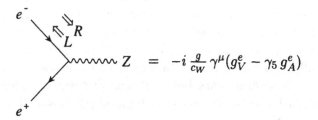

Figure 5. Neutral-current coupling of an electron to the Z boson. A left-handed electron has negative helicity, a right-handed electron has positive helicity.

Let's consider a simple and very relevant example, the left-right asymmetry in e^+e^- annihilation near the Z mass, shown in Fig. 5. Left and right refer to the helicity of the incident electron, either negative (left) or positive (right). The asymmetry is defined in terms of the total cross section for a negative-helicity or positive-helicity electron to annihilate with an unpolarized positron and produce a Z boson,

$$A_{LR} \equiv \frac{\sigma_L - \sigma_R}{\sigma_L + \sigma_R}$$
$$= \frac{2g_V^e g_A^e}{g_V^{e2} + g_A^{e2}} , \tag{29}$$

where

$$g_V^e = \sqrt{\rho_e}\left(-\frac{1}{2} + 2\kappa_e s_W^2\right)$$
$$g_A^e = \sqrt{\rho_e}\left(-\frac{1}{2}\right) \tag{30}$$

are the vector and axial-vector couplings of the electron to the Z boson. At tree level, $\rho_e = \kappa_e = 1$, but there are one-loop corrections. The correction quadratic in the top-quark mass is

$$(\rho_e)_{\text{top}} \approx 1 + \frac{3G_F m_t^2}{8\sqrt{2}\pi^2}$$
$$(\kappa_e)_{\text{top}} \approx 1 + \frac{3G_F m_t^2}{8\sqrt{2}\pi^2}\frac{1}{t_W^2} . \tag{31}$$

Different neutral-current measurements have different dependencies on m_t and m_h, so by combining two or more measurements one can extract both m_t and m_h. The solid ellipse in Fig. 4 represents the 68%

CL constraint from all neutral-current measurements combined. It is in good agreement with the direct measurements of M_W and m_t, and strengthens the case for a light Higgs boson. Combining all precision electroweak data, one finds 45 GeV $\leq m_h \leq$ 191 GeV [2].

Historically, neutral-current data were used to successfully predict the top-quark mass several years before it was discovered. This is a good reason to trust the prediction of a light Higgs boson from precision electroweak analyses.

It is also significant that the two ellipses in Fig. 4 lie on or near the lines of constant Higgs mass (within the allowed range of the Higgs mass). These measurements could have ended up far from those lines, thereby disproving the existence of the hypothetical Higgs boson. Instead, these measurements bolster our belief in the standard model in general, and in the Higgs boson in particular.

$\overline{\text{MS}}$ *scheme* – Before we leave this topic, let's discuss the other most often-used definition of $\sin^2 \theta_W$. This is the minimal-subtraction-bar ($\overline{\text{MS}}$) scheme, so-called due to the simple way in which ultraviolet divergences in loop diagrams are subtracted.

Exercise 2.4 (*) - Show that

$$\sin^2 \theta_W = \frac{g'^2}{g^2 + g'^2} \tag{32}$$

at tree level.

The $\overline{\text{MS}}$ scheme promotes this to the definition of $\sin^2 \theta_W$:

$$\hat{s}_Z^2 \equiv \frac{g'^2(M_Z)}{g^2(M_Z) + g'^2(M_Z)} \tag{33}$$

where the gauge couplings are evaluated at the Z mass. Its numerical value is $\hat{s}_Z^2 = 0.23113(15)$.

The analogues of Eqs. (26) and (24) in the $\overline{\text{MS}}$ scheme are

$$M_W^2 = \frac{\frac{\pi \alpha}{\sqrt{2} G_F}}{\hat{s}_Z^2 (1 - \Delta \hat{r}_W)} \tag{34}$$

$$M_Z^2 = \frac{M_W^2}{\hat{c}_Z^2 \hat{\rho}} . \tag{35}$$

Unlike its on-shell analogue Δr, the one-loop quantity $\Delta \hat{r}_W$ has no quadratic dependence on the top-quark mass. This appears instead in the quantity $\hat{\rho}$ (which is unity in the on-shell scheme):

$$\hat{\rho} \approx 1 + \frac{3 G_F m_t^2}{8 \sqrt{2} \pi^2} . \tag{36}$$

Although the quadratic dependence on the top-quark mass has been shifted from one relation to another, the physical predictions, such as the constraint on the Higgs mass, remain unchanged.

Exercise 2.5 ($* * *$) - Repeat Exercise 2.3 in the $\overline{MS}$ scheme. [Note that $\hat{s}_Z^2$ depends on M_W via Eq. (35).]

4. Top Strong Interactions

We now begin to discuss the study of the top quark itself. In the introduction we listed several reasons why the top quark is an interesting object to study. The strategy that follows from these motivations is to get to know the top quark by measuring everything we can about it, and comparing with the predictions of the standard model. This program will occupy a large portion of our efforts at the Fermilab Tevatron and the CERN Large Hadron Collider (LHC). In this section I discuss some of the measurements that can be made at these machines related to the strong interactions of the top quark, and in the next section I turn to its weak interactions.

The top quark is produced at hadron colliders primarily via the strong interaction. The Feynman diagrams for the two contributing subprocesses, quark-antiquark annihilation and gluon fusion, are shown in Fig. 6. In Table 3 I give the predicted cross sections, at next-to-leading-order (NLO) in QCD, for $m_t = 175$ GeV. I also show the percentage

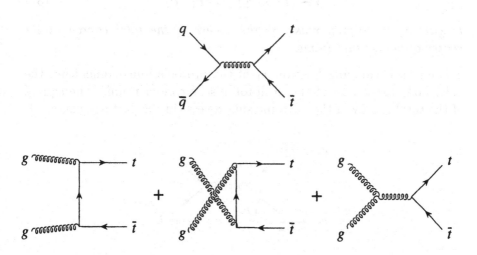

Figure 6. Top-quark production via the strong interaction at hadron colliders proceeds through quark-antiquark annihilation (upper diagram) and gluon fusion (lower diagrams).

Table 3. Cross sections, at next-to-leading-order in QCD, for top-quark production via the strong interaction at the Tevatron and the LHC [5]. Also shown is the percentage of the total cross section from the quark-antiquark-annihilation and gluon-fusion subprocesses.

	σ_{NLO} (pb)	$q\bar{q} \to t\bar{t}$	$gg \to t\bar{t}$
Tevatron ($\sqrt{s} = 1.8$ TeV $p\bar{p}$)	$4.87 \pm 10\%$	90%	10%
Tevatron ($\sqrt{s} = 2.0$ TeV $p\bar{p}$)	$6.70 \pm 10\%$	85%	15%
LHC ($\sqrt{s} = 14$ TeV pp)	$803 \pm 15\%$	10%	90%

of the cross section that results from each of the two subprocesses. At the Tevatron, the quark-antiquark-annihilation subprocess dominates; at the LHC, gluon fusion reigns. To understand why this is, we need to discuss the parton model of the proton.

The parton model is shown schematically in Fig. 7, where I illustrate how a proton-antiproton collision results in a $t\bar{t}$ pair produced via the quark-antiquark-annihilation subprocess. The proton is regarded as a collection of quarks, antiquarks, and gluons (collectively called partons), each carrying some fraction x of the proton's four-momentum. Figure 7 shows a proton of four-momentum P_1 colliding with an antiproton of four-momentum P_2.

Exercise 3.1 (*) - Show that

$$S \equiv (P_1 + P_2)^2 \approx 2P_1 \cdot P_2 \tag{37}$$

(neglecting the proton mass) is the square of the total energy in the center-of-momentum frame.

The quark is carrying fraction x_1 of the proton's four-momentum, the antiquark fraction x_2 of the antiproton's four-momentum. The square of the total energy of the partonic subprocess (in the partonic center-of-

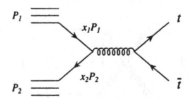

Figure 7. The parton-model description of top-quark pair production. A quark carrying fraction x_1 of the proton's momentum P_1 annihilates with an antiquark carrying fraction x_2 of the antiproton's momentum P_2.

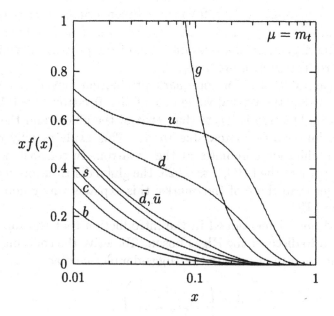

Figure 8. Parton distribution functions at the scale $\mu = m_t$, relevant for top-quark production.

momentum frame) is similarly

$$\hat{s} \equiv (x_1 P_1 + x_2 P_2)^2 \approx 2 x_1 x_2 P_1 \cdot P_2 = x_1 x_2 S . \tag{38}$$

Since there has to be at least enough energy to produce a $t\bar{t}$ pair at rest, $\hat{s} \geq 4 m_t^2$. It follows from Eq. (38) that

$$x_1 x_2 = \frac{\hat{s}}{S} \geq \frac{4 m_t^2}{S} . \tag{39}$$

Since the probability of finding a quark of momentum-fraction x in the proton falls off with increasing x, the typical value of $x_1 x_2$ is near the threshold for $t\bar{t}$ production. Setting $x_1 \approx x_2 = x$ in Eq. (39) gives

$$x \approx \frac{2 m_t}{\sqrt{S}} \tag{40}$$

as the typical value of x for $t\bar{t}$ production.

Figure 8 shows the parton distribution functions in the proton for all the different species of partons.[7] The probability of finding a given

---

[7] I will explain in Section 5 the presence of antiquarks in the proton, as well as strange, charm, and bottom quarks.

parton species with momentum fraction between x and $x+dx$ is $f(x)dx$. [What is plotted in Fig. 8 is actually $xf(x)$]. The parton distribution functions also depend on the relevant scale of the process, μ, which for top-quark production is of order m_t.

The typical value of x for top-quark production may be computed from Eq. (40). For the typical value of x at the Tevatron, $x \approx 0.18$, the up distribution function is larger than that of the gluon, and the down distribution function is comparable to it. This explains why quark-antiquark annihilation dominates at the Tevatron. In contrast, for the typical value of x at the LHC, $x \approx 0.025$, the gluon distribution function is much larger than those of the quarks; this explains why gluon fusion reigns at the LHC.

Higgs and top – I mentioned in the introduction that the top quark could be used to discover the Higgs boson. To derive the coupling of the Higgs boson to fermions, write the Higgs-doublet field as

$$\phi = \begin{pmatrix} 0 \\ \frac{1}{\sqrt{2}}(v+h) \end{pmatrix} \tag{41}$$

where h is the Higgs boson, which corresponds to oscillations about the vacuum-expectation value of the field, Eq. (7). Inserting this expression for ϕ into the Yukawa Lagrangian, Eq. (9), yields the desired coupling, shown in Fig. 9.

Exercise 3.2 (* * *) - Show that the coupling of the Higgs boson to fermions is as given in Fig. 9. [Hint: Recall $M^{ij} = \Gamma^{ij}v/\sqrt{2}$, Eq. (13).]

The Feynman diagrams for Higgs-boson production in association with a $t\bar{t}$ pair are the same as those of Fig. 6, but with a Higgs boson attached to the top quark or antiquark. The Higgs boson can also be produced by itself via its coupling to a virtual top-quark loop, as shown

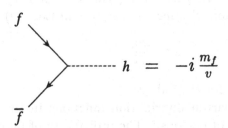

Figure 9. The coupling of the Higgs boson to fermions.

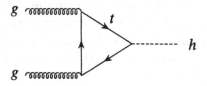

Figure 10.   Higgs-boson production via gluon fusion through a top-quark loop.

in Fig. 10. Remarkably, this is the largest source of Higgs bosons at the Tevatron or the LHC. It is amusing that the virtual top quark points to the existence of a light Higgs boson, as discussed in the previous section, and may also help us discover the Higgs boson.

Top-quark spin – One of the remarkable features of the top quark is that it is the only quark whose spin is directly observable. This is a consequence of its very short lifetime, $\Gamma_t^{-1} \approx (1.5 \text{ GeV})^{-1}$. Figure 11 shows an example of the evolution of a heavy quark of a definite spin after it is produced in a hard-scattering collision. On a time scale of order $\Lambda_{QCD}^{-1} \approx (200 \text{ MeV})^{-1}$, the heavy quark picks up a light antiquark of the opposite spin from the vacuum and hadronizes into a meson. Some time later, on the order of $(\Lambda_{QCD}^2/m_Q)^{-1} \approx (1 \text{ MeV})^{-1}$ (for $m_Q = m_t$), the spin-spin interaction between the heavy quark and the light quark[8] cause the meson to evolve into a spin-zero state, $(|\uparrow\downarrow\rangle - |\downarrow\uparrow\rangle)/\sqrt{2}$, thereby depolarizing the heavy quark [6]. The top quark is the only quark that decays before it has a chance to depolarize (or even hadronize), so its spin is observable in the angular distribution of its decay products.[9]

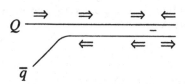

Figure 11.   A heavy quark hadronizes with a light quark of the opposite spin, then evolves into a spin-zero meson.

---

[8]This is the QCD analogue of the spin-spin interaction that produces the hyperfine splitting in atomic physics.
[9]Actually, the spin of a long-lived heavy quark is observable if it hadronizes into a baryon, such as a Λ_b.

Let's discuss the spin of a fermion in some detail. For a moving fermion, it is conventional to use the helicity basis, in which the spin quantization axis is the direction of motion of the fermion. The free fermion field may be decomposed into states of definite four-momentum,

$$\psi(x) = \int \frac{d^3p}{(2\pi)^3\sqrt{2E}} \sum_{\lambda=\pm} (a_p^\lambda u_\lambda(p)e^{-ip\cdot x} + b_p^{\lambda\dagger} v_\lambda(p)e^{ip\cdot x}), \qquad (42)$$

where the sum is over positive and negative helicity, a_p^λ and $b_p^{\lambda\dagger}$ are the annihilation and creation operators for a fermion and an antifermion, and $u_\lambda(p)$ and $v_\lambda(p)$ are the momentum-space spinors for a fermion and an antifermion. These spinors are given explicitly in Table 4, in the representation where the Dirac matrices are [7]

$$\gamma^0 = \begin{pmatrix} 0 & 1 \\ 1 & 0 \end{pmatrix} \quad \gamma^i = \begin{pmatrix} 0 & \sigma^i \\ -\sigma^i & 0 \end{pmatrix} \quad \gamma_5 = \begin{pmatrix} -1 & 0 \\ 0 & 1 \end{pmatrix}, \qquad (43)$$

where each entry in the above matrices is itself a 2×2 matrix.

We used the concept of chirality when formulating the standard model in Section 2. In the representation of the Dirac matrices given above,

$$\psi_L \equiv \frac{1-\gamma_5}{2}\psi = \begin{pmatrix} 1 & 0 \\ 0 & 0 \end{pmatrix}\psi \qquad (44)$$

$$\psi_R \equiv \frac{1+\gamma_5}{2}\psi = \begin{pmatrix} 0 & 0 \\ 0 & 1 \end{pmatrix}\psi \qquad (45)$$

so a left-chiral spinor has nonzero upper components and a right-chiral spinor has nonzero lower components. Chirality is conserved in gauge

Table 4. Spinors for a fermion of energy E and three-momentum of magnitude p pointing in the (θ,ϕ) direction. The spinors $u_\lambda(p)$ and $v_\lambda(p)$ correspond to fermions and antifermions of helicity $\lambda\frac{1}{2}$.

$$u_+(p) = \begin{pmatrix} \sqrt{E-p}\begin{pmatrix}\cos\frac{\theta}{2} \\ e^{i\phi}\sin\frac{\theta}{2}\end{pmatrix} \\ \sqrt{E+p}\begin{pmatrix}\cos\frac{\theta}{2} \\ e^{i\phi}\sin\frac{\theta}{2}\end{pmatrix} \end{pmatrix} \qquad u_-(p) = \begin{pmatrix} \sqrt{E+p}\begin{pmatrix}\sin\frac{\theta}{2} \\ -e^{i\phi}\cos\frac{\theta}{2}\end{pmatrix} \\ \sqrt{E-p}\begin{pmatrix}\sin\frac{\theta}{2} \\ -e^{i\phi}\cos\frac{\theta}{2}\end{pmatrix} \end{pmatrix}$$

$$v_+(p) = \begin{pmatrix} \sqrt{E+p}\begin{pmatrix}-e^{-i\phi}\sin\frac{\theta}{2} \\ \cos\frac{\theta}{2}\end{pmatrix} \\ -\sqrt{E-p}\begin{pmatrix}-e^{-i\phi}\sin\frac{\theta}{2} \\ \cos\frac{\theta}{2}\end{pmatrix} \end{pmatrix} \qquad v_-(p) = \begin{pmatrix} \sqrt{E-p}\begin{pmatrix}e^{-i\phi}\cos\frac{\theta}{2} \\ \sin\frac{\theta}{2}\end{pmatrix} \\ -\sqrt{E+p}\begin{pmatrix}e^{-i\phi}\cos\frac{\theta}{2} \\ \sin\frac{\theta}{2}\end{pmatrix} \end{pmatrix}$$

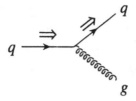

Figure 12.   Helicity is conserved for massless quarks involved in a gauge interaction.

interactions because the matter Lagrangian, Eq. (2), connects fields of the same chirality. In the massless limit, helicity and chirality are related, because the factor $\sqrt{E-p}$ vanishes in the expressions for the spinors in Table 4, causing either the upper or lower components to vanish:

$$u_+(p) = u_R(p) \quad u_-(p) = u_L(p)$$
$$v_+(p) = v_L(p) \quad v_-(p) = v_R(p) \tag{46}$$

Note that the relationship between helicity and chirality is reversed for fermions and antifermions.

For massless fermions, chirality conservation implies helicity conservation, as shown in Fig. 12. For massive fermions, helicity is no longer related to chirality, so although chirality is conserved, helicity is not. This is illustrated in Fig. 13. Both helicity-conserving and helicity-nonconserving gauge interactions occur; the latter are proportional to the fermion mass, since they are forbidden in the massless limit.

Exercise 3.3 (∗) - Do the quark mass terms in $\mathcal{L}_M$, Eq. (12), conserve

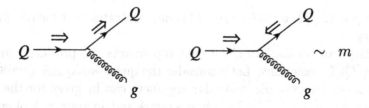

Figure 13.   For massive quarks, there are helicity-conserving and nonconserving gauge interactions. The amplitude for the latter is proportional to the quark mass.

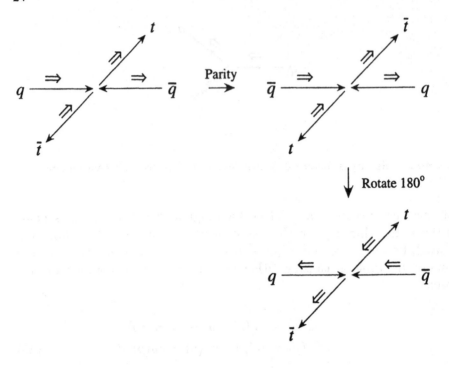

Figure 14. Parity and rotational symmetry are used to show that the top quark is produced unpolarized in (unpolarized) $p\bar{p}$ collisions.

chirality?

A useful discrete symmetry of QCD is parity, $\mathbf{x} \to -\mathbf{x}$, $\mathbf{p} \to -\mathbf{p}$. Helicity flips under parity, because although spin does not flip,[10] the direction of motion of the fermion does. One can show that the spinors of Table 4 are related to each other under parity as follows:

$$u_-(p) = \gamma^0 u_+(\tilde{p})$$
$$v_-(p) = \gamma^0 v_+(\tilde{p}) \tag{47}$$

where $p = (E, \mathbf{p})$, $\tilde{p} = (E, -\mathbf{p})$. This demonstrates that parity flips the helicity.

Parity can be used to show that top quarks are produced unpolarized in QCD reactions. Let's consider the quark-antiquark-annihilation subprocess, for example; a similar argument can be given for the gluon-fusion subprocess. In Fig. 14 I show a quark and an antiquark of opposite

---

[10]Spin angular momentum, like orbital angular momentum ($\mathbf{L} = \mathbf{x} \times \mathbf{p}$), does not change sign under parity.

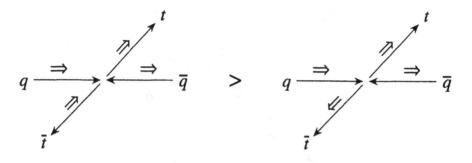

Figure 15. The cross section for opposite-helicity $t\bar{t}$ production is greater than that for same-helicity $t\bar{t}$ production.

helicity annihilating to produce a top quark and a top antiquark of opposite helicity. (Due to helicity conservation in the massless limit, the helicities of the light quark and antiquark must be opposite; this is not true of the top quark and antiquark.) Applying a parity transformation to this reaction yields the second diagram in Fig. 14. Rotating this figure by 180° in the scattering plane yields the third diagram, which is the same as the first diagram but with all helicities reversed. Since parity is a symmetry of QCD, the rates for the first and third reactions are the same. The light quarks are unpolarized in (unpolarized) $p\bar{p}$ collisions, so the first and third reactions will occur with equal probabilities. The first reaction produces positive-helicity top quarks, the second negative-helicity top quarks. Thus top quarks are produced with positive and negative helicity with equal probability, *i.e.*, they are produced unpolarized.

However, there is another avenue open to observe the spin of the top quark. Although the top quark is produced unpolarized, the spin of the top quark is correlated with that of the top antiquark. This is shown in Fig. 15; the rate for opposite-helicity $t\bar{t}$ production is greater than that of same-helicity $t\bar{t}$ production.

Exercise 3.4 (*) - Argue that in the limit $E \gg m$, the correlation between the helicities of the top quark and antiquark is 100%.

There is a special basis in which the correlation is 100% for all energies, dubbed the "off-diagonal" basis [8]. This basis is shown in Fig. 16. Rather than using the direction of motion of the quarks as the spin quantization axis, one uses another direction, which makes an angle ψ

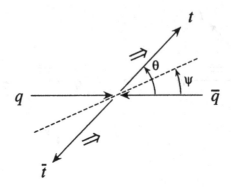

Figure 16. The "off-diagonal" basis. The spins of the top quark and antiquark point in the same direction.

with respect to the beam, related to the scattering angle θ by

$$\tan \psi = \frac{\beta^2 \sin \theta \cos \theta}{1 - \beta^2 \sin^2 \theta} , \tag{48}$$

where β is the velocity of the top quark and antiquark in the center-of-momentum frame. When the spin is projected along this axis, the correlation is 100%; the spins of the top quark and antiquark point in the same direction along this axis.

The moral of this story is that, for massive fermions, there is nothing special about the helicity basis. We will see this again in the next section on the weak interaction. The spin correlation between top quarks and antiquarks should be observed for the first time in Run II of the Tevatron.

Exercise 3.5 (*) - Use Eq. (48) to show that in the limit $E \gg m$, the off-diagonal basis is identical to the helicity basis. Argue that this had to be the case. [Hint: See Exercise 3.4.]

Exercise 3.6 (***) - What is the off-diagonal basis at threshold ($E = m$)? Give a physics argument for this basis at threshold.

5. Top Weak Interactions

In this section I discuss the charged-current weak interaction of the top quark, shown in Fig. 17. This interaction connects the top quark with a down-type quark, with an amplitude proportional to the CKM matrix element V_{tq} ($q = d, s, b$). The interaction has a vector-minus-

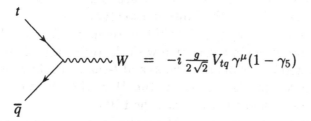

Figure 17. Charged-current weak interaction of the top quark.

axial-vector $(V - A)$ structure because only the left-chiral component of the top quark participates in the $SU(2)$ gauge interaction (see Table 1).

The charged-current weak interaction is responsible for the rapid decay of the top quark, as shown in Fig. 18. The partial width into the final state Wq is proportional to $|V_{tq}|^2$.[11] The CDF Collaboration has measured [9]

$$\frac{BR(t \to Wb)}{BR(t \to Wq)} = 0.94^{+0.31}_{-0.24}$$

$$= \frac{|V_{tb}|^2}{|V_{td}|^2 + |V_{ts}|^2 + |V_{tb}|^2} \tag{49}$$

This implies that $|V_{tb}| \gg |V_{td}|, |V_{ts}|$, but it does not tell us the absolute magnitude of V_{tb}.

Exercise 4.1 (*) - Show that the denominator of the last expression in Eq. (49) is unity if one assumes that there are just three generations.

Thus, if we assume three generations, Eq. (49) implies $|V_{tb}| = 0.97^{+0.16}_{-0.12}$. However, we already know $V_{tb} = 0.9990 - 0.9993$ if there are just three generations [2].

Single top – The magnitude of V_{tb} can be extracted directly by measuring the cross section for top-quark production via the weak interaction. There are three such processes, depicted in Fig. 19, all of which result in a single top quark rather than a $t\bar{t}$ pair [10]. The cross sections for these single-top processes are proportional to $|V_{tb}|^2$.

The first subprocess in Fig. 19, which is mediated by the exchange of an s-channel W boson, is analogous to the Drell-Yan subprocess. The

---

[11]The W boson then goes on to decay to a fermion-antifermion pair.

28

second subprocess is simply the first subprocess turned on its side, so the W boson is in the t channel. The b quark is now in the initial state, so this subprocess relies on the b distribution function in the proton, which we will discuss momentarily.[12] In the third subprocess, the W boson is real, and is produced in association with the top quark. This subprocess is also initiated by a b quark. The s- and t-channel subprocesses should be observed for the first time in Run II of the Tevatron; associated production of W and t must await the LHC.

The cross sections for these three single-top processes are given in Table 5 at the Tevatron and the LHC. The largest cross section at both machines is from the t-channel subprocess; it is nearly one third of the cross section for $t\bar{t}$ pair production via the strong interaction (see Table 3). The next largest cross section at the Tevatron is from the s-channel subprocess. This is the smallest of the three at the LHC, because it is initiated by a quark-antiquark collision. As is evident from Fig. 8, the light-quark distribution functions grow with decreasing x more slowly than the gluon or b distribution functions, so quark-antiquark annihilation is relatively suppressed at the LHC. For a similar reason, associated production of W and t (which is initiated by a gluon-b collision) is relatively large at the LHC, while it is very small at the Tevatron.

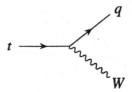

Figure 18. Top-quark decay via the charged-current weak interaction.

Table 5. Cross sections (pb), at next-to-leading-order in QCD, for top-quark production via the weak interaction at the Tevatron and the LHC [11, 12, 13].

	s channel	t channel	Wt
Tevatron ($\sqrt{s} = 2.0$ TeV $p\bar{p}$)	$0.90 \pm 5\%$	$2.1 \pm 5\%$	$0.1 \pm 10\%$
LHC ($\sqrt{s} = 14$ TeV pp)	$10.6 \pm 5\%$	$250 \pm 5\%$	$75 \pm 10\%$

---

[12]If one instead uses a d or s quark in the initial state, the cross section is much less due to the CKM suppression.

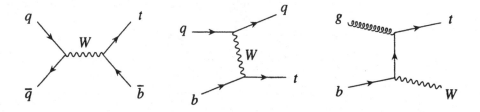

Figure 19. Single-top-quark production via the weak interaction. The first diagram corresponds to the *s*-channel subprocess, the second to the *t*-channel subprocess, and the third to *Wt* associated production (only one of the two contributing diagrams is shown).

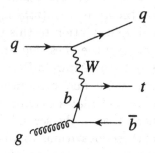

Figure 20. When the $\bar{b}$ is produced at high transverse momentum, the leading-order process for *t*-channel single-top production is W-gluon fusion.

Let's consider the largest of the three processes, *t*-channel single-top production, in more detail. This process was originally dubbed W-gluon fusion [14], because it was thought of as a virtual W striking a gluon to produce a $t\bar{b}$ pair, as shown in Fig. 20. If the $\bar{b}$ in the final state is at high transverse momentum (p_T), this is indeed the leading-order diagram for this process. If we instead integrate over the p_T of the $\bar{b}$, we obtain an enhancement from the region where the $\bar{b}$ is at low p_T, nearly collinear with the incident gluon.

Exercise 4.2 (**) - Show that a massless quark propagator blows up in the collinear limit, as shown in Fig. 21.

The b mass regulates the collinear divergence, such that the resulting cross section is proportional to $\alpha_S \ln(m_t^2/m_b^2)$, where the weak couplings are tacit.

This collinear enhancement is desirable — it yields a larger cross section — but it also makes perturbation theory less convergent. Each emission of a collinear gluon off the internal b quark produces another

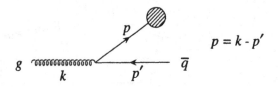

Figure 21. When a gluon splits into a real antiquark and a virtual quark, the quark propagator becomes singular when the kinematics are collinear.

power of $\alpha_S \ln(m_t^2/m_b^2)$, because it yields another b propagator that is nearly on-shell, as shown in Fig. 22. The result is that the expansion parameter for perturbation theory is $\alpha_S \ln(m_t^2/m_b^2)$, rather than α_S [12].

Fortunately, there is a simple solution to this problem. The collinear logarithms that arise are exactly the ones that are summed to all orders by the Dokshitzer-Gribov-Lipatov-Altarelli-Parisi (DGLAP) equations. In order to sum these logarithms, one introduces a b distribution function in the proton. When one calculates t-channel single-top production using a b distribution function, as in the second diagram in Fig. 19, one is automatically summing these logarithms to all orders. The expansion parameter for perturbation theory is now simply α_S [15].

Figure 23 shows how the b distribution function in the proton arises from a gluon splitting into a (virtual) $b\bar{b}$ pair. The strange and charm distributions arise in the same way; this also explains the presence of up and down antiquarks in the proton (see Fig. 8). Unlike the other "sea" quark distributions, which are extracted from experiment, the b distribution function is calculated from the initial condition $b(x) = 0$ at

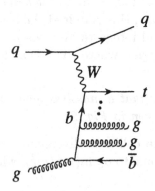

Figure 22. The emission of collinear gluons is suppressed only by $\alpha_S \ln(m_t^2/m_b^2)$, rather than α_S.

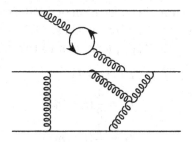

Figure 23. The quark "sea" in the proton arises from loops of virtual quarks.

$\mu = m_b$, and is evolved to higher μ via the DGLAP equations.

Exercise 4.3 (**) - Draw the leading-order Feynman diagrams for the subprocesses that contribute to

$$\text{(a) } p\bar{p} \to W + X$$
$$\text{(b) } p\bar{p} \to W + 1 \text{ jet} + X$$

where X denotes the remnants of the proton and antiproton.

Top-quark spin – In the previous section we studied the top-quark spin in the context of the strong interaction. Let's now consider this topic in relation to the weak interaction, beginning with the decay of the top quark.

The top-quark decay to the final state $b\bar{\ell}\nu$ is depicted in Fig. 24.

Exercise 4.4 (**) Determine the helicities of all final-state particles in top decay (neglecting their masses).

The partial width for this decay, summed over the two spin states of the top quark, is given by a very simple formula:

$$d\Gamma \sim \sum_{\text{spin}} |\mathcal{M}|^2 \sim t \cdot \ell b \cdot \nu \, , \tag{50}$$

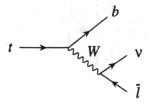

Figure 24. Semileptonic top-quark decay.

where the four-momentum of the fermion or antifermion is denoted by its label. To undo the sum over the top-quark spin, it is useful to decompose the four-momentum of the top quark, t, into two lightlike four vectors,

$$t = t_1 + t_2 \tag{51}$$

$$t_1 = \frac{1}{2}(t + ms) \tag{52}$$

$$t_1 = \frac{1}{2}(t - ms) \tag{53}$$

where s is the spin four-vector. In the top-quark rest frame, the spin four-vector is $s = (0, \hat{s})$, where $\hat{s}$ is a unit vector that defines the spin quantization axis of the top quark.

Exercise 4.5 (*) - Show that t_1 and t_2 are lightlike four-vectors, $t_1^2 = t_2^2 = 0$.

In the top-quark rest frame, the spatial components of t_1 point in the spin-up direction, while the spatial components of t_2 point in the spin-down direction. The partial widths for the decay of these two spin states are

$$d\Gamma_\uparrow \sim t_2 \cdot \ell b \cdot \nu$$
$$d\Gamma_\downarrow \sim t_1 \cdot \ell b \cdot \nu . \tag{54}$$

Note that Eq. (50) is the sum of these two partial widths, as expected.

Let's consider the decay of a top quark with spin up along the $\hat{s}$ direction in its rest frame, as depicted in Fig. 25. In this frame, the spatial components of t_2 point in the $-\hat{s}$ direction. Hence

$$d\Gamma_\uparrow \sim t_2 \cdot \ell \sim 1 + \cos\theta , \tag{55}$$

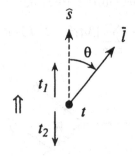

Figure 25. Semileptonic top-quark decay in the top rest frame. The vector $\hat{s}$ indicates the spin-quantization axis. The four-vectors t_1 and t_2 have spatial components that point in the spin up and down directions, respectively.

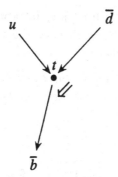

Figure 26. In single-top production, the top-quark spin is polarized along the direction of the $\bar{d}$ quark in the top rest frame.

where θ is the angle between the spin direction and the charged-lepton three-momentum (see Fig. 25).

Exercise 4.6 (∗) - Confirm Eq. (55).

Thus

$$\frac{d\Gamma_\uparrow}{d\cos\theta} \sim 1 + \cos\theta \,, \tag{56}$$

which means that the charged lepton in top decay tends to go in the direction of the top-quark spin. In fact, the charged lepton is the most efficient analyzer of the top-quark spin, via the angular distribution of Eq. (56) [16].

We can use these same formulas to analyze the top-quark spin in single-top production [17]. The Feynman diagram for the s-channel subprocess, Fig. 19, is the same as that for top-quark decay, Fig. 24, with the replacement $\nu \to u$, $\bar{\ell} \to \bar{d}$. Thus

$$d\sigma_\uparrow \sim t_2 \cdot db \cdot u$$
$$d\sigma_\downarrow \sim t_1 \cdot db \cdot u \tag{57}$$

from Eqs. (54). If we choose the spin-quantization axis to point in the direction of the $\bar{d}$ (in the top-quark rest frame), then $t_1 \sim d$, and the latter cross section above vanishes. Thus the top-quark is 100% polarized in the direction of the $\bar{d}$ (in the top-quark rest frame) in s-channel single-top production, as depicted in Fig. 26. This result holds true for t-channel single-top production as well, since it proceeds via the same Feynman diagram, just turned on its side.

Although the top quark is 100% polarized when produced via the weak interaction, it is not in a state of definite helicity. Just as we saw in the previous section, there is nothing special about helicity for massive fermions. It may be possible to observe the polarization of single top quarks in Run II of the Tevatron.

Exercise 4.7 (* * *) - Show that in the limit $E \gg m$, the top quark has negative helicity when produced via the s- or t-channel subprocesses, as expected.

Acknowledgments

I would like to thank Harrison Prosper for organizing a very memorable summer school. I am also grateful to Reinhard Schwienhorst and Martin Hennecke for reading the manuscript, and to Kevin Paul for preparing the figures. This work was supported in part by the U. S. Department of Energy under contract No. DOE DE-FG02-91ER40677.

Solutions to the exercises

Section 2

Exercise 1.1 – It is easiest to show this using index-free notation. Write the first term in the Lagrangian of Eq. (2) as

$$i\bar{Q}_L^i \, \slashed{D} Q_L^i = iQ_L^\dagger \gamma^0 \, \slashed{D} Q_L \,, \tag{58}$$

where Q_L is a 3-component vector in generation space. This term is invariant under the transformation $Q_L \to U_{Q_L} Q_L$, because the 3×3 unitary matrix U_{Q_L} commutes with the Dirac matrices (which are the same for all three generations):

$$iQ_L^\dagger \gamma^0 \, \slashed{D} Q_L \to iQ_L^\dagger U_{Q_L}^\dagger \gamma^0 \, \slashed{D} U_{Q_L} Q_L = iQ_L^\dagger \gamma^0 \, \slashed{D} Q_L \,, \tag{59}$$

where I have used $U_{Q_L}^\dagger U_{Q_L} = 1$. The same argument applies to the other terms in the matter Lagrangian and their corresponding symmetries.

Exercise 1.2 – Consider the transformation of the first term in the Yukawa Lagrangian, Eq. (9), under the symmetry U_{Q_L} of Eq. (5):

$$\bar{Q}_L^i \epsilon \phi^* u_R^j \to \bar{Q}_L^i U_{Q_L}^\dagger \epsilon \phi^* u_R^j \,. \tag{60}$$

This is not invariant under the symmetry transformation, so U_{Q_L} is violated. In contrast, baryon number symmetry, Eq. (10), is respected:

$$\bar{Q}_L^i \epsilon \phi^* u_R^j \to \bar{Q}_L^i e^{-i\theta/3} \epsilon \phi^* e^{i\theta/3} u_R^j = \bar{Q}_L^i \epsilon \phi^* u_R^j \,. \tag{61}$$

The same applies to the other terms in the Yukawa Lagrangian, and also to lepton number, Eq. (11).

Exercise 1.3 – Consider the transformation of the Lagrangian of Eq. (16) under the first field redefinition of Eq. (15) (using the index-free notation introduced in the solution to Exercise 1.1):

$$\mathcal{L}_{KE} = i\bar{u}_L \not{\partial} u_L \to i\bar{u}_L A_{u_L}^\dagger \not{\partial} A_{u_L} u_L = i\bar{u}_L \not{\partial} u_L \ . \tag{62}$$

The last step requires that A_{u_L} be unitary, $A_{u_L}^\dagger A_{u_L} = 1$. The same argument applies to the other fermionic kinetic-energy terms in the Lagrangian.

Exercise 1.4 – If $A_{u_L} = A_{d_L}$, then we may combine the first two field redefinitions in Eq. (15) into one equation:

$$Q_L^i = A_{Q_L}^{ij} Q_L'^j \ , \tag{63}$$

where $A_{Q_L} = A_{u_L} = A_{d_L}$. This is exactly the symmetry U_{Q_L} of Eq. (5). The field redefinitions of u_R^i and d_R^i in Eq. (15) are the symmetries U_{u_R} and U_{d_R} of Eq. (5).

Exercise 1.5 – This follows from the definition of the CKM matrix, $V \equiv A_{d_L}^\dagger A_{u_L}$:

$$V^\dagger V = (A_{d_L}^\dagger A_{u_L})^\dagger A_{d_L}^\dagger A_{u_L} = A_{u_L}^\dagger A_{d_L} A_{d_L}^\dagger A_{u_L} = 1 \ , \tag{64}$$

where I have used the unitarity of the A matrices.

Exercise 1.6 – A useful equation to remember is $\hbar c = 197$ MeV fm. Using this, one can convert length to mass^{-1}:

$$\text{length} = \hbar c / \text{mass} \, c^2 = \text{mass}^{-1} \tag{65}$$

using $\hbar = c = 1$.

Exercise 1.7 – Such a term is not invariant under SU(3) gauge symmetry, $Q_L \to U Q_L$, where U acts on the (suppressed) color indices of the quarks:

$$Q_L^{iT} \epsilon \phi C \phi^T \epsilon Q_L^j \to Q_L^{iT} U^T \epsilon \phi C \phi^T \epsilon U Q_L^j \ . \tag{66}$$

This involves $U^T U$, which is not equal to unity (rather, $U^\dagger U = 1$). This term is also not invariant under $U(1)_Y$, as the total hypercharge is nonzero (1/6+1/6+1/2+1/2).

Exercise 1.8 – Lepton number, Eq. (11), is violated because

$$L_L^{iT} \epsilon \phi C \phi^T \epsilon L_L^j \to L_L^{iT} e^{i\phi} \epsilon \phi C \phi^T \epsilon e^{i\phi} L_L^j \tag{67}$$

is not invariant. Recall that lepton number is an accidental symmetry of the standard model. Once you go beyond the standard model by including higher-dimension operators, there is no reason for lepton number (and baryon number) to be conserved.

Exercise 1.9 – We'll follow a similar argument as the one made to count the number of parameters in the CKM matrix. The Yukawa matrix Γ_e has $2 \times 3 \times 3$ parameters, and the complex, symmetric matrix c^{ij} has 2×6 parameters. The symmetries U_{L_L} and U_{e_R} contain $2 \times 3 \times 3$ degrees of freedom, so the number of physically-relevant parameters is

$$2 \times 3 \times 3 + 2 \times 6 - 2 \times 3 \times 3 = 12 . \tag{68}$$

[Note that we did not remove lepton number from the symmetries, because lepton number is violated by $\mathcal{L}_5$, Eq. (21)]. Of these parameters, six are the charged-lepton and neutrino masses, leaving six parameters for the MNS matrix. Three are mixing angles, and three are CP-violating phases.

Section 3

Exercise 2.1 – Plug the expressions for α, G_F, and M_Z in terms of g, g', and v, given at beginning of Section 3, into Eq. (23) and carry through the algebra to obtain $M_W^2 = (1/4)g^2 v^2$.

Exercise 2.2 – Using Eq. (24), we can write Eq. (25) as

$$M_W^2 \left(1 - \frac{M_W^2}{M_Z^2} \right) = \frac{\pi \alpha}{\sqrt{2} G_F} . \tag{69}$$

Solving this quadratic equation for M_W^2 yields Eq. (23). Alternatively, one could plug the expressions for α, G_F, and M_Z in terms of g, g', and v, given at beginning of Section 3, as well as $M_W^2 = (1/4)g^2 v^2$, into the above equation to check its veracity.

Exercise 2.3 – Starting from Eq. (26), the one-loop analogue of Eq. (69) is

$$M_W^2 \left(1 - \frac{M_W^2}{M_Z^2} \right) = \frac{\frac{\pi \alpha}{\sqrt{2} G_F}}{(1 - \Delta r)} . \tag{70}$$

The differential of this equation (with respect to M_W^2 and m_t^2, keeping everything else fixed) is

$$dM_W^2 - 2 \frac{M_W^2}{M_Z^2} dM_W^2 = - \frac{\frac{\pi \alpha}{\sqrt{2} G_F}}{(1 - \Delta r)^2} \frac{3 G_F dm_t^2}{8 \sqrt{2} \pi^2} \frac{1}{t_W^2} , \tag{71}$$

where I have used Eq. (27) for Δr. We can now set $\Delta r = 0$ to leading-order accuracy, and solve for dM_W^2/dm_t^2:

$$\frac{dM_W^2}{dm_t^2} = \frac{3\alpha}{16\pi} \frac{1}{(2c_W^2 - 1)t_W^2} , \tag{72}$$

where I've used Eq. (24). Using $dM_W/dm_t = (m_t/M_W)dM_W^2/dm_t^2$ and evaluating numerically (for $M_W = 80$ GeV, $m_t = 175$ GeV) gives a slope of 0.0060, in good agreement with the slope of the lines of constant Higgs mass in Fig. 4.

Exercise 2.4 – The desired result follow from inserting $M_W^2 = (1/4)g^2v^2$ and $M_Z^2 = (1/4)(g^2 + g'^2)v^2$ [Eq. (8)] into the on-shell definition of $\sin^2\theta_W$, Eq. (24).

Exercise 2.5 – Combining Eqs. (34) and (35) to eliminate $\hat{s}_Z^2$ gives

$$M_W^2 \left(1 - \frac{M_W^2}{M_Z^2 \hat{\rho}}\right) = \frac{\frac{\pi\alpha}{\sqrt{2}G_F}}{(1 - \Delta\hat{r}_W)} . \tag{73}$$

The differential of this equation is

$$dM_W^2 - 2\frac{M_W^2}{M_Z^2 \hat{\rho}}dM_W^2 + \frac{M_W^4}{M_Z^2 \hat{\rho}^2}\frac{3G_F dm_t^2}{8\sqrt{2}\pi^2} = 0 , \tag{74}$$

where I have used Eq. (35) for $\hat{\rho}$ (there is no m_t dependence in $\Delta\hat{r}_W$). We can now set $\hat{\rho} = 1$ to leading-order accuracy. Using the leading-order expressions of Eq. (25) and $M_W^2/M_Z^2 = c_W^2$, it is easy to show that Eq. (74) is identical Eq. (71) at leading order.

Section 4

Exercise 3.1 – The four-momenta of the quark and antiquark in the center-of-momentum frame are

$$P_1 = (E, 0, 0, p)$$
$$P_2 = (E, 0, 0, -p) .$$

Thus $S \equiv (P_1 + P_2)^2 = (2E, 0, 0, 0)^2 = (2E)^2$, which is the square of the total energy of the collision. The last expression in Eq. (37) follows from $(P_1 + P_2)^2 = P_1^2 + P_2^2 + 2P_1 \cdot P_2 \approx 2P_1 \cdot P_2$, if we neglect the proton mass, $P_1^2 = P_2^2 = m_p^2$.

Exercise 3.2 – Inserting Eq. (41) into the second term in the Yukawa Lagrangian, Eq. (9), yields

$$\mathcal{L}_Y = -\Gamma_d^{ij}\frac{1}{\sqrt{2}}(v + h)\bar{d}_L^i d_R^j + h.c. \tag{75}$$

(analogous results are obtained for the other terms in the Lagrangian). Using Eq. (13), this can be written

$$\mathcal{L}_Y = -M_d^{ij}\left(1 + \frac{h}{v}\right)\bar{d}_L^i d_R^j + h.c. \tag{76}$$

The field redefinitions that diagonalize the mass matrix, Eq. (15), will therefore also diagonalize the couplings of the fermions to the Higgs boson. The coupling to a given fermion is thus given by $-m_f/v$ (times a factor i since the Feynman rules come from $i\mathcal{L}$), as shown in Fig. 9.

Exercise 3.3 – The answer is evidently no, since these terms connect fields of different chirality.

Exercise 3.4 – In the ultrarelativistic limit, $E \gg m$, the mass of the top quark is negligible. Since helicity is conserved for massless quarks, the top quark and antiquark must be produced with opposite helicities.

Exercise 3.5 – In the limit $E \gg m$ ($\beta \to 1$), Eq. (48) implies $\psi = \theta$, which means that the off-diagonal and helicity bases are the same. This is as expected, because in the massless limit the helicities of the top quark and antiquark are 100% correlated (see Exercise 3.4), which is the defining characteristic of the off-diagonal basis.

Exercise 3.6 – At threshold ($\beta \to 0$), Eq. (48) implies $\psi = 0$, which means that the top quark and antiquark spins are 100% correlated along the beam direction. This is a consequence of angular-momentum conservation. At threshold, the top quark and antiquark are produced at rest with no orbital angular momentum. The colliding light quark and antiquark have no orbital angular momentum along the beam direction. Thus spin angular momentum along the beam direction must be conserved. The light quark and antiquark have opposite helicity (due to helicity conservation in the massless limit), so the top quark and antiquark are produced with their spins pointing in the same direction along the beam.

Section 5

Exercise 4.1 – This follows from the unitarity of the CKM matrix, $VV^\dagger = 1$. Displaying indices, this may be written

$$V_{ik}V_{kj}^\dagger = V_{ik}V_{jk}^* = \delta_{ij} . \tag{77}$$

For $i = j$, this implies

$$\sum_{k=d,s,b} |V_{ik}|^2 = 1 , \tag{78}$$

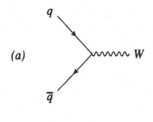

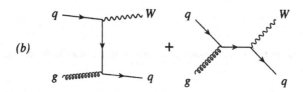

Figure 27. (a) Leading-order subprocess for W production. (b) Leading-order subprocess for $W + 1$ jet production.

which yields the desired result for $i = t$.

Exercise 4.2 – The square of the four-momentum of the quark propagator in Fig. 21 is

$$p^2 = (k - p')^2 = -2k \cdot p' \tag{79}$$

This vanishes for collinear kinematics,

$$k = (E, 0, 0, E)$$
$$p' = (E', 0, 0, E') \ .$$

Thus the denominator of the quark propagator vanishes in the collinear limit (if we neglect the quark mass).

Exercise 4.3 – (a) There is just one diagram, shown in Fig. 27(a). (b) There are two contributing subprocesses, $gq \to Wq$ and $q\bar{q} \to Wg$; each consists of two Feynman diagrams, shown in Fig. 27(b) for $gq \to Wq$. The two diagrams for $q\bar{q} \to Wg$ may be obtained by radiating a gluon off either fermion line in Fig. 27(a).

Exercise 4.4 – The charged-current weak interaction couples only to left-chiral fields. Thus the fermions in the final state (b, ν) have negative helicity, and the antifermion $(\bar{\ell})$ has positive helicity, due to the relationship between chirality and helicity for massless particles (discussed in Section 4).

Exercise 4.5 – In the top-quark rest frame, $s^2 = (0, \hat{s})^2 = -1$, since $\hat{s}$

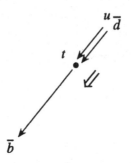

Figure 28. Single-top production in the ultrarelativistic limit, as viewed from the top rest frame.

is a unit vector. Because s^2 is Lorentz invariant, this is true in all reference frames. Similarly, $t \cdot s = 0$, because $t = (m, 0, 0, 0)$ in the top-quark rest frame. Thus

$$t_1^2 = \frac{1}{4}(t + ms)^2 = \frac{1}{4}(m^2 - m^2 + 2mt \cdot s) = 0 , \qquad (80)$$

and similarly for t_2^2.

Exercise 4.6 – The spatial part of the lightlike four-vector t_2 is pointing in the $-\hat{s}$ direction. Thus $t_2 \cdot \ell \sim 1 - \cos \alpha$, where α is the angle between $-\hat{s}$ and the direction of the charged lepton. This angle is supplementary to θ ($\alpha + \theta = \pi$), so $t_2 \cdot \ell \sim 1 - \cos \alpha = 1 + \cos \theta$.

Exercise 4.7 – The s-channel subprocess, in the top-quark rest frame, looks like Fig. 26. In the limit $E \gg m$, this figure looks like Fig. 28; the u and $\bar{d}$ approach each other along a line and annihilate to make a top quark at rest and a $\bar{b}$ that carries off the incoming momentum. As always, the top-quark spin points in the direction of the $\bar{d}$. To view this event from the center-of-momentum frame, one boosts opposite the direction of motion of the u and $\bar{d}$. This boosts the top quark in the direction opposite its spin, so it is in a state of negative helicity. This is as expected; in the limit $E \gg m$, the top quark acts like a massless quark, and is therefore produced in a negative-helicity state by the weak interaction (see Exercise 4.4).

References

[1] A. F. Falk, "The CKM matrix and the heavy quark expansion," in *Flavor Physics for the Millennium*, TASI 2000, ed. J. Rosner (World Scientific, Singapore, 2001), p. 379 [arXiv:hep-ph/0007339].

[2] K. Hagiwara *et al.* [Particle Data Group Collaboration], Phys. Rev. D **66**, 010001 (2002).

[3] S. Weinberg, "Conceptual Foundations Of The Unified Theory Of Weak And Electromagnetic Interactions," Rev. Mod. Phys. **52**, 515 (1980) [Science **210**, 1212 (1980)].

[4] W. Buchmüller and D. Wyler, "Effective Lagrangian Analysis Of New Interactions And Flavor Conservation," Nucl. Phys. B **268**, 621 (1986).

[5] R. Bonciani, S. Catani, M. L. Mangano and P. Nason, "NLL resummation of the heavy-quark hadroproduction cross-section," Nucl. Phys. B **529**, 424 (1998) [arXiv:hep-ph/9801375].

[6] A. F. Falk and M. E. Peskin, "Production, decay, and polarization of excited heavy hadrons," Phys. Rev. D **49**, 3320 (1994) [arXiv:hep-ph/9308241].

[7] M. Peskin and D. Schroeder, *An Introduction to Quantum Field Theory* (Addison-Wesley, Reading, 1995).

[8] G. Mahlon and S. Parke, "Maximizing spin correlations in top quark pair production at the Tevatron," Phys. Lett. B **411**, 173 (1997) [arXiv:hep-ph/9706304].

[9] T. Affolder *et al.* [CDF Collaboration], "First measurement of the ratio $B(t \to Wb)/B(t \to Wq)$ and associated limit on the CKM element $|V_{tb}|$," Phys. Rev. Lett. **86**, 3233 (2001) [arXiv:hep-ex/0012029].

[10] T. Stelzer, Z. Sullivan and S. Willenbrock, "Single top quark production at hadron colliders," Phys. Rev. D **58**, 094021 (1998) [arXiv:hep-ph/9807340].

[11] M. C. Smith and S. Willenbrock, "QCD and Yukawa Corrections to Single-Top-Quark Production via $q\bar{q} \to t\bar{b}$," Phys. Rev. D **54**, 6696 (1996) [arXiv:hep-ph/9604223].

[12] T. Stelzer, Z. Sullivan and S. Willenbrock, "Single-top-quark production via W-gluon fusion at next-to-leading order," Phys. Rev. D **56**, 5919 (1997) [arXiv:hep-ph/9705398].

[13] S. Zhu, "Next-To-Leading Order QCD Corrections to $bg \to tW^-$ at the CERN Large Hadron Collider," Phys. Lett. B **524**, 283 (2002) [Erratum-ibid. B **537**, 351 (2002)].

[14] S. S. Willenbrock and D. A. Dicus, "Production Of Heavy Quarks From W-Gluon Fusion," Phys. Rev. D **34**, 155 (1986).

[15] M. A. Aivazis, J. C. Collins, F. I. Olness and W. K. Tung, "Leptoproduction of heavy quarks. 2. A Unified QCD formulation of charged and neutral current processes from fixed target to collider energies," Phys. Rev. D **50**, 3102 (1994) [arXiv:hep-ph/9312319].

[16] M. Jeżabek and J. H. Kühn, "Lepton Spectra From Heavy Quark Decay," Nucl. Phys. B **320**, 20 (1989).

[17] G. Mahlon and S. Parke, "Improved spin basis for angular correlation studies in single top quark production at the Tevatron," Phys. Rev. D **55**, 7249 (1997) [arXiv:hep-ph/9611367].

NEUTRINO PHYSICS

M.C. Gonzalez-Garcia

Theory Division, CERN, CH-1211, Geneva 23, Switzerland
and

Instituto de Física Corpuscular, Universitat de València – C.S.I.C
Edificio Institutos de Paterna, Apt 22085, 46071 València, Spain
and

C.N. Yang Institute for Theoretical Physics
State University of New York at Stony Brook
Stony Brook,NY 11794-3840, USA
concha.gonzalez-garcia@cern.ch

Abstract   These lectures discuss some aspects of neutrino physics making special emphasis on the phenomenology of neutrino oscillations in vacuum and in matter. The existing evidence from solar and atmospheric neutrinos as well as the results from laboratory searches are also reviewed.

1.   Lecture 1: Neutrino Masses

1.1   Introduction

In 1930 Wolfgang Pauli postulated the existence of a new particle in order to reconcile the observed continuous spectrum of nuclear beta decay with energy conservation. The postulated particle, which he called the neutron, had no electric charge and, in fact, Pauli himself pointed out that in order to do the job it had to weigh less than one percent of the proton mass, thus establishing the first limit on the *neutrino* mass. It was Fermi, who, in 1934 [1], named the new particle the neutrino and first proposed the four-fermion theory of beta decay. The neutrino was first observed by Cowans and Reines [2] in 1956 in a reactor experiment.

Neutrinos are copiously produced in natural sources: in the burning of the stars, in the interaction of cosmic rays, even as relics of the Big Bang. Starting in the 1960s, neutrinos produced in the Sun and in the atmosphere have been observed. In 1987, neutrinos from a supernova in the Large Magellanic Cloud were also detected. These observations play an important role in our understanding of the detailed features of

H.B. Prosper and M. Danilov (eds.), Techniques and Concepts of High-Energy Physics XII, 43–87.

neutrinos. Neutrinos are also produced in "man-made" sources, such as nuclear reactors or in beam-dump experiments at accelerators.

The properties of the neutrino and, in particular, the question of its mass have intrigued physicists ever since the particle was proposed. In the laboratory, kinematic effects of neutrino masses have been searched for without any positive result. Experiments achieved higher and higher precision, reaching upper limits for the electron-neutrino mass of 10^{-9} the proton mass, rather than the 10^{-2} originally obtained by Pauli. This raised the question of whether neutrinos, like photons, are truly massless.

It is clear that the answer to this question is limited by our capability of detecting the effect of a non-zero neutrino mass. This is a very difficult task in direct kinematic measurements. In 1957, however, Bruno Pontecorvo [3] realized that the existence of neutrino masses may not only reveal itself in kinematic effects but it implies also the possibility of neutrino oscillations. Flavor oscillations of neutrinos have been searched for using either neutrino beams from reactors or accelerators, or natural neutrinos generated at astrophysical sources (the Sun giving the largest flux) or in the atmosphere. The longer the distance that the neutrinos travel from their production point to the detector, the smaller the masses that can be probed by their oscillation. Indeed, solar neutrinos allow us to search for masses as small as 10^{-5} eV, that is 10^{-14} times the proton mass!

In fact, in recent years, experiments studying natural neutrino fluxes have provided us with the strongest evidence of neutrino masses and mixing. Experiments that measure the flux of atmospheric neutrinos have found results suggesting the disappearance of muon-neutrinos when propagating over distances of order hundreds (or more) kilometers. Experiments that measure the flux of solar neutrinos found results that suggest the disappearance of electron-neutrinos while propagating within the Sun or between the Sun and the Earth. The disappearance of both atmospheric ν_μ's and solar ν_e's is most easily explained in terms of neutrino oscillations. Regarding experiments performed with laboratory beams, most have given no evidence of oscillations. One exception is the LSND experiment, which has observed the appearance of electron anti-neutrinos in a muon anti-neutrino beam.

In these lectures I first discuss the low energy formalism for adding neutrino masses to the SM and the induced leptonic mixing. In the second lecture I describe the phenomenology associated with neutrino oscillations in vacuum and in matter. The third lecture is devoted to the evidence from solar and atmospheric neutrinos.

In preparing these lectures I have benefited from the many excellent books and reviews on the subject existing in the literature. In particular, the ones by Bahcall [4], Bilenky, Giunti and Grimus [5], Boehm and Vogel [6], Gaisser [7], Kayser, Gibrat-Debu and Perrier [8], Kim and Pevsner [9], and Mohapatra and Pal [10]. In the writing of these notes I have used material from our recent review article [11] to which I refer for details and references.

1.2 Standard Model of Massless Neutrinos

The Standard Model (SM) is based on the gauge group

$$G_{\rm SM} = SU(3)_{\rm C} \times SU(2)_{\rm L} \times U(1)_{\rm Y}, \tag{1}$$

with three fermion generations, where a single generation consists of five different representations of the gauge group,

$$Q_L(3,2,\tfrac{1}{6})\,,\ U_R(3,1,\tfrac{2}{3})\,,\ D_R(3,1,-\tfrac{1}{3})\,,\ L_L(1,2,-\tfrac{1}{2})\,,\ E_R(1,1,-1), \tag{2}$$

where the numbers in parenthesis represent the corresponding charges under the group (1).

The model contains a single Higgs boson doublet, $\phi(1,2,1/2)$, whose vacuum expectation value breaks the gauge symmetry,

$$\langle\phi\rangle = \begin{pmatrix} 0 \\ \frac{v}{\sqrt{2}} \end{pmatrix} \implies G_{\rm SM} \to SU(3)_{\rm C} \times U(1)_{\rm EM}. \tag{3}$$

Neutrinos are fermions that have neither strong nor electromagnetic interactions, i.e., they are singlets of $SU(3)_{\rm C} \times U(1)_{\rm EM}$. Active neutrinos have weak interactions, that is, they are not singlets of $SU(2)_{\rm L}$. They reside in the lepton doublets L_L. Sterile neutrinos are defined as having no SM gauge interactions; they are singlets of the SM gauge group.

The SM has three active neutrinos accompanying the charged lepton mass eigenstates, e, μ and τ:

$$L_{L\ell} = \begin{pmatrix} \nu_{L\ell} \\ \ell_L^- \end{pmatrix}, \quad \ell = e, \mu, \tau. \tag{4}$$

Thus the charged current (CC) interaction terms for leptons read

$$-\mathcal{L}_{\rm CC} = \frac{g}{\sqrt{2}} \sum_\ell \overline{\nu_{L\ell}}\gamma^\mu \ell_L^- W_\mu^+ + {\rm h.c.}. \tag{5}$$

In addition, the SM neutrinos have neutral current (NC) interactions,

$$-\mathcal{L}_{\rm NC} = \frac{g}{2\cos\theta_W} \sum_\ell \overline{\nu_{L\ell}}\gamma^\mu \nu_{L\ell} Z_\mu^0. \tag{6}$$

Equations (5) and (6) give all the neutrino interactions within the SM. In particular, Eq. (6) determines the decay width of the Z^0 boson into neutrinos, which is proportional to the number of light active neutrinos. At present, the measurement of the invisible Z width yields $N_\nu = 3.00 \pm 0.06$ making the existence of three, and only three, light (that is, $m_\nu \leq m_Z/2$) active neutrinos an experimental fact.

In the SM fermions masses arise from the Yukawa interactions,

$$-\mathcal{L}_{\text{Yukawa}} = Y_{ij}^d \overline{Q_{Li}} \phi D_{Rj} + Y_{ij}^u \overline{Q_{Li}} \tilde{\phi} U_{Rj} + Y_{ij}^\ell \overline{L_{Li}} \phi E_{Rj} + \text{h.c.}, \qquad (7)$$

(where $\tilde{\phi} = i\tau_2\phi^\star$) which after spontaneous symmetry breaking generate fermion masses $m_{ij}^f = Y_{ij}^f v/\sqrt{2}$. However, since no right-handed neutrinos exist in the model, the Yukawa interactions of Eq. (7) leave the neutrinos massless. In other words, the SM predicts that neutrinos are precisely massless and consequently, there is neither mixing nor CP violation in the leptonic sector.

1.3 Introducing Massive Neutrinos

It is very likely that the SM is not the ultimate theory of Nature and there is new physics (NP) that will appear at higher energies. In this case, the SM would be an effective low energy theory valid up to the scale Λ_{NP} which characterizes the NP. In this approach, the gauge group Eq. (1), the fermionic spectrum Eq. (2) and the pattern of spontaneous symmetry breaking Eq. (3) are still valid ingredients to describe Nature at energies $E \ll \Lambda_{\text{NP}}$. The difference between the SM as a complete description of Nature and as a low energy effective theory is that in the latter case we must consider also non-renormalizable (dim> 4) terms whose effect will be suppressed by powers $1/\Lambda_{\text{NP}}^{\text{dim}-4}$. In this approach the largest effects at low energy are expected to come from dim= 5 operators.

There is a single set of dimension-five terms that is made of SM fields and is consistent with the gauge symmetry given by

$$\frac{Z_{ij}^\nu}{\Lambda_{\text{NP}}} \left(\bar{L}_{Li}\tilde{\phi}\right) \left(\phi^+ L_{Lj}^C\right) + \text{h.c.}, \qquad (8)$$

where $L_{Li}^C = C\bar{L}_{Li}^T$. Equation (8) contains two lepton doublet vector fields and therefore it violates total lepton number by two units. Upon spontaneous symmetry breaking it leads to neutrino masses:

$$(M_\nu)_{ij} = \frac{Z_{ij}^\nu}{2} \frac{v^2}{\Lambda_{\text{NP}}}. \qquad (9)$$

This is a Majorana mass term.

Equation (9) arises in a generic extension of the SM, which means that neutrino masses are very likely to appear if there is new physics. Furthermore, from Eq. (9) we find that the scale of neutrino masses is suppressed by v/Λ_{NP} when compared to the scale of charged fermion masses, thus providing an explanation not only for the existence of neutrino masses but also for their smallness. Finally, Eq.(9) breaks not only total lepton number but also the lepton flavor symmetry $U(1)_e \times U(1)_\mu \times U(1)_\tau$. Therefore we should expect lepton mixing and CP violation.

The best known scenario that leads to Eq. (8) is the *see-saw mechanism*[12], which we discuss next.

1.3.1   Dirac and Majorana Neutrino Mass Terms.

With the fermionic content and gauge symmetry of the SM neutrinos can only gain masses through non-renormalizable terms of the form Eq. (8), which give a Majorana ($\Delta L = 2$) mass to the neutrino.

Other possibilities can be open if one adds to the SM an arbitrary number m of sterile neutrinos $\nu_{si}(1,1,0)$. In this case there are, in general, two types of mass terms that arise from *renormalizable* terms:

$$-L_{M_\nu} = M_{Dij}\overline{\nu_{Li}}\nu_{sj} + \frac{1}{2}M_{Nij}\overline{\nu_{si}^c}\nu_{sj} + \text{h.c.}. \qquad (10)$$

Here ν^c indicates a charge conjugated field, $\nu^c = C\bar{\nu}^T$ and C is the charge conjugation matrix. M_D is a complex $3 \times m$ matrix and M_N is a symmetric matrix (as follows from simple Dirac algebra) of dimension $m \times m$.

The first term is a Dirac mass term. It is generated after spontaneous electroweak symmetry breaking from the Yukawa interactions $Y_{ij}^\nu \overline{L_{Li}}\tilde{\phi}\nu_{sj}$, similarly to the charged fermion masses discussed in Sec. 1.2. It conserves total lepton number.

The second term in Eq. (10) is a Majorana mass term. It is different from the Dirac mass terms in many important aspects. It is a singlet of the SM gauge group. Therefore, it can appear as a bare mass term. Furthermore, since it involves two neutrino fields, it breaks lepton number conservation by two units. More generally, such a term is allowed only if the neutrinos carry no additive conserved charge. This is the reason that such terms are not allowed for any charged fermions which, by definition, carry $U(1)_{EM}$ charges.

Equation (10) can be written as:

$$-L_{M_\nu} = \frac{1}{2}\overline{\vec{\nu}^c}M_\nu\vec{\nu} + \text{h.c.} , \qquad (11)$$

where

$$M_\nu = \begin{pmatrix} 0 & M_D \\ M_D{}^T & M_N \end{pmatrix}, \tag{12}$$

and $\vec{\nu} = \begin{pmatrix} \nu_{Li} \\ \nu_{sj} \end{pmatrix}$ is a $(3+m)$-dimensional vector. The matrix M_ν is complex and symmetric. It can be diagonalized by a unitary matrix of dimension $(3+m)$. The resulting mass eigenstates, ν_k, obey the Majorana condition, $\nu_k^c = \nu_k$.

There are three interesting cases, differing in the hierarchy of scales between M_N and M_D:

(1) The scale of the mass eigenvalues of M_N is much higher than the scale of electroweak symmetry breaking $\langle \phi \rangle$. In this case the diagonalization of M_ν leads to three light mass eigenstates with a mass matrix of the form Eq. (9). In particular, the scale Λ_{NP} is identified with the mass scale of the heavy sterile neutrinos, that is the typical scale of the eigenvalues of M_N. This is the natural situation in various extensions of the SM that are characterized by a high energy scale. This is the *see-saw mechanism*.

(2) The scale of some eigenvalues of M_N is not higher than the electroweak scale. Now the SM is not even a good low energy effective theory: there are more than three light neutrinos, and they are mixtures of doublet and singlet fields. These light fields are all of the Majorana-type.

(3) $M_N = 0$. This is equivalent to imposing lepton number symmetry on this model. Again, the SM is not a good low energy theory: both the fermionic content and the assumed symmetries are different. (Recall that within the SM lepton number conservation is an accidental symmetry.) Now only the first term in Eq. (10) is allowed, which is a Dirac mass term. It is generated by the Higgs mechanism in the same way that charged fermions masses are generated. If indeed it is the only neutrino mass term present and $m = 3$, we can identify the three sterile neutrinos with the right handed component of a four-component spinor neutrino field (actually with its charge conjugate). In this way, the six massive Majorana neutrinos combine to form three massive neutrino Dirac states, equivalently to the charged fermions. In this particular case the 6×6 diagonalizing matrix is block diagonal and it can be written in terms of a 3×3 unitary matrix.

As we will see the analysis of neutrino oscillations is the same whether the light neutrinos are of the Majorana- or Dirac-type. Only in the discussion of neutrinoless double beta decay will the question of Majorana versus Dirac neutrinos be crucial.

1.4   Lepton Mixing

The possibility of arbitrary mixing between two massive neutrino states was first introduced in Ref. [13]. In the general case, we denote the neutrino mass eigenstates by $(\nu_1, \nu_2, \nu_3, \ldots, \nu_n)$ where $n = 3 + m$, and the charged lepton mass eigenstates by (e, μ, τ). The corresponding interaction eigenstates are denoted by (e^I, μ^I, τ^I) and $\vec{\nu} = (\nu_{Le}, \nu_{L\mu}, \nu_{L\tau}, \nu_{s1}, \ldots, \nu_{sm})$. In the mass basis, leptonic CC interactions are given by

$$-\mathcal{L}_{CC} = \frac{g}{\sqrt{2}} (\overline{e_L}\ \overline{\mu_L}\ \overline{\tau_L}) \gamma^\mu U \begin{pmatrix} \nu_1 \\ \nu_2 \\ \nu_3 \\ \cdot \\ \cdot \\ \cdot \\ \nu_n \end{pmatrix} W_\mu^+ - \text{h.c.}. \qquad (13)$$

Here U is a $3 \times n$ matrix.

Given the charged lepton mass matrix M_ℓ and the neutrino mass matrix M_ν in some interaction basis,

$$-\mathcal{L}_M = (\overline{e_L^I}\ \overline{\mu_L^I}\ \overline{\tau_L^I})\, M_\ell \begin{pmatrix} e_R^I \\ \mu_R^I \\ \tau_R^I \end{pmatrix} + \frac{1}{2}\overline{\vec{\nu}^c}M_\nu\vec{\nu} + \text{h.c.} , \qquad (14)$$

we can find the diagonalizing matrices V^ℓ and V^ν:

$$V^{\ell\dagger} M_\ell M_\ell^\dagger V^\ell = \text{diag}(m_e^2, m_\mu^2, m_\tau^2),$$
$$V^{\nu\dagger} M_\nu^\dagger M_\nu V^\nu = \text{diag}(m_1^2, m_2^2, m_3^2, \ldots, m_n^2). \qquad (15)$$

Here V^ℓ is a unitary 3×3 matrix while V^ν is a unitary $n \times n$ matrix. The $3 \times n$ mixing matrix U can be found from these diagonalizing matrices:

$$U_{ij} = P_{\ell,ii}\, V_{ik}^{\ell\,\dagger}\, V_{kj}^\nu\, (P_{\nu,jj}). \qquad (16)$$

P_ℓ is a diagonal 3×3 phase matrix, that is conventionally used to reduce by three the number of phases in U. P_ν is a diagonal matrix with additional arbitrary phases (chosen to reduce the number of phases in U) only for Dirac states. For Majorana neutrinos, this matrix is simply a unit matrix. The reason for that is that if one rotates a Majorana neutrino by a phase, this phase will appear in its mass term which will no longer be real. Thus, the number of phases that can be absorbed by redefining the mass eigenstates depends on whether the neutrinos are Dirac or Majorana particles. In particular, if there are only three Majorana neutrinos, U is a 3×3 matrix analogous to the CKM matrix for the quarks [14] but due to the Majorana nature of the neutrinos it depends on six independent parameters: three mixing angles and three phases. This is to be compared to the case of three Dirac neutrinos, where the number of physical phases is one, similarly to the CKM matrix. Note, however, that the two extra Majorana phases affect only lepton number violating processes and are very hard to measure.

If no new interactions for the charged leptons are present we can identify their interaction eigenstates with the mass eigenstates after phase redefinitions. In this case the CC lepton mixing matrix U is simply given by a $3 \times n$ sub-matrix of the unitary matrix V^ν.

1.5 Direct Determination of m_ν

It was Fermi who first proposed a kinematic search for the neutrino mass from the hard part of the beta spectra in ^{3}H beta decay ^{3}H $\rightarrow ^3$He $+ e^- + \bar{\nu}_e$ In the absence of leptonic mixing this search provides a measurement of the electron neutrino mass.

^{3}H beta decay is a superallowed transition, which means that the nuclear matrix elements do not generate any energy dependence, so that the electron spectrum is given by the phase space alone

$$\frac{dN}{dT} = CpE(Q - T)\sqrt{(Q - T)^2 - m_\nu^2}F(E) , \qquad (17)$$

where $E = T + m_e$, Q is the maximum energy and $F(E)$ is the Fermi function which incorporates final state Coulomb interactions.

Plotted in terms of the Kurie function $K(T) \equiv \sqrt{\frac{dN}{dT}\frac{1}{pEF(E)}}$ a non-vanishing neutrino mass m_ν provokes a distortion from the straight-line T-dependence at the end point: for $m_\nu = 0 \rightarrow T_{max} = Q$ whereas for $m_\nu \neq 0 \rightarrow T_{max} = Q - m_\nu$ as illustrated in Fig. 1. ^{3}H beta decay has

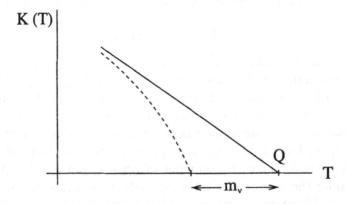

Figure 1. Kinematic determination of m_ν

a very small energy release $Q = 18.6$ KeV which makes it particularly sensitive to this kinematic effect.

At present the most precise determination from the Troitsk and Mainz experiments [15] give no indication in favour of $m_\nu \neq 0$ and one sets an upper limit $m_\nu < 2.2$ eV at 95% CL.

For the other flavours the present limits are [15]

$$m_{\nu_\tau} < 18.2 \text{ MeV} \quad (95\% \text{ CL}) \qquad \text{from} \qquad \tau^- \to n\pi + \nu_\tau \qquad (18)$$

$$m_{\nu_\mu} < 190 \text{ keV} \quad (90\% \text{ CL}) \qquad \text{from} \qquad \pi^- \to \mu^- + \bar{\nu}_\mu \qquad (19)$$

In the presence of mixing these limits have to be modified and in general they involve more than one flavor parameter. For example for neutrinos with small mass differences the distortion of the beta spectrum can be described by the single parameter substituting m_{ν_e} by

$$m_\beta = \frac{\sum_i m_i |U_{ei}|^2}{\sum_i |U_{ei}|^2}.$$

Direct information on neutrino masses can also be obtained from neutrinoless double beta decay ($2\beta0\nu$) searches:

$$(A, Z) \to (A, Z + 2) + e^- + e^-. \qquad (20)$$

Notice that in this process lepton number is violated by 2 units, thus in order to induce the $2\beta0\nu$ decay, ν_e must be a Majorana particle. Thus ($2\beta0\nu$) decay can give us the answer to one of the outstanding questions in neutrino physics: are neutrinos Dirac or Majorana particles?

The ($2\beta0\nu$) rate is proportional to the *effective Majorana mass of ν_e*,

$$m_{ee} = \left| \sum_i m_i U_{ei}^2 \right| \qquad (21)$$

which, in addition to five parameters that affect ^{3}H beta decay spectrum, depends also on the three leptonic CP violating phases. The strongest bound from $2\beta0\nu$-decay, from the Heidelberg-Moscow group [15], is

$$m_{ee} < 0.34 \ (0.26) \text{ eV}, \qquad 90 \% \ (68\%) \text{ C.L.}. \qquad (22)$$

Taking into account systematic errors related to nuclear matrix elements, the bound may be weaker by a factor of about 3.

Summarizing, in the SM neutrinos are strictly massless. Neutrino masses can be introduced in the model at the expense of adding new sterile states and/or breaking total lepton number. Depending on the way the mass term is introduced, neutrinos may be Dirac particles, as are other fermions of the SM for which neutrinos and antineutrinos are different states, or they may be Majorana particles, being their own antiparticles. In this second case one may gain an understanding of why neutrino masses are smaller than other fermion masses. Massive neutrinos open up the possibility of flavour mixing and CP violation in the lepton sector similar to the quark sector. So far direct searches for neutrino masses have resulted only in limits, the strongest of which is $\sim$ few eV.

2.   Lecture 2: Neutrino Oscillations

2.1   Neutrino Oscillations in Vacuum

If neutrinos have mass, the weak eigenstates, ν_α, produced in a CC interaction are, in general, linear combinations of the mass eigenstates ν_i

$$|\nu_\alpha\rangle = \sum_{i=1}^{n} U_{\alpha i}^* |\nu_i\rangle \tag{23}$$

where n is the number of light neutrino species and U is the the mixing matrix. (Implicit in our definition of the state $|\nu\rangle$ is its energy-momentum and space-time dependence). After traveling a distance L (or, equivalently for relativistic neutrinos, time t), a neutrino originally produced with a flavor α evolves as follows:

$$|\nu_\alpha(t)\rangle = \sum_{i=1}^{n} U_{\alpha i}^* |\nu_i(t)\rangle \ . \tag{24}$$

It can be detected in the CC interaction $\nu_\alpha(t) N' \to \ell_\beta N$ with a probability

$$P_{\alpha\beta} = |\langle \nu_\beta | \nu_\alpha(t) \rangle|^2 = |\sum_{i=1}^{n} \sum_{j=1}^{n} U_{\alpha i}^* U_{\beta j} \langle \nu_j(0) | \nu_i(t) \rangle|^2 \ , \tag{25}$$

where E_i and m_i are, respectively, the energy and the mass of the neutrino mass eigenstate ν_i.

Using the standard approximation that $|\nu\rangle$ is a plane wave $|\nu_i(t)\rangle = e^{-iE_i t}|\nu_i(0)\rangle$, that neutrinos are relativistic with $p_i \simeq p_j \equiv p \simeq E$

$$E_i = \sqrt{p_i^2 + m_i^2} \simeq p + \frac{m_i^2}{2E} \ , \tag{26}$$

and the orthogonality relation $\langle \nu_j(0)|\nu_i(0)\rangle = \delta_{ij}$, we get the following transition probability

$$P_{\alpha\beta} = \delta_{\alpha\beta} - 4\sum_{i<j}^{n} \text{Re}[U_{\alpha i} U_{\beta i}^* U_{\alpha j}^* U_{\beta j}] \sin^2 x_{ij}$$

$$+2\sum_{i<j}^{n} \text{Im}[U_{\alpha i}U_{\beta i}^{*}U_{\alpha j}^{*}U_{\beta j}]\sin^{2}\frac{x_{ij}}{2} \,, \tag{27}$$

where in convenient units:

$$x_{ij} = 1.27\frac{\Delta m_{ij}^{2}}{\text{eV}^{2}}\frac{L/E}{\text{m/MeV}} \,. \tag{28}$$

with $\Delta m_{ij}^{2} \equiv m_{i}^{2} - m_{j}^{2}$. $L = t$ is the distance between the production point of ν_{α} and the detection point of ν_{β}. The first line in Eq. (27) is CP conserving while the second one is CP violating and has opposite sign for neutrinos and antineutrinos. In what follows we will only discuss CP conserving effects.

The transition probability [Eq. (27)] has an oscillatory behavior, with oscillation lengths

$$L_{0,ij}^{\text{osc}} = \frac{4\pi E}{\Delta m_{ij}^{2}} \tag{29}$$

and amplitude that is proportional to elements in the mixing matrix. Thus, in order to have oscillations, neutrinos must have different masses ($\Delta m_{ij}^{2} \neq 0$) and they must mix ($U_{\alpha i}U_{\beta i} \neq 0$).

An experiment is characterized by the typical neutrino energy E and by the source-detector distance L. But, in general, neutrino beams are not monoenergetic and, rather than measuring $P_{\alpha\beta}$, the experiments are sensitive to the average probability

$$\langle P_{\alpha\beta}\rangle = \frac{\int dE_{\nu}\frac{d\Phi}{dE_{\nu}}\sigma_{CC}(E_{\nu})P_{\alpha\beta}(E_{\nu})\epsilon(E_{\nu})}{\int dE_{\nu}\frac{d\Phi}{dE_{\nu}}\sigma_{CC}(E_{\nu})\epsilon(E_{\nu})}$$

$$= \delta_{\alpha\beta} - 4\sum_{i=1}^{n-1}\sum_{j=i+1}^{n} \text{Re}[U_{\alpha i}U_{\beta i}^{*}U_{\alpha j}^{*}U_{\beta j}]\langle\sin^{2}x_{ij}\rangle, \tag{30}$$

where Φ is the neutrino energy spectrum, σ_{CC} is the cross section for the process in which the neutrino is detected (in general, a CC interaction), and $\epsilon(E_{\nu})$ is the detection efficiency.

In order to be sensitive to a given value of Δm_{ij}^{2}, the experiment has to be set up with $E/L \approx \Delta m_{ij}^{2}$ ($L \sim L_{0,ij}^{\text{osc}}$). The typical values of L/E for different types of neutrino sources and experiments are summarized in Table 1. If $(E/L) \gg \Delta m_{ij}^{2}$ ($L \ll L_{0,ij}^{\text{osc}}$), the oscillation does not have time to give an appreciable effect because $\sin^{2}x_{ij} \ll 1$. If $L \gg L_{0,ij}^{\text{osc}}$, the oscillating phase goes through many cycles before the detection and is averaged to $\langle\sin^{2}x_{ij}\rangle = 1/2$.

Table 1. Characteristic values of L and E.

Experiment	L (m)	E (MeV)	Δm^2 (eV^2)
Solar	10^{10}	1	10^{-10}
Atmospheric	$10^4 - 10^7$	$10^2 - 10^5$	$10^{-1} - 10^{-4}$
Reactor	$10^2 - 10^3$	1	$10^{-2} - 10^{-3}$
Accelerator	10^2	$10^3 - 10^4$	> 0.1
LBL Accelerator	$10^5 - 10^6$	10^4	$10^{-2} - 10^{-3}$

For a two-neutrino case, the mixing matrix depends on a single parameter,

$$U = \begin{pmatrix} \cos\theta & \sin\theta \\ -\sin\theta & \cos\theta \end{pmatrix}, \qquad (31)$$

and there is a single mass-squared difference Δm^2. Then $P_{\alpha\beta}$ of Eq. (27) takes the well known form

$$P_{\alpha\beta} = \delta_{\alpha\beta} - (2\delta_{\alpha\beta} - 1) \sin^2 2\theta \sin^2 x . \qquad (32)$$

The physical parameter space is covered with $\Delta m^2 \geq 0$ and $0 \leq \theta \leq \frac{\pi}{2}$ (or, alternatively, $0 \leq \theta \leq \frac{\pi}{4}$ and either sign for Δm^2).

Changing the sign of the mass difference, $\Delta m^2 \to -\Delta m^2$, and changing the octant of the mixing angle, $\theta \to \frac{\pi}{2} - \theta$, amounts to redefining the mass eigenstates, $\nu_1 \leftrightarrow \nu_2$: $P_{\alpha\beta}$ must be invariant under such transformation. Equation (32) reveals, however, that $P_{\alpha\beta}$ is actually invariant under each of these transformations separately. This situation implies that there is a two-fold discrete ambiguity in the interpretation of $P_{\alpha\beta}$ in terms of two-neutrino mixing: the two different sets of physical parameters, $(\Delta m^2, \theta)$ and $(\Delta m^2, \frac{\pi}{2} - \theta)$, give the same transition probability in vacuum. One cannot tell from a measurement of, say, $P_{e\mu}$ in vacuum whether the larger component of ν_e resides in the heavier or in the lighter neutrino mass eigenstate.

Neutrino oscillation experiments measure $P_{\alpha\beta}$. It is common practice for the experiments to interpret their results in the two-neutrino framework. In other words, the constraints on $P_{\alpha\beta}$ are translated into allowed or excluded regions in the plane $(\Delta m^2, \sin^2 2\theta)$ by using Eq. (32). An example is given in Fig. 2.

When an experiment is taking data at fixed $\langle L \rangle$ and $\langle E \rangle$, as is the case for most laboratory searches, its result can always be accounted for by Δm^2 that is large enough to be in the region of averaged oscillations, $\langle \sin^2 x_{ij} \rangle = 1/2$. Consequently, no upper bound on Δm^2 can be achieved by such an experiment. So, for negative searches that set an upper bound on the oscillation probability, $\langle P_{\alpha\beta} \rangle \leq P_L$, the excluded region

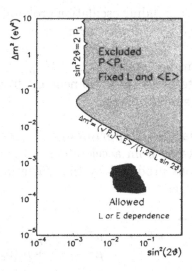

Figure 2. The characteristic form of an excluded region from a negative search with fixed L/E and of an allowed region from a positive search with varying L/E in the two-neutrino oscillation parameter plane.

lies always on the upper-right side of the $(\Delta m^2, \sin^2 2\theta)$ plane, limited by the following asymptotic lines:

- For $\Delta m^2 \gg 1/\langle L/E\rangle$, a vertical line at $\sin^2 2\theta = 2\,P_L$.

- For $\Delta m^2 \ll 1/\langle L/E\rangle$, the oscillating phase can be expanded and the limiting curve takes the form $\Delta m^2 \sin 2\theta = 4\sqrt{P_L}/\langle L/E\rangle$, which in a log-log plot gives a straight line of slope $-1/2$.

If, instead, data are taken at several values of $\langle L\rangle$ and/or $\langle E\rangle$, the corresponding region may be closed as it is possible to have direct information on the characteristic oscillation wavelength.

2.2 Laboratory Searches for Oscillations

Laboratory experiments to search for neutrino oscillations are performed with neutrino beams produced at either accelerators or nuclear reactors. In *disappearance* experiments, one looks for the attenuation of a neutrino beam primarily composed of a single flavor due to the mixing with other flavors. In *appearance* experiments, one searches for interactions by neutrinos of a flavor not present in the original neutrino beam.

Most of the past and present laboratory experiments did not have an oscillation signal. In such a case, as discussed in Sec. 2.1, the experiment sets a limit on the corresponding oscillation probability. Appearance experiments set limits $\langle P_{\alpha\beta} \rangle < P_L$ for given flavors $\alpha \neq \beta$. Disappearance experiments set limits $\langle P_{\alpha\alpha} \rangle > 1 - P_L$ for a given flavor α which, in the two neutrino case, can be translated into $\langle P_{\alpha\beta} \rangle < P_L$ for $\beta \neq \alpha$. The results are usually interpreted in a two neutrino framework as exclusion regions in the $(\Delta m^2, \sin^2 2\theta)$ plane.

2.2.1   Short Baseline Experiments at Accelerators.   Conventional neutrino beams from accelerators are mostly produced by π decays, with the pions produced by the scattering of the accelerated protons on a fixed target:

$$
\begin{aligned}
p + \text{target} &\to \quad \pi^\pm + X \\
\pi^\pm &\to \quad \mu^\pm + \nu_\mu(\bar{\nu}_\mu) \\
\mu^\pm &\to e^\pm + \nu_e(\bar{\nu}_e) + \bar{\nu}_\mu(\nu_\mu)
\end{aligned}
\tag{33}
$$

Thus the beam can contain both μ- and e-neutrinos and antineutrinos. The final composition and energy spectrum of the neutrino beam is determined by selecting the sign of the decaying π and by stopping the produced μ in the beam line.

Most oscillation experiments performed so far with neutrino beams from accelerators have characteristic distances of the order of hundreds of meters. We call them *short baseline (SBL) experiments*. With the exception of the LSND experiment, which we discuss below, all searches have been negative. Owing to the short path length, these experiments are not sensitive to the low values of Δm^2, which we will invoke in the last lecture to explain either the solar or the atmospheric neutrino data.

The only positive signature of oscillations at a laboratory experiment comes from the Liquid Scintillator Neutrino Detector (LSND) [16] running at Los Alamos Meson Physics Facility. Its primary neutrino flux comes from π^+'s produced in a 30-cm-long water target when hit by protons from the LAMPF linac with 800 MeV kinetic energy. The detector is a tank filled with 167 metric tons of dilute liquid scintillator, located about 30 m from the neutrino source. The experiment observed an excess of events as compared to the expected background while the excess was consistent with $\bar{\nu}_\mu \to \bar{\nu}_e$ oscillations. In the latest results the total fitted excess is of $87.9 \pm 22.4 \pm 6$ events, corresponding to an oscillation probability of $P_{\mu e} = (2.64 \pm 0.67 \pm 0.45) \times 10^{-3}$. In the two-family formalism these results lead to the oscillation parameters shown in Fig. 3 (from Ref. [17]). The shaded regions are the 90 % and 99 % likelihood regions from LSND. The best fit point corresponds to $\Delta m^2 = 1.2$ eV^2 and $\sin^2 2\theta = 0.003$. The region of parameter space which is favoured by the LSND observations has been partly tested by other experiments in par-

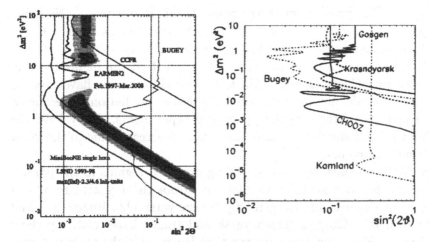

Figure 3. (Left) Allowed regions (at 90 and 99 % CL) for $\nu_e \rightarrow \nu_\mu$ oscillations from the LSND experiment compared with the exclusion regions (at 90% CL) from KARMEN2 and other experiments. The 90 % CL expected sensitivity curve for MinimBoNE is also shown. (Right) Excluded regions at 90% for ν_e oscillations from reactors experiments and the expected sensitivity from the KamLAND experiment.

ticular by the KARMEN [18] experiment. The KARMEN experiment is performed at the neutron spallation facility ISIS of the Rutherford Appleton Laboratory. They find a number of events in good agreement with the total background expectation.The corresponding exclusion curve in the two-neutrino parameter space is also given in Fig. 3.

At large Δm^2, KARMEN results exclude the region favored by LSND. At low Δm^2, KARMEN leaves some allowed space, but the reactor experiment at Bugey add stringent limits for the larger mixing angles. This figure represents the final status of the LSND oscillation signal.

The MiniBooNE experiment [19] searches for $\nu_\mu \rightarrow \nu_e$ oscillations and is specifically designed to make a conclusive statement about the LSND's neutrino oscillation evidence. MiniBooNE uses a ν_μ beam of energy 0.5 – 1.0 GeV initiated by a primary beam of 8 GeV protons from the Fermilab Booster, which contains only a small intrinsic ν_e component (less than 0.3%). They search for an excess of electron neutrino events in a detector located approximately 500 m from the neutrino source. The L/E ratio is similar to that of LSND, giving MiniBooNE sensitivity to the same mode of oscillations. In Fig. 3 we show the 90% CL limits that MiniBooNE can achieve.

2.2.2 Disappearance Experiments at Reactors.

Neutrino oscillations are also searched for using neutrino beams from nuclear reactors. Nuclear reactors produce $\bar{\nu}_e$ beams with $E_\nu \sim$ MeV. Owing to the low energy, e's are the only charged leptons that can be produced in the neutrino CC interaction. If the $\bar{\nu}_e$ oscillated to another flavor, its CC interaction could not be observed. Therefore, oscillation experiments performed at reactors are disappearance experiments. They have the advantage that smaller values of Δm^2 can be accessed due to the lower neutrino beam energy.

In Fig. 3 we show the corresponding excluded regions in the parameter space for two neutrino oscillations from the negative results of the reactor experiments Gosgen [20], Krasnoyarsk [22], Bugey [21], and CHOOZ [23]. Gosgen, Krasnoyarsk and Bugey have relatively short baselines. From the figure we see that Bugey sets the strongest constraint on the allowed mixing in the Δm^2 range that is interesting for the LSND signal. CHOOZ, which can be considered the first long baseline (LBL) reactor experiment ($L \simeq 1$ km), is sensitive to lower values of Δm^2. Its 90% CL limits include $\Delta m^2 < 7 \times 10^{-4}$ eV^2 for maximal mixing, and $\sin^2 2\theta < 0.10$ for large Δm^2. The CHOOZ results are significant in excluding part of the region that corresponds to the LMA solution of the solar neutrino problem and rules out with high significance the possibility that $\nu_\mu \to \nu_e$ oscillations explain the atmospheric neutrino deficit (see Sec. 3). The CHOOZ constraint is also relevant to the interpretation of the solar and atmospheric neutrino data in the framework of three-neutrino mixing.

Smaller values of Δm^2 can be accessed at future reactor experiments using longer baseline. Pursuing this idea, the KamLAND experiment [24], a 1000 ton liquid scintillation detector, is currently in operation in the Kamioka mine in Japan. This underground site is located at a distance of 150-210 km from several Japanese nuclear power stations. The measurement of the flux and energy spectrum of the $\bar{\nu}_e$'s emitted by these reactors will provide a test to the LMA solution of the solar neutrino anomaly. In Fig. 3 we plot the expected 90% sensitivity for the KamLAND experiment after 3 years of data taking. The experiment will, for the first time, provide a completely solar model independent test of this particle physics solution of the solar neutrino problem. After a few years of data taking, it should be capable of either excluding the entire LMA region or, not only establishing $\nu_e \leftrightarrow \nu_{\text{other}}$ oscillations, but also measuring the oscillation parameters with unprecedented precision.

2.2.3 Long Baseline Experiments at Accelerators.

Smaller values of Δm^2 can also be accessed using accelerator beams at long baseline (LBL) experiments. In these experiments the intense neutrino beam from an accelerator is aimed at a detector located underground at a distance of several hundred kilometers. The main goal of these experiments is to test the presently allowed solution for the atmospheric neutrino problem by searching for either ν_μ disappearance or ν_τ appearance.

At present there are three such projects approved: K2K [25] which runs with a baseline of about 235 km from KEK to SuperKamiokande (SK), MINOS [26] under construction with a baseline of 730 km from Fermilab to the Soudan mine where the detector will be placed, and OPERA [27], under construction with a baseline of 730 km from CERN to Gran Sasso.

2.3 Neutrinos in Matter: Effective Potentials

When neutrinos propagate in dense matter, the interactions with the medium affect their properties. These effects are either coherent or incoherent. For purely incoherent inelastic ν-p scattering, the characteristic cross section is very small:

$$\sigma \sim \frac{G_F^2 s}{\pi} \sim 10^{-43} \text{cm}^2 \left(\frac{E}{1\ \text{MeV}}\right)^2 . \tag{34}$$

This cross section is too small to give any effect in neutrino propagation. But Eq. (34) does not contain the contribution from forward elastic coherent interactions. In coherent interactions, the medium remains unchanged and it is possible to have interference of scattered and unscattered neutrino waves which enhances the effect. Coherence further allows one to decouple the evolution equation of the neutrinos from the equations of the medium. In this approximation, the effect of the medium is described by an effective potential which depends on the density and composition of the matter [28].

For example, the effective potential for the evolution of ν_e in a medium with electrons, protons and neutrons due to its CC interactions is given by (a detailed derivation of this result can be found, for instance, in Refs.[9, 11])

$$V_C = \sqrt{2} G_F N_e . \tag{35}$$

where N_e is the electron number density. For $\overline{\nu_e}$ the sign of V_C is reversed. This potential can also be expressed in terms of the matter density ρ:

$$V_C = \sqrt{2} G_F N_e \simeq 7.6\, Y_e \frac{\rho}{10^{14} \text{g/cm}^3}\ \text{eV} , \tag{36}$$

where $Y_e = \frac{N_e}{N_p + N_n}$ is the relative electron number density. Three examples that are relevant to observations are the following:

• At the Earth core $\rho \sim 10$ g/cm^3 and $V_C \sim 10^{-13}$ eV;
• At the solar core $\rho \sim 100$ g/cm^3 and $V_C \sim 10^{-12}$ eV

In the same way we can obtain the effective potentials for any flavour neutrino or antineutrino due to interactions with different particles in the medium. For ν_μ and ν_τ, $V_C = 0$ for most media, while for any active neutrino the effective potential due to NC interactions in a neutral medium is $V_N = -1/\sqrt{2}G_F N_n$, where N_n is the number density of neutrons.

2.4 Evolution in Matter: Effective Mass and Mixing

There are several derivations in the literature of the evolution equation of a neutrino system in matter (see, for instance, Refs. [29, 30, 31]).

Consider a state which is an admixture of two neutrino species $|\nu_e\rangle$ and $|\nu_X\rangle$ or, equivalently, of $|\nu_1\rangle$ and $|\nu_2\rangle$:

$$|\Phi(x)\rangle = \Phi_e(x)|\nu_e\rangle + \Phi_X(x)|\nu_X\rangle = \Phi_1(x)|\nu_1\rangle + \Phi_2(x)|\nu_2\rangle \qquad (37)$$

We decompose the neutrino state: $\Phi_i(x) = \nu_i(x)\phi_i(x)$ where $\phi_i(x)$ is the Dirac spinor part.

The evolution of Φ in a medium is described by a system of coupled Dirac equations, but after several approximations the spinoral part can be dropped out and we end up with an equation which can be written in matrix form as [28]:

$$-i\frac{\partial}{\partial x}\begin{pmatrix} \nu_e \\ \nu_X \end{pmatrix} = \left(-\frac{M_w^2}{2E}\right)\begin{pmatrix} \nu_e \\ \nu_X \end{pmatrix}, \qquad (38)$$

We have defined an effective mass matrix in matter (for simplicity, we do not display explicitly the x [or t] of the neutrino components, matter potentials and functions depending on those):

$$M_w^2 = \begin{pmatrix} \frac{m_1^2+m_2^2}{2} + 2EV_e - \frac{\Delta m^2}{2}\cos 2\theta & \frac{\Delta m^2}{2}\sin 2\theta \\ \frac{\Delta m^2}{2}\sin 2\theta & \frac{m_1^2+m_2^2}{2} + 2EV_X + \frac{\Delta m^2}{2}\cos 2\theta \end{pmatrix}. \qquad (39)$$

Here $\Delta m^2 = m_2^2 - m_1^2$.

We define the instantaneous mass eigenstates in matter, ν_i^m, as the eigenstates of M_w for a fixed value of x (or t). They are related to the interaction eigenstates through a unitary rotation,

$$\begin{pmatrix} \nu_e \\ \nu_X \end{pmatrix} = U(\theta_m)\begin{pmatrix} \nu_1^m \\ \nu_2^m \end{pmatrix} = \begin{pmatrix} \cos\theta_m & \sin\theta_m \\ -\sin\theta_m & \cos\theta_m \end{pmatrix}\begin{pmatrix} \nu_1^m \\ \nu_2^m \end{pmatrix}. \qquad (40)$$

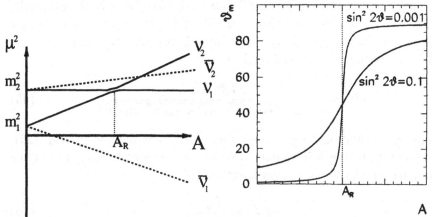

Figure 4. Effective masses (left) and mixing(right) acquired in the medium by a system of two massive neutrinos as a function of the potential A [see Eq. (41)].

The eigenvalues of M_w, that is, the effective masses in matter are given by [28, 32]:

$$\mu^2_{1,2} = \frac{m_1^2 + m_2^2}{2} + E(V_e + V_X) \mp \frac{1}{2}\sqrt{(\Delta m^2 \cos 2\theta - A)^2 + (\Delta m^2 \sin 2\theta)^2} ,$$

(41)

while the mixing angle in matter is given by

$$\tan 2\theta_m = \frac{\Delta m^2 \sin 2\theta}{\Delta m^2 \cos 2\theta - A}.$$

(42)

The quantity A is defined by

$$A \equiv 2E(V_e - V_X).$$

(43)

In Fig. 4 we plot the effective masses and the mixing angle in matter as functions of the potential A, for $A > 0$ and $\Delta m^2 \cos 2\theta > 0$. Notice that even massless neutrinos acquire non-vanishing effective masses in matter. The resonant density (or potential) A_R is defined as the value of A for which the difference between the effective masses is minimal:

$$A_R = \Delta m^2 \cos 2\theta .$$

(44)

The mixing angle $\tan \theta_m$ changes sign at A_R. As can be seen in Fig. 4, for $A > A_R$ we have $\theta_m \gg \theta$.

Notice that once the sign of $V_e - V_X$ (which depends on the composition of the medium and on the state X) is known, this resonance condition can only be achieved for a given sign of $\Delta m^2 \cos 2\theta$, *i.e.* for mixing angles in only one of the two possible octants. We learn that the symmetry present in vacuum oscillations is broken by matter potentials.

Also if the resonant condition is achieved for two neutrinos it cannot be achieved for antineutrinos of the same flavor and vice versa.

We define the oscillation length in matter:

$$L^{osc} = \frac{L_0^{osc} \Delta m^2}{\sqrt{(\Delta m^2 \cos 2\theta - A)^2 + (\Delta m^2 \sin 2\theta)^2}}, \tag{45}$$

where the oscillation length in vacuum, L_0^{osc}, was defined in Eq. (29). The oscillation length in matter displays a resonant behaviour. At the resonance point the oscillation length is

$$L_R^{osc} = \frac{L_0^{osc}}{\sin 2\theta}. \tag{46}$$

The width (in distance) of the resonance, δr_R, corresponding to $\delta A_R = 2\Delta m^2 \sin^2 2\theta$, and the resonance height h_R are given by

$$\delta r_R = \frac{\delta A_R}{|\frac{dA}{dr}|_R} = \frac{2 \tan 2\theta}{h_R}, \qquad h_R \equiv \left|\frac{1}{A}\frac{dA}{dr}\right|_R. \tag{47}$$

For constant A, *i.e.*, for constant matter density, the evolution of the neutrino system is described just in terms of the masses and mixing in matter. But for varying A, this is in general not the case.

2.5   Adiabatic and Non-adiabatic Transitions

Taking the time derivative of Eq. (40), we find:

$$\frac{\partial}{\partial t}\begin{pmatrix} \nu_e \\ \nu_X \end{pmatrix} = \dot{U}(\theta_m)\begin{pmatrix} \nu_1^m \\ \nu_2^m \end{pmatrix} + U(\theta_m)\begin{pmatrix} \dot{\nu}_1^m \\ \dot{\nu}_2^m \end{pmatrix}. \tag{48}$$

Using the evolution equation in the flavor basis, Eq. (38), we get

$$i\begin{pmatrix} \dot{\nu}_1^m \\ \dot{\nu}_2^m \end{pmatrix} = \frac{1}{2E}U^\dagger(\theta_m)M_w^2 U(\theta_m)\begin{pmatrix} \nu_1^m \\ \nu_2^m \end{pmatrix} - i\,U^\dagger \dot{U}(\theta_m)\begin{pmatrix} \nu_1^m \\ \nu_2^m \end{pmatrix}. \tag{49}$$

For constant matter density, θ_m is constant and the second term vanishes. In general, using the definition of the effective masses $\mu_i(t)$ in Eq. (41), and subtracting a diagonal piece $(\mu_1^2 + \mu_2^2)/2E \times I$, we can rewrite the evolution equation as:

$$i\begin{pmatrix} \dot{\nu}_1^m \\ \dot{\nu}_2^m \end{pmatrix} = \frac{1}{4E}\begin{pmatrix} -\Delta(t) & -4iE\dot{\theta}_m(t) \\ 4iE\dot{\theta}_m(t) & \Delta(t) \end{pmatrix}\begin{pmatrix} \nu_1^m \\ \nu_2^m \end{pmatrix}, \tag{50}$$

where we defined $\Delta(t) \equiv \mu_2^2(t) - \mu_1^2(t)$.

The evolution equations (50) constitute a system of coupled equations: the instantaneous mass eigenstates, ν_i^m, mix in the evolution and are not energy eigenstates. The importance of this effect is controlled by the relative size of the off-diagonal piece $4\,E\,\dot{\theta}_m(t)$ with respect to the diagonal one $\Delta(t)$. When $\Delta(t) \gg 4\,E\,\dot{\theta}_m(t)$, the instantaneous mass eigenstates, ν_i^m, behave approximately as energy eigenstates and they do not mix in the evolution. This is the *adiabatic* transition approximation. From the definition of θ_m in Eq. (42) we find that the adiabaticity condition can be expressed in terms of the adiabaticity parameter Q as

$$\frac{Q}{2} \equiv \frac{\Delta(t)}{4E\dot{\theta}_m(t)} = \frac{\Delta(t)^3}{2EA\Delta m^2 \sin 2\theta}\left|\frac{A}{\dot{A}}\right| \gg 1 \,. \tag{51}$$

For small mixing angles the maximum of θ_m occurs at the resonance point (as seen in Fig. 4) and the strongest adiabaticity condition is obtained when Eq. (51) is evaluated at the resonance

$$Q = \frac{\Delta m^2 \sin^2 2\theta}{E\cos 2\theta\, h_R} = \frac{2\,\pi\,\delta r_R}{L_R^{osc}} \,, \tag{52}$$

where we used the definitions of A_R, δr_R, and h_R in Eqs. (44) and (47). Written in this form, we see that the adiabaticity condition, $Q \gg 1$, implies that many oscillations take place in the resonant region. Conversely, when $Q \leq 1$ the transition is non-adiabatic. The generalization of the condition of maximum adiabaticity violation to large mixings can be found in Refs. [33, 34].

The survival amplitude of a ν_e produced in matter at t_0 and exiting the matter at $t > t_0$ can be written as follows:

$$A(\nu_e \rightarrow \nu_e; t) = \sum_{i,j} A(\nu_e(t_0) \rightarrow \nu_i(t_0))\, A(\nu_i(t_0) \rightarrow \nu_j(t))\, A(\nu_j(t) \rightarrow \nu_e(t)),$$

$$A(\nu_e(t_0) \rightarrow \nu_i(t_0)) = \langle \nu_i(t_0)|\nu_e(t_0)\rangle = U_{ei}^*(\theta_{m,0}),$$

$$A(\nu_j(t) \rightarrow \nu_e(t)) = \langle \nu_e(t)|\nu_j(t)\rangle = U_{ej}(\theta),$$

$$\tag{53}$$

where $U_{ei}^*(\theta_{m,0})$ [$U_{ej}(\theta)$] is the (ei) [(ej)] element of the mixing matrix in matter at the production point [in vacuum].

In the adiabatic approximation the mass eigenstates do not mix so

$$A(\nu_i(t_0) \rightarrow \nu_j(t)) = \delta_{ij}\,\langle \nu_i(t)|\nu_i(t_0)\rangle = \delta_{ij}\,\exp\left\{i\int_{t_0}^t E_i(t')dt'\right\} \,. \tag{54}$$

Note that E_i is a function of time because the effective mass μ_i is a function of time,

$$E_i(t') \simeq p + \frac{\mu_i^2(t')}{2p} \,. \tag{55}$$

Thus the transition probability for the adiabatic case is given by

$$P(\nu_e \to \nu_e; t) = \left| \sum_i U_{ei}(\theta) U_{ei}^{\star}(\theta_{m,0}) \exp\left(-\frac{i}{2E}\int_{t_0}^{t} \mu_i^2(t')dt'\right) \right|^2 . \quad (56)$$

For the case of two-neutrino mixing, Eq. (56) takes the form

$$P(\nu_e \to \nu_e; t) = \cos^2\theta_m \cos^2\theta + \sin^2\theta_m \sin^2\theta + \frac{1}{2}\sin 2\theta_m \sin 2\theta \cos\left(\frac{\delta(t)}{2E}\right) , \quad (57)$$

where

$$\delta(t) = \int_{t_0}^{t} \Delta(t')dt' = \int_{t_0}^{t} \sqrt{(\Delta m^2 \cos 2\theta - A(t'))^2 + (\Delta m^2 \sin 2\theta)^2}\, dt' ,$$

which, in general, has to be evaluated numerically. There are some analytical approximations for specific forms of $A(t')$: exponential, linear ... (see, for instance, Ref. [35]). For $\delta(t) \gg E$ the last term in Eq. (57) is averaged out and the survival probability takes the form

$$P(\nu_e \to \nu_e; t) = \frac{1}{2}\left[1 + \cos 2\theta_m \cos 2\theta\right]. \quad (58)$$

In Fig. 5 we plot isocontours of constant survival probability in the parameter plane $(\Delta m^2, \tan^2\theta)$ for the particular case of the Sun density for which $A > 0$. Notice that, unlike $\sin^2 2\theta$, $\tan^2\theta$ is a single valued function in the full parameter range $0 \le \theta \le \pi/2$. Therefore it is a more appropriate variable once matter effects are included and the symmetry of the survival probability with respect to the change of octant for the mixing angle is lost. As seen in the figure, for $\theta < \pi/4$, $P(\nu_e \to \nu_e)$ in matter can be larger or smaller than $1/2$, in contrast to the case of vacuum oscillations where, in the averaged regime, $P_{ee}^{vac} = 1 - \frac{1}{2}\sin^2 2\theta > \frac{1}{2}$. In Fig. 5 we also plot the limiting curve for $Q = 1$. To the left and below of this curve the adiabatic approximation breaks down and the isocontours in Fig. 5 deviate from the expression in Eq. (58). In this region, the off-diagonal term $\dot{\theta}_m$ cannot be neglected and the mixing between instantaneous mass eigenstates is important. In this case we can write

$$A(\nu_i(t_0) \to \nu_j(t)) = \langle\nu_j(t)|\nu_j(t_R)\rangle\langle\nu_j(t_R)|\nu_i(t_R)\rangle\langle\nu_i(t_R)|\nu_i(t_0)\rangle, \quad (59)$$

where t_R is the point of maximum adiabaticity violation, which, for small mixing angles, corresponds to the resonant point. The possibility of this *level crossing* can be described in terms of the Landau-Zener probability [36]:

$$P_{LZ} = |\langle\nu_j(t_R)|\nu_i(t_R)\rangle|^2 \quad (i \ne j) . \quad (60)$$

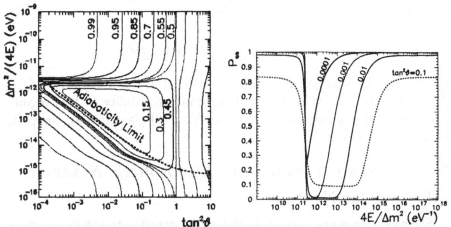

Figure 5. (Left) Isocontours of the survival probability P_{ee} in the Sun. Also shown is the limit of applicability of the adiabatic approximation $Q = 1$ (dashed line). (Right) The survival probability for a ν_e state produced in the center of the Sun as a function of $E/\Delta m^2$ for various values of the mixing angle.

Introducing this transition probability in Eq. (53) we find that in the non-adiabatic regime (after averaging out the oscillatory term), the survival probability can be written as

$$P(\nu_e \to \nu_e; t) = \frac{1}{2}\left[1 + (1 - 2P_{LZ})\cos 2\theta_m \cos 2\theta\right] . \qquad (61)$$

The physical interpretation of this expression is as follows. An electron neutrino produced at $A > A_R$ consists of an admixture of ν_1 with fraction $\cos^2 \theta_m$ and ν_2 with fraction $\sin^2 \theta_m$. In particular, for very small mixing angles in vacuum, $\theta_m \sim \pi/2$ (see Fig. 4) so ν_e is almost a pure $\nu_2(t_0)$ state. When the neutrino state reaches the resonance, ν_2 (ν_1) can jump to ν_1 (ν_2) with probability P_{LZ} or remain as ν_2 (ν_1) with probability $[1 - P_{LZ}]$. So after passing the resonance, the ν_e flux contains a fraction of ν_1: $P_{e1} = \sin^2 \theta_m P_{LZ} + \cos^2 \theta_m (1 - P_{LZ})$, and a fraction of ν_2: $P_{e2} = \cos^2 \theta_m P_{LZ} + \sin^2 \theta_m (1 - P_{LZ})$. At the exit ν_1 consists of ν_e with fraction $\cos^2 \theta$ and ν_2 consists of ν_e with fraction $\sin^2 \theta$ so [37, 38, 39] $P_{ee} = \cos^2 \theta P_{e1} + \sin^2 \theta P_{e2}$, which reproduces Eq. (61).

The Landau-Zener probability can be evaluated in the WKB approximation. The general form of the Landau-Zener probability for an exponential density can be written as [40, 41]:

$$P_{LZ} = \frac{\exp(-\gamma\sin^2 \theta) - \exp(-\gamma)}{1 - \exp(-\gamma)}, \quad \gamma \equiv \pi \frac{\Delta m^2}{E|\dot{A}/A|_R} = \pi Q \frac{\cos 2\theta}{\sin^2 2\theta}. \qquad (62)$$

When ν_e is produced at $A \gg A_R$ and θ is small, $\theta_m \sim 90°$. In this case γ is very large and $P_{LZ} \simeq \exp(-\gamma\sin^2 \theta) \simeq \exp\left(-\frac{\pi}{4}Q\right)$ and the

66

survival probability is simply given by

$$P(\nu_e \to \nu_e; t) \simeq P_{LZ} \simeq \exp\left(-\frac{\pi}{4}Q\right)$$

where Q is the adiabaticity parameter defined in Eq. (52). Since $Q \sim \Delta m^2 \sin^2 2\theta / E$, the isocontours of constant probability in this regime correspond to diagonal lines in the $(\Delta m^2, \tan^2\theta)$ plane in a log-log plot, as illustrated in Fig. 5.

3.   Lecture 3: Solar and Atmospheric Neutrinos

3.1   Solar Neutrinos

Solar neutrinos are electron neutrinos produced in the thermonuclear reactions that generate the solar energy. These reactions occur via two main chains, the pp-chain and the CNO cycle. There are five reactions which produce ν_e in the pp chain and three in the CNO cycle.   In Fig. 6 we show the energy spectrum of the fluxes from the five pp chain reactions. Both chains result in the overall fusion of protons into ^{4}He:

$$4p \;\to\; {}^4\text{He} + 2e^+ + 2\nu_e + \gamma, \tag{63}$$

where the energy released in the reaction, $Q = 4m_p - m_{^4\text{He}} - 2m_e \simeq 26$ MeV, is mostly radiated through the photons with only a small fraction carried by the neutrinos, $\langle E_{2\nu_e} \rangle = 0.59$ MeV.

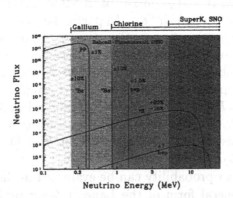

Figure 6.   Energy dependence of neutrino fluxes from the pp chain reactions.

In what follows we refer to the neutrino fluxes by the corresponding source reaction, so, for instance, the neutrinos produced from ^{8}B decay are called ^{8}B neutrinos. Most reactions produce a neutrino spectrum characteristic of β decay. For ^{8}B neutrinos the energy distribution presents deviations with respect to the maximum allowed energy because the final state, ^{8}Be, is a wide resonance. On the other hand, the ^{7}Be neutrinos are almost monochromatic, with an energy width of about 2 keV which is characteristic of the temperature in the core of the Sun.

In order to precisely determine the rates of the different reactions in the two chains which would give us the final neutrino fluxes and their energy spectrum, a detailed knowledge of the Sun and its evolution is needed. Solar Models describe the properties of the Sun and its evolution after entering the main sequence. The models are based on a set of observational parameters: the surface luminosity (3.844×10^{26} W), the age (4.5×10^9 years), the radius (6.961×10^8 m) and the mass (1.989×10^{30} kg), and on several basic assumptions: spherical symmetry, hydrostatic and thermal equilibrium, equation of state of an ideal gas, and present surface abundances of elements similar to the primordial composition. We use as Standard Solar Model (SSM) the most updated version of the model developed by Bahcall, Pinsonneault and Basu [53].

To describe the evolution of neutrinos in the solar matter, one needs to know other quantities that are predicted by the SSM, such as the density and composition of solar matter. Furthermore, in order to precisely determine the evolution of the neutrino system one also needs to know the production point distribution for the different neutrino fluxes.

3.1.1   Experiments.

• **Chlorine experiment: Homestake** The first result on the detection of solar neutrinos was announced by Ray Davis Jr and his collaborators from Brookhaven in 1968 [54]. In the gold mine of Homestake in Lead, South Dakota, they installed a detector consisting of $\sim$ 615 Tons of C_2Cl_4. Solar ν_e's are captured via

$$^{37}\text{Cl} \, (\nu, e^-) \, ^{37}\text{Ar}.$$

The energy threshold for this reaction is 0.814 MeV, so the relevant fluxes are the ^{7}Be and ^{8}B neutrinos. For the SSM fluxes, 78% of the expected number of events are due to ^{8}B neutrinos while 13% arise from ^{7}Be neutrinos. The produced ^{37}Ar is extracted radiochemically about every three months and the number of ^{37}Ar decays ($t_{\frac{1}{2}}$=34.8 days) is measured in a proportional counter.

The average event rate measured during the more than 20 years of operation is

$$R_{\text{Cl}} = 2.56 \pm 0.16 \pm 0.16 \, \text{SNU} \Rightarrow \frac{R_{\text{Cl}}}{\text{BP00}} = 0.337 \pm 0.065 \qquad (64)$$

(1 SNU $= 10^{-36}$ captures/atom/sec).

• Gallium experiments: SAGE and GALLEX/GNO: In January 1990 and May 1991, two new radiochemical experiments using a ^{71}Ga target started taking data, SAGE [56] and GALLEX [55]. The SAGE detector is located in Baksan, Kaberdino-Balkaria, Russia, with 30 Tons (increased to 57 Tons from July 1991) of liquid metallic Ga. GALLEX is located in Gran Sasso, Italy, and consists of 30 Tons of GaCl$_3$-HCl. In these experiments the solar neutrinos are captured via

$$^{71}\text{Ga}(\nu, e^-)^{71}\text{Ge}.$$

This target has a low threshold (0.233 MeV) and a strong transition to the ground level of ^{71}Ge, which gives a large cross section for the lower energy pp neutrinos. According to the SSM, approximately 54% of the events are due to pp neutrinos, while 26% and 11% arise from ^{7}Be and ^{8}B neutrinos, respectively. The extraction of ^{71}Ge takes places every 3–4 weeks and the number of ^{71}Ge decays ($t_{\frac{1}{2}}$=11.4 days) is measured in a proportional counter. The GALLEX program was completed in fall 1997 and its successor GNO started taking data in spring 1998.

The event rates measured by SAGE and GALLEX+GNO are

$$\begin{aligned}
R_{\text{GALLEX+GNO}} &= 70.8 \pm 5.9\,\text{SNU}\,, \\
R_{\text{SAGE}} &= 70.8^{+5.3}_{-5.2}\,{}^{+3.7}_{-3.2}\,\text{SNU}\,,
\end{aligned} \qquad \Rightarrow \frac{R_{\text{Ga}}}{\text{BP00}} = 0.55 \pm 0.048\,. \quad (65)$$

The pp flux is directly constrained by the solar luminosity, and in all stationary solar models there is a theoretical minimum of the expected number of events of 79 SNU.

• Water Cerenkov: Kamiokande and SuperKamiokande:

Kamiokande [57] and its successor SuperKamiokande (SK) [58] in Japan are water Cerenkov detectors able to detect in real time the electrons scattered from the water by elastic interaction of the solar neutrinos,

$$\nu_a + e^- \to \nu_a + e^-\,, \qquad (66)$$

The scattered electrons produce Cerenkov light which is detected by photomultipliers. Notice that, while the detection process in radiochemical experiments is purely a CC (W-exchange) interaction, the detection process of Eq. (66) goes through both CC and NC (Z-exchange) interactions. Consequently, the detection process (66) is sensitive to all active neutrino flavors, although ν_e's (which are the only ones to scatter via W-exchange) give a contribution that is about 6 times larger than that of ν_μ's or ν_τ's.

Kamiokande, with 2140 tons of water, started taking data in January 1987 and was terminated in February 1995. SuperKamiokande, with 45000 tons of water (of which 22500 are usable in solar neutrino measurements) started in May 1996 and it has analyzed so far the events corresponding to 1258 days. The detection threshold in Kamiokande was 7.5 MeV while SK's later runs were at 5 MeV. This means that these experiments are able to measure only the ^{8}B neutrinos (and the very small hep neutrino flux). Their results are presented in terms of the measured ^{8}B flux.

$$\Phi_{Kam} = (2.80 \pm 0.19 \pm 0.33) \times 10^6 \text{ cm}^{-2}\text{s}^{-1},$$

$$\Phi_{SK} = (2.35 \pm 0.02 \pm 0.08) \times 10^6 \text{ cm}^{-2}\text{s}^{-1} \Rightarrow \frac{\Phi_{SK}}{\Phi_{BP00}} = 0.465 \pm 0.094.$$

There are three features unique to the water Cerenkov detectors. First, they are real time experiments. Each event is individually recorded. Second, for each event the scattered electron keeps the neutrino direction within an angular interval which depends on the neutrino energy as $\sqrt{2m_e/E_\nu}$. Thus, it is possible, for example, to correlate the neutrino detection with the position of the Sun. Third, the amount of Cerenkov light produced by the scattered electron allows a measurement of its energy. In summary, the experiment provides information on the time, direction and energy for each event. Signatures of neutrino oscillations might include distortion of the recoil electron energy spectrum, difference between the night-time solar neutrino flux and the day-time flux, or a seasonal variation in the neutrino flux. Observation of these effects would be strong evidence in support of solar neutrino oscillations independent of absolute flux calculations. Conversely, non-observation of these effects can constrain oscillation solutions to the solar neutrino problem.

Over the years the SK collaboration has presented the information on the energy and time dependence of their event rates in different forms. In Fig. 7 we show their spectrum, corresponding to 1258 days of data, relative to the predicted spectrum. As seen in the figure they do not observe any significant energy dependence of their event rates beyond the expected one in the SSM. No time dependence is also observed. For example they measure a day-night asymmetry,

$$A_{N-D} \equiv 2\frac{N - D}{D + N} = 0.021 \pm 0.020(\text{stat.}) \pm 0.013(\text{syst.}), \qquad (67)$$

where D (N) is the event rate during the day (night) period. The SK results show a small excess of events during the night but only at the 0.8 σ level.

• **SNO:** The Sudbury Neutrino Observatory (SNO) was first proposed in 1987 and it started taking data in November 1999 [59]. The detector, a great sphere surrounded by photomultipliers, contains approximately 1000 Tons of heavy water, D_2O, and is located at the Creighton mine,

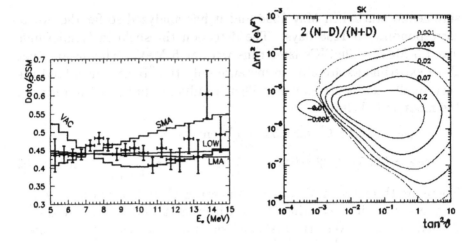

Figure 7. (Left) The electron recoil energy spectrum measured in SK normalized to the SSM prediction, and the expectations for the best fit points for the LMA, SMA, LOW and VAC solutions from Fig. 8. (Right) Isocontours of the day-night asymmetry at SK.

near Sudbury in Canada. SNO was designed to give a model independent test of the possible explanations of the observed deficit in the solar neutrino flux by having sensitivity to all flavors of active neutrinos and not just to ν_e. This sensitivity is achieved because energetic neutrinos can interact in the D_2O of SNO via three different reactions. Electron neutrinos may interact via the CC reaction

$$\nu_e + d \rightarrow p + p + e^- , \qquad (68)$$

and can be detected above an energy threshold of a few MeV (presently $T_e > 5$ MeV). All active neutrinos $(\nu_a = \nu_e, \nu_\mu, \nu_\tau)$ interact via the NC reaction

$$\nu_a + d \rightarrow n + p + \nu_a , \qquad (69)$$

with an energy threshold of 2.225 MeV. The non-sterile neutrinos can also interact via Elastic Scattering (ES), $\nu_a + e^- \rightarrow \nu_a + e^-$, but with smaller cross section.

SNO can also perform measurements of the energy spectrum and time variation of the event rates. But the uniqueness of SNO lies in its ability to directly test if the deficit of solar ν_e is due to changes in the flavor composition of the solar neutrino beam, since the ratio CC/NC compares the number of ν_e interactions with those from all active flavors. This comparison is independent of the overall flux normalization.

In June 2001, SNO published their first results on the CC measurement and in April 2002 [59] they published their results from the first phase of the experiment which include the day-night spectrum data in the full energy range above a threshold $T_e > 5$ MeV. Their spectrum is the result of the combination of the three possible signals. Assuming an undistorted energy spectrum they extract the individual rates:

$$\Phi_{SNO}^{CC} = (1.76_{-0.05}^{+0.06} \pm 0.09) \times 10^6 \text{ cm}^{-2}\text{s}^{-1} \Rightarrow \frac{\Phi_{SNO}^{CC}}{\Phi_{BP00}} = 0.348 \pm 0.073,$$

$$\Phi_{SNO}^{ES} = (2.39 \pm 0.24 \pm 0.12) \times 10^6 \text{ cm}^{-2}\text{s}^{-1},$$

$$\Phi_{SNO}^{NC} = (5.09_{-0.43}^{+0.44}\,_{-0.43}^{+0.46}) \times 10^6 \text{ cm}^{-2}\text{s}^{-1} \Rightarrow \frac{\Phi_{SNO}^{NC}}{\Phi_{BP00}} = 1.01 \pm 0.23$$

and a day–night asymmetry $A_{N-D} = 0.07 \pm 0.049$(stat.) ± 0.013(syst.). Notice that the NC measured flux Φ_{SNO}^{NC} is in excellent agreement with the prediction of the SSM.

3.1.2   The Solar Neutrino Problem.

From the experimental results we can conclude that:

- Before the NC measurement at SNO, all experiments observed a flux that was smaller than the SSM predictions, $\Phi^{obs}/\Phi^{SSM} \sim 0.3 - 0.6$.

- The deficit is not the same for the various experiments, which may indicate that the effect is energy dependent.

These two statements constitute the solar neutrino problem [60].

The results of SNO have provided further model independent evidence of the problem. Both SNO and SK are sensitive mainly to the ^{8}B flux. Without NP, the measured fluxes in any reaction at these two experiments should be equal. Schematically, in presence of flavour conversion

$$\begin{aligned}
\Phi^{CC} &= \Phi_e, \\
\Phi^{ES} &= \Phi_e + r\,\Phi_{\mu\tau}, \\
\Phi^{NC} &= \Phi_e + \Phi_{\mu\tau},
\end{aligned} \tag{70}$$

where $r \equiv \sigma_\mu/\sigma_e \simeq 0.15$ is the ratio of the the $\nu_e - e$ and $\nu_\mu - e$ elastic scattering cross-sections. The flux $\Phi_{\mu\tau}$ of active non-electron neutrinos is zero in the SSM. Thus, the three observed rates should be equal.

The first reported SNO CC result compared with the ES rate from SK showed that the hypothesis of no flavour conversion was excluded at $\sim 3\sigma$. Finally, with the NC measurement at SNO one finds that

$$\Phi_{SNO,\mu\tau} \equiv \Phi_{SNO}^{NC} - \Phi_{SNO}^{CC} = (3.41 \pm 0.45^{+0.48}_{-0.45}) \times 10^6 \text{ cm}^{-2}\text{s}^{-1}. \quad (71)$$

This result provides evidence for neutrino flavour transition (from ν_e to $\nu_{\mu,\tau}$) at the level of 5.3σ. This evidence is independent of the solar model.

3.1.3   Oscillations in Vacuum and in Matter: MSW Effect.

The most generic explanation of the solar neutrino anomaly is oscillations of ν_e into an active (ν_μ and/or ν_τ) or a sterile (ν_s) neutrino.

Solar neutrinos can oscillate "in vaccum" on their way from the Sun to the Earth. In this case the corresponding survival probability is given by Eq. (32). Given the distance between the Sun and the Earth ($\sim 10^{10}$ m) the explanation of the observed deficit requires $\Delta m^2 \sim 10^{-10}$ eV^2 and large mixing angle $\sin^2(2\theta) \sim 1$.

The propagation of solar neutrinos can also be affected by matter effects on their way out from the production point inside the core of Sun as we describe next.

Let's consider the propagation of a $\nu_e - \nu_X$ neutrino system in the matter density of the Sun where X is some superposition of μ and τ. The solar density distribution decreases monotonically with the distance R from the center of the Sun . For $R < 0.9 R_\odot$ the density can be approximated by an exponential $N_e(R) = N_e(0) \exp(-R/r_0)$, with $r_0 = R_\odot/10.54 = 6.6 \times 10^7$ m $= 3.3 \times 10^{14}$ eV^{-1}. Given the characteristic solar neutrino energies and the solar density, matter effects in the Sun are most important for $\Delta m^2 \sim 10^{-4}$–10^{-8} eV^2, for which the interference effects leading to vaccum oscillations between the Sun and the Earth are averaged out.

Thus after traversing this density the dominant component of the exiting neutrino state depends on the value of the mixing angle in vacuum and the relative size of $\Delta m^2 \cos 2\theta$ versus $A_0 = 2 E G_F N_{e,0}$ (at the neutrino production point)

(i) $\Delta m^2 \cos 2\theta \gg A_0$: matter effects are negligible and the propagation occurs as in vacuum. The survival probability at the sunny surface of the Earth is

$$P_{ee}(\Delta m^2 \cos 2\theta \gg A_0) = 1 - \frac{1}{2}\sin^2 2\theta > \frac{1}{2}, \quad (72)$$

which corresponds to Eq. (32) with the oscillatory term averaged to $1/2$.

(ii) $\Delta m^2 \cos 2\theta > A_0$: the neutrino does not pass the resonance but its mixing is affected by the matter. This effect is well described by an adiabatic propagation:

$$P_{ee}(\Delta m^2 \cos 2\theta \geq A_0) = \frac{1}{2}[1 + \cos 2\theta_m \cos 2\theta] \ . \qquad (73)$$

Since the resonance is not crossed, $\cos 2\theta_m$ has the same sign as $\cos 2\theta$ and the corresponding survival probability is also larger than $1/2$.

(iii) $\Delta m^2 \cos 2\theta < A_0$: the neutrino can cross the resonance on its way out. In this case for small mixing angle in vacuum, $\nu_e \sim \nu_2^m$ at the production point and remains ν_2^m till the resonance point (for larger mixing but still in the first octant ν_e is a combination of ν_1^m and ν_2^m with larger ν_2^m component). It is important in this case to find whether the transition is adiabatic. For the solar density, $Q \sim 1$ corresponds to

$$\frac{(\Delta m^2/\text{eV}^2)\sin^2 2\theta}{(E/\text{MeV})\cos 2\theta} \sim 3 \times 10^{-9} \ . \qquad (74)$$

For $Q \gg 1$ the transition is adiabatic and the neutrino state remains in the same linear combination of mass eigenstates after the resonance determined by θ_m. As seen in Fig. 4, θ_m (that is, the ν_e component of the state) decreases after crossing the resonance and, consequently, so does the survival probability P_{ee}. In particular, for small mixing angle, ν_2 at the exit point is almost a pure ν_X and, consequently, P_{ee} can be very small. Explicitly,

$$P_{ee}(\Delta m^2 \cos 2\theta < A_0, Q \gg 1) = \frac{1}{2}[1 + \cos 2\theta_{m,0} \cos 2\theta] \qquad (75)$$

can be much smaller than $1/2$ because $\cos 2\theta_{m,0}$ and $\cos 2\theta$ can have opposite signs. Note that the smaller the mixing angle in vacuum the larger is the deficit of electron neutrinos in the outgoing state. This is the MSW effect [28, 32]. This behaviour is illustrated in Fig. 5 where we plot the electron survival probability as a function of $\Delta m^2/E$ for different values of the mixing angle.

For smaller values of $\Delta m^2/E$ (right side of Fig. 5) we approach the regime where $Q < 1$ and non-adiabatic effects start playing a role. In this case the state can jump from ν_2 into ν_1 (or vice versa) with probability P_{LZ}. For small mixing angle, at the surface $\nu_1 \sim \nu_e$ and the ν_e component of the exiting neutrino increases. This can be seen from the expression for P_{ee},

$$P_{ee}(\Delta m^2 \cos 2\theta < A_0, Q \ll 1) = \frac{1}{2}[1 + (1 - 2P_{LZ})\cos 2\theta_m \cos 2\theta] , \qquad (76)$$

and from Fig. 5. For large mixing angles this expression is still valid.

For $\Delta m^2 \sim 10^{-8}$–10^{-10} eV^2 both the propagation inside the Sun and between the Sun and the Earth play a role and must be taken into account in the computation of the ν_e survival probability.

Finally one must also take into account the effect of propagation within the Earth when the neutrinos are detected at night. Given the Earth's matter potential (see Sec. 2.3) those effects are important for $\cos 2\theta \Delta m^2 \sim 10^{-6}$ eV$^2 \frac{E}{\text{MeV}}$. For most of the relevant oscillation parameters, Earth matter effects lead to the regeneration of the ν_e component resulting in an increase of the survival probability.

3.1.4   Two-Neutrino Oscillation Analysis.

The goal of the analysis of the solar neutrino data in terms of neutrino oscillation is to determine which range of mass-squared difference and mixing angle can be responsible for the observed deficit [61]. In order to answer this question in a statistically meaningful way one must compare the predictions in the different oscillation regimes with the observations, including all the sources of uncertainties and their correlations.

The expected event rate in the presence of oscillations in the experiment i, R_i^{th}, can be written as follows:

$$R_i^{\text{th}} = \sum_{k=1,8} \phi_k \int dE_\nu \, \lambda_k(E_\nu) \times [\sigma_{e,i} \langle P_{ee}(E_\nu, t) \rangle + \sigma_{x,i}(1 - \langle P_{ee}(E_\nu, t) \rangle)],$$

(77)

where E_ν is the neutrino energy, ϕ_k is the total neutrino flux and λ_k is the neutrino energy spectrum (normalized to 1) from the solar nuclear reaction k. $\sigma_{e,i}$ ($\sigma_{x,i}$) is the ν_e (ν_x, $x = \mu$, τ) interaction cross section in the SM with the target corresponding to experiment i, and $\langle P_{ee}(E_\nu, t) \rangle$ is the time-averaged ν_e survival probability. The expected signal in the absence of oscillations, R_i^{BP00}, can be obtained from Eq. (77) by substituting $P_{ee} = 1$. For the Chlorine, Gallium and SNO(CC) measurements, only the electron neutrino contributes and the $\sigma_{x,i}$-term in Eq. (77) vanishes. For ES at SK or SNO there is a possible contribution from the NC interaction of the other active neutrino flavors present in the beam. For the NC rate at SNO all active flavours contribute equally.

The main sources of uncertainty are the theoretical errors in the prediction of the solar neutrino fluxes for the different reactions. These errors are due to uncertainties in the twelve basic ingredients of the solar model, which include the nuclear reaction rates (parametrized in terms of the astrophysical factors S_{11}, S_{33}, S_{34}, $S_{1,14}$ and S_{17}), the solar luminosity, the metalicity Z/X, the Sun age, the opacity, the diffusion, and the electronic capture of ^{7}Be, C_{Be}. Another source of theoretical error arises from the uncertainties in the neutrino interaction cross section for the different detection processes. For a detailed description of the way to include all these uncertainties and correlations see Ref. [11] and references therein.

The results of the analysis of the total event rates are shown in Fig. 8 where we plot the allowed regions, which correspond to 90%, 95%, 99% and 99.73% (3σ) CL for ν_e oscillations into active neutrinos. As seen in the figure, for oscillations into active neutrinos there are several oscillation regimes which are compatible within errors with the experimental data. These allowed parameter regions can be classified according to the regions for which oscillations occur in vaccum (VAC) and regions for which the flavour conversion occurs due to matter effects in the Sun (MSW regions). There are three allowed MSW regions:

- *MSW small mixing angle* (SMA): within this regime the resonant MSW effect is most relevant and leads to the almost total suppression of the ^{7}Be neutrinos. The survival probability for ^{8}B neutrinos is a steeply increasing function of the energy.

- *MSW large mixing angle* (LMA): oscillations for the ^{8}B neutrinos occur in the adiabatic regime and the survival probability increases for the ^{7}Be and pp neutrinos. This fits well the higher rate observed at gallium experiments. Matter effects in the Earth are important for ^{8}B neutrinos if $\Delta m^2 \leq$ few 10^{-5} eV^2.

- *MSW low mass* (LOW): in this regime the survival probability is larger for the ^{8}B neutrinos and smaller the lower energy ^{7}Be and pp neutrinos. Matter effects in the Earth for pp and ^{7}Be neutrinos enhance the average annual survival probability allowing to fit the rate observed at gallium experiments.

Before including the SNO(CC) data, the best fit corresponded to SMA solution, but after SNO the best fit corresponds to LMA.

Oscillations into pure sterile neutrinos are strongly disfavoured by the SNO data since if the beam is comprised of only ν_e's and ν_s's, the three observed CC, ES and NC rates should be equal (up to effects due to spectral distortions), an hypothesis which is now ruled out at $\sim$ 5-sigma by the SNO data. Oscillations into an admixture of active and sterile states are still allowed provided that the ^{8}B neutrino flux is allowed to be larger than the SSM expectation [62].

Further information on the different oscillation regimes can be obtained from the analysis of the energy and time dependence data from real time experiments. In Fig.7 we show the predicted energy spectrum at SK for the different allowed solutions from the analysis of the total rates. For LMA and LOW, the expected spectrum is very little distorted.

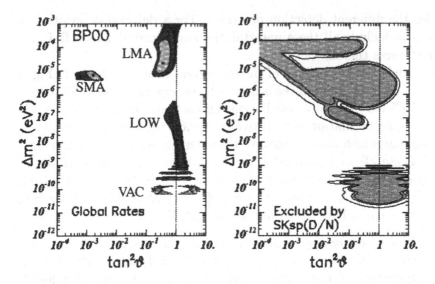

Figure 8. (Left) Allowed oscillation parameters (at 90, 95, 99 and 99.7% CL) from the analysis of the total event rates of the Chlorine, Gallium, SK and SNO CC experiments. The best fit point is marked with a star. (Right) Excluded oscillation parameters at 95, 99 (shadowed region) and 99.7% CL from the analysis of the day-night spectrum data.

For SMA, a positive slope is expected, with larger slope for larger mixing angle within SMA. For VAC, large distortions are expected. The details are dependent on the precise values of the oscillation parameters. Also shown in Fig.7 are the predicted values of the day-night asymmetry at SK in the oscillation parameter plane.

The observed day-night spectrum in SK is essentially undistorted in comparison to the SSM expectation and shows no significant differences between the day and the night periods. Consequently, a large region of the oscillation parameter space where these variations are expected to be large can be excluded. In Fig. 8 we show the excluded regions at the 99% CL from the analysis of SK day-night spectrum data, together with the contours corresponding to the 95% and 99.73% (3σ) CL. In particular, the central region ($2 \times 10^{-5} < \Delta m^2 < 3 \times 10^{-7}$, $\tan^2\theta > 3 \times 10^{-3}$) is excluded due to the small observed day-night variation (compare with Fig. 7). The rest of the excluded region is due to the absence of any observed distortion of the energy spectrum.

From Fig. 8 we can deduce the main consequences of adding the day-night spectrum information to the analysis of the total event rates:

- SMA: within this region, the part with larger mixing angle fails to comply with the observed energy spectrum, while the part with smaller mixing angles gives a bad fit to the total rates.

- VAC: the observed flat spectrum cannot be accommodated.

- LMA and LOW: the small Δm^2 part of LMA and the large Δm^2 part of LOW are reduced because they predict a day-night variation that is larger than observed. Both active LMA and active LOW solutions predict a flat spectrum in agreement with the observation.

We show in Fig. 9 the allowed regions which correspond to 90%, 95%, 99% and 99.73% (3σ) CL for ν_e oscillations into active neutrinos from the global analysis of Ref. [63].

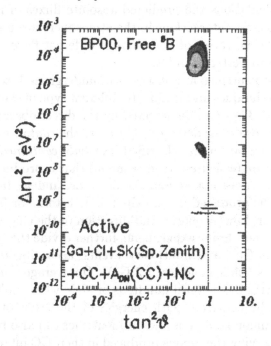

Figure 9. Allowed oscillation parameters (at 90, 95, 99 and 99.7% CL) from the global analysis of the solar neutrino data. The best fit point (LMA active) is marked with a star.

The results show that at present the most favoured solution is the LMA oscillation while the LOW solution provides a worse fit. There are some small allowed *islands* for VAC oscillations. The active SMA solution does not appear at 3σ as a consequence of the incompatibility between the observed small CC rate at SNO, which would favour larger mixing, and the flat spectrum which prefers smaller mixing. Oscillations of solar neutrinos into pure sterile state are also disfavoured at the $\sim 5\sigma$ level.

3.2   Atmospheric Neutrinos

Cosmic rays interacting with the nitrogen and oxygen in the Earth's atmosphere at an average height of 15 kilometers produce mostly pions and some kaons that decay into electron and muon neutrinos and antineutrinos. Since ν_e is produced mainly from the decay chain $\pi \to \mu\nu_\mu$ followed by $\mu \to e\nu_\mu\nu_e$, one naively expects a 2 : 1 ratio of ν_μ to ν_e. (For higher energy events the expected ratio is smaller because some of the muons arrive at the Earth's surface before they decay.) In practice, however, the theoretical calculation of the ratio of muon-like interactions to electron-like interactions in each experiment is more complicated. In the present calculations the predicted absolute fluxes of neutrinos produced by cosmic-ray interactions in the atmosphere are uncertain at the 20% level while the ratios of neutrinos of different flavor are, expected to be accurate to better than 5%.

Atmospheric neutrinos are observed in underground experiments using different techniques and leading to different type of events depending on their energy. They can be detected by the direct observation of their CC interaction inside the detector. These are the *contained* events. Contained events can be further classified into *fully contained* events, when the charged lepton (either electron or muon) that is produced in the neutrino interaction does not escape the detector, and *partially contained* muons, when the produced muon exits the detector. For fully contained events the flavor, kinetic energy and direction of the charged lepton can be best determined. Some experiments further divide the contained data sample into sub-GeV and multi-GeV events, according to whether the visible energy is below or above 1.2 GeV. On average, sub-GeV events arise from neutrinos of several hundreds of MeV while multi-GeV events are originated by neutrinos with energies of the order of several GeV. Higher energy muon neutrinos and antineutrinos can also be detected indirectly by observing the muons produced in their CC interactions in the vicinity of the detector. These are the so called *upgoing muons*. Should the muon stop inside the detector, it is classified as a *stopping muon* (which arises from neutrinos $E_\nu \sim 10$ GeV), while if the muon track crosses the full detector the event is classified as a *through-going muon* (which is originated by neutrinos with energies of the order of hundreds of GeV). Downgoing muons from ν_μ interactions above the detector cannot be distinguished from the background of cosmic ray muons. Higher energy ν_e's cannot be detected this way as the produced e showers immediately in the rock. In Fig. 10 we display the characteristic neutrino energy distribution for these different type of events.

Atmospheric neutrinos were first detected in the 1960's by the underground experiments in South Africa [42] and the Kolar Gold Field experiment in India [43]. These experiments measured the flux of horizontal muons (they could not discriminate between downgoing and upgoing directions) and although the observed total rate was not in full agreement with theoretical predictions the effect was not statistically significant.

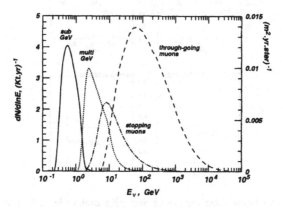

Figure 10. Event rates as a function of neutrino energy for fully contained events, stopping muons, and through-going muons at SuperKamiokande.

A set of modern experiments were proposed and built in the 1970's and 1980's. The original purpose was to search for nucleon decay, for which atmospheric neutrinos constitute background. Two different detection techniques were employed. In water Cerenkov detectors the target is a large volume of water surrounded by photomultipliers which detect the Cerenkov-ring produced by the charged leptons. The event is classified as an electron-like (muon-like) event if the ring is diffuse (sharp). In iron calorimeters, the detector is composed of a set of alternating layers of iron which act as a target and some tracking element (such as plastic drift tubes) which allows the reconstruction of the shower produced by the electrons or the tracks produced by muons. Both types of detectors allow for flavor classification of the events as well as the measurement of the scattering angle of the outgoig charged lepton and some determination of its energy.

The two oldest iron calorimeter experiments, Fréjus [44] and NU-SEX [45], found atmospheric neutrino fluxes in agreement with the theoretical predictions. On the other hand, two water Cerenkov detec-

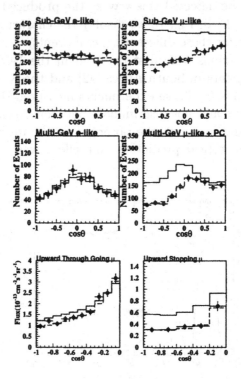

Figure 11. Zenith angle distribution of SuperKamiokande 1289 days data samples. Dots, solid line and dashed line correspond to data, MC with no oscillation and MC with best fit oscillation parameters, respectively.

tors, IMB [46]and Kamiokande, detected a ratio of ν_μ-induced events to ν_e-induced events smaller than the expected one by a factor of about 0.6. Kamiokande performed separate analyses for both sub-GeV neutrinos and multi-GeV neutrinos [47], which showed the same deficit. This was the original formulation of the atmospheric neutrino anomaly. Whether $R_{\mu/e}/R_{\mu/e}^{MC}$ is small because there is ν_μ disappearance or ν_e appearance or a combination of both could not be determined. Furthermore, the fact that the anomaly appeared only in water Cerenkov and not in iron calorimeters left the window open for the suspicion of a possible systematic problem as the origin of the effect.

Kamiokande also presented the zenith angular dependence of the deficit for the multi-GeV neutrinos. The zenith angle, parametrized in terms of $\cos\theta$, measures the direction of the reconstructed charged lepton with respect to the vertical of the detector. Vertically downgoing (upgoing) particles correspond to $\cos\theta = +1(-1)$. Horizontally arriving particles come at $\cos\theta = 0$. Kamiokande results seemed to indicate that the deficit was mainly due to the neutrinos coming from below the horizon. Atmospheric neutrinos are produced isotropically at a distance of about 15 km above the surface of the Earth. Therefore neutrinos coming from the top of the detector have traveled approximately those 15 kilometers before interacting while those coming from the bottom of the detector have traversed the full diameter of the Earth, $\sim 10^4$ Km before reaching the detector. The Kamiokande distribution suggested that the deficit increases with the distance between the neutrino production and interaction points.

In the last five years, the case for the atmospheric neutrino anomaly has become much stronger with the high precision and large statistics data from SK [48] and it has received important confirmation from the iron calorimeter detectors Soudan2 [49] and MACRO [50]. In June 1998, in the Neutrino98 conference, SK presented *evidence* of ν_μ oscillations [48] based on the angular distribution for their contained event data sample. Since then SK accumulated more statistics and has also studied the angular dependence of the upgoing muon sample. In Fig. 11 we show their data corresponding to 79 kiloton year (1289 days) exposure. In the figure we show the angular zenith distribution of the different samples. Comparing the observed and the expected (MC) distributions, we can make the following statements:

(i) ν_e distributions are well described by the MC while ν_μ presents a deficit. Thus the atmospheric neutrino deficit is mainly due to disappearance of ν_μ and not the appearance of ν_e.

(ii) The suppression of contained μ-like events is stronger for larger $\cos\theta$, which implies that the deficit grows with the distance traveled by the neutrino from its production point to the detector. This effect is more obvious for multi-GeV events because at higher energy the direction of the charged lepton is more aligned with the direction of the neutrino. It can also be described in terms of an up-down asymmetry:

$$A_\mu \equiv \frac{U - D}{U + D} = -0.316 \pm 0.042(\text{stat.}) \pm 0.005(\text{syst.}) \qquad (78)$$

where U (D) are the contained μ-like events with zenith angle in the range $-1 < \cos\theta < -0.2$ ($0.2 < \cos\theta < 1$). It deviates from the SM value, $A_\mu = 0$, by 7.5 standard deviations.

(iii) The overall suppression of the flux of stopping-muons, Φ_{ST}, is by a factor of about 0.6, similar to contained events. However, for the flux of through-going muons, Φ_{TH}, the suppression is weaker, which implies that the effect is smaller at larger neutrino energy. This effect is also parametrized in terms of the double flux ratio:

$$\frac{\Phi_{ST}/\Phi_{TH}|_{obs}}{\Phi_{ST}/\Phi_{TH}|_{MC}} = 0.635 \pm 0.049(\text{stat.}) \pm 0.035(\text{syst.}) \pm 0.084(\text{theo.}) \quad (79)$$

which deviates from the SM value of 1 by about 3 standard deviations.

These effects have been confirmed by the results of the iron calorimeters Soudan2 and MACRO which removed the suspicion that the atmospheric neutrino anomaly is simply a systematic effect in the water detectors.

To analyze the atmospheric neutrino data in terms of oscillations one needs to have a good understanding of the different elements entering into the theoretical predictions of the event rates: the atmospheric neutrino fluxes and their interaction cross section (see, Refs.[7, 11] and references therein).

3.2.1 Two-Neutrino Oscillation Analysis.

The simplest and most direct interpretation of the atmospheric neutrino anomaly is that of muon neutrino oscillations [51]. The estimated value of the oscillation parameters can be easily derived in the following way:

- The angular distribution of contained events shows that, for $E \sim$ 1 GeV, the deficit comes mainly from $L \sim 10^2 - 10^4$ km. The corresponding oscillation phase must be maximal, $\frac{\Delta m^2 (\text{eV}^2) L(\text{km})}{2E(\text{GeV})} \sim$ 1, which requires $\Delta m^2 \sim 10^{-4} - 10^{-2}$ eV^2.

- Assuming that all upgoing ν_μ's which would lead to multi-GeV events oscillate into a different flavor while none of the downgoing ones do, the up-down asymmetry is given by $|A_\mu| = \sin^2 2\theta/(4 - \sin^2 2\theta)$. The present one sigma bound reads $|A_\mu| > 0.27$ [see Eq. (78)], which requires that the mixing angle is close to maximal, $\sin^2 2\theta > 0.85$.

In order to go beyond these rough estimates, one must compare in a statistically meaningful way the experimental data with the detailed theoretical expectations.

• **Expected Event Rates**: For a given neutrino oscillation channel, the expected number of μ-like and e-like contained events, N_α ($\alpha = \mu, e$), can be computed as:

$$N_\mu = N_{\mu\mu} + N_{e\mu}, \qquad N_e = N_{ee} + N_{\mu e}, \quad (80)$$

where

$$N_{\alpha\beta} = n_t T \int \frac{d^2\Phi_\alpha}{dE_\nu d(\cos\theta_\nu)} \kappa_\alpha(h) P_{\alpha\beta} \frac{d\sigma}{dE_\beta} \varepsilon(E_\beta) dE_\nu dE_\beta d(\cos\theta_\nu) dh .$$

$$(81)$$

Here $P_{\alpha\beta}$ is the conversion probability of $\nu_\alpha \to \nu_\beta$ for given values of E_ν, $\cos\theta_\nu$ and h, i.e., $P_{\alpha\beta} \equiv P(\nu_\alpha \to \nu_\beta; E_\nu, \cos\theta_\nu, h)$. In the SM, the only non-zero elements are the diagonal ones, i.e. $P_{\alpha\alpha} = 1$ for all α. In Eq. (81), n_t denotes the number of target particles, T is the experiment running time, E_ν is the neutrino energy, Φ_α is the flux of atmospheric ν_α's, E_β is the final charged lepton energy, $\varepsilon(E_\beta)$ is the detection efficiency for such a charged lepton, σ is the neutrino-nucleon interaction cross section, and θ_ν is the angle between the vertical direction and the incoming neutrinos ($\cos\theta_\nu=1$ corresponds to the down-coming neutrinos). In Eq. (81), h is the slant distance from the production point to the sea level for α-type neutrinos with energy E_ν and zenith angle θ_ν, and κ_α is the slant distance distribution, normalized to one.

To obtain the expectation for the angular distribution of contained events one must integrate the corresponding bins for $\cos\theta_\beta$, where θ_β is the angle of the detected lepton, taking into account the opening angle between the neutrino and the charged lepton directions as determined by the kinematics of the neutrino interaction.

Experimental results on upgoing muons are presented in the form of measured muon fluxes. To obtain the effective muon fluxes for both stopping and through-going muons, one must convolute the survival probabilities for ν_μ's with the corresponding muon fluxes produced by the neutrino interactions with the Earth. One must further take into account the muon energy loss during propagation both in the rock and in the detector, and also the effective detector area for both types of events, stopping and through-going.

• **Oscillation Probabilities**: The oscillation probabilities are obtained by solving the evolution equation of the $\nu_\mu - \nu_X$ system in the matter background of the Earth (see section 2.4):

$$i\frac{d}{dt}\begin{pmatrix} \nu_\mu \\ \nu_X \end{pmatrix} = \begin{pmatrix} H_\mu & H_{\mu X} \\ H_{\mu X} & H_X \end{pmatrix}\begin{pmatrix} \nu_\mu \\ \nu_X \end{pmatrix}, \tag{82}$$

where

$$H_\mu = V_\mu - \frac{\Delta m^2}{4E_\nu}\cos 2\theta, \quad H_X = V_X + \frac{\Delta m^2}{4E_\nu}\cos 2\theta, \quad H_{\mu X} = \frac{\Delta m^2}{4E_\nu}\sin 2\theta.$$

The various neutrino potentials in matter are given by $V_e = V_C + V_N$, $V_\mu = V_\tau = V_N$, and $V_s = 0$ where the CC and NC potentials V_C and V_N are proportional to the electron and neutron number density (see Sec. 2.3), or equivalently, to the matter density in the Earth [52], For anti-neutrinos, the signs of the potentials are reversed.

For $X = \tau$, we have $V_\mu = V_\tau$ and consequently these potentials can be removed from the evolution equation. The solution of Eq. (82) is then

84

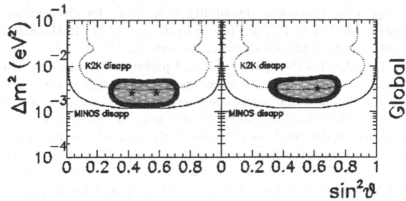

Figure 12. Allowed regions (at 90, 95 and 99 % CL) from the analysis of the full data sample of atmospheric neutrinos for the oscillation channels $\nu_\mu \to \nu_\tau$ and $\nu_\mu \to \nu_s$. The best fit points are marked with a star (see text for details). Also shown are the expected sensitivities from LBL experiments.

straightforward and the probability takes the well-known vacuum form [Eq. (32)], which is equal for neutrinos and anti-neutrinos. For $X = e$ or s, the effect of the matter potentials requires a numerical solution of the evolution equations in order to obtain $P_{\alpha\beta}$ which, furthermore, is different for neutrinos and anti-neutrinos.

3.2.2 $\nu_\mu \to \nu_e$.   At present $\nu_\mu \to \nu_e$ is excluded with high CL as the explanation of the atmospheric neutrino anomaly for two different reasons:

(i) SK high precision data show that the ν_e contained events are very well described by the SM prediction both in normalization and in their zenith angular dependence [see Fig. 11]. The ν_μ distribution, however, shows an angle-dependent deficit. $\nu_\mu \to \nu_e$ oscillations can explain the angular dependence of the ν_μ flux only at the price of introducing angular dependence of the ν_e flux, in contradiction with the data.

(ii) Explaining the atmospheric data with a $\nu_\mu \to \nu_e$ transition has direct implications for the $\bar{\nu}_e \to \bar{\nu}_\mu$ transition. In particular, there should be a $\bar{\nu}_e$ deficit in the CHOOZ reactor experiment. Thus the neutrino parameters not only give a poor fit to the atmospheric data but are actually excluded by the negative results from the CHOOZ reactor experiment.

3.2.3 $\nu_\mu \to \nu_\tau$ and $\nu_\mu \to \nu_s$.   In Fig. 12 we plot the allowed regions from the global analysis, including all the atmospheric neutrino data. Also shown are the expected sensitivity from ν_μ disappearance at the long baseline (LBL) experiments K2K and MINOS discussed in 2.2.3. We emphasize the following points:

(i) The allowed regions for $\nu_\mu \to \nu_\tau$ transition are symmetric with respect to maximal mixing. This must be the case because the corresponding probabilities take the vacuum expression and therefore depend on $\sin^2 2\theta$.

(ii) The allowed regions for $\nu_\mu \to \nu_s$ transition are asymmetric due to Earth matter effects. These effects are more pronounced when the condition for maximal matter effect, $\Delta m^2 \cos 2\theta \sim 2EV_s$, is fulfilled. Since in the chosen convention the potential difference $A = 2E(V_\mu - V_s) < 0$, the matter effects enhance neutrino oscillations for $\cos 2\theta < 0$ ($\sin^2 \theta > 0.5$). The opposite situation holds for antineutrinos, but neutrino fluxes are larger and dominate in the resulting effect. As a result the allowed regions are *wider* in the $\sin^2 \theta > 0.5$ side of the plot.

(iii) The best fit to the full data for $\nu_\mu \to \nu_\tau$ corresponds to $\Delta m^2 = 2.6 \times 10^{-3}$ eV^2 and $\sin^2 2\theta = 0.97$.

In order to discriminate between the $\nu_\mu \to \nu_\tau$ and $\nu_\mu \to \nu_s$ options, one can use the difference in the survival probabilities due to the presence of matter effects for oscillations into sterile neutrinos. SK has also used other methods to distinguish between the ν_τ and ν_s hypotheses for explaining atmospheric ν_μ disappearance. One is to examine events likely to have been caused by NC interactions. While ν_τ's readily undergo such interactions, ν_s's do not, resulting in a relative suppression of the NC signal. Another method attempts to observe appearance of the newly created ν_τ, even if only on a statistical basis, by selecting enriched samples. All methods favour $\nu_\mu \leftrightarrow \nu_\tau$ oscillations over $\nu_\mu \leftrightarrow \nu_s$.

Acknowledgments

I wish to thank Harrison Prosper for creating a pleasant and enjoyable atmosphere during this school.

References

[1] E. Fermi, Nuovo Cimento **11**, 1 (1934); Z. Phys. **88**, 161 (1934).

[2] F. Reines, C. Cowan, Phys. Rev. **113**, 273 (1959).

[3] B. Pontecorvo, B., J. Exptl. Theoret. Phys. **33**, 549 (1957) [Sov. Phys. JETP **6**, 429 (1958)].

[4] J.N. Bahcall, *Neutrino Astrophysics*, (Cambridge University Press, (1998).

[5] S. M. Bilenky, C. Giunti and W. Grimus, *Phenomenology of neutrino oscillations*, Prog. Part. Nucl. Phys. **43**, 1 (1999)

[6] F. Boehm, and P. Vogel, *Physics of Massive Neutrinos*, Cambridge University Press, (1987).

[7] T. K. Gaisser, "Cosmic Rays And Particle Physics," Cambridge University Press (1990).

[8] B. Kayser, F. Gibrat-Debu and F. Perrie, *The Physics Of Massive Neutrinos*, World Sci. Lect. Notes Phys. **25** (1989).

[9] C.W. Kim and A. Pevsner *Neutrinos in Physics and Astrophysics*, Contemporary Concepts in Physics, Vol. 8, Harwood Academic Press, Chur, Switzerland (1993).

[10] R. N. Mohapatra and P. B. Pal, *Massive Neutrinos In Physics And Astrophysics* World Sci. Lect. Notes Phys. **60** (1998) 1.

[11] M.C. Gonzalez-Garcia and Y. Nir, hep-ph/0202058, to appear in Review of Modern Physics.

[12] M.P. Gell-Mann, P. Ramond and R. Slansky, 1979, in *Supergravity*, edited by P. van Nieuwenhuizen and D.Z. Freedman (North Holland); T. Yanagida, 1979, in *Proceedings of Workshop on Unified Theory and Baryon Number in the Universe*, edited by O. Sawada and A. Sugamoto (KEK).

[13] Z. Maki, M. Nakagawa, and S. Sakata, Prog. Theor. Phys. **28**, 870 (1962).

[14] Kobayashi, M., and T. Maskawa, 1973, Prog. Theor. Phys. **49**, 652.

[15] Groom, D.E., *et al.*, 2000, Particle Data Group, Eur. Phys. J.**C15**, 1.

[16] C. Athanassopoulos *et al.*, Phys. Rev. Lett. **75**, 2650 (1995); A. Aguilar *et al.*, LSND Coll., Phys. Rev. D **64**, 112007 (2001).

[17] J. Wolf *et al.*, KARMEN Coll., Proceedings of the EPS HEP 2001 Conference, Budapest, Hungary.

[18] H. Gemmeke *et al.*, Nucl. Instrum. Meth. A **289**, 490 (1990).

[19] A. Bazarko *et al.*, MiniBooNE Coll. , Nucl. Phys. Proc. Suppl. **91**, 210 (2000).

[20] G. Zacek, *et al.*, Phys. Rev. D **34**, 2621 (1986).

[21] B. Achkar *et al.*, Bugey Coll., Nucl. Phys. B**434**, 503 (1995).

[22] G.S. Vidyakin *et al.*, JETP Lett. **59**, 390 (1994).

[23] M. Apollonio *et al.*, CHOOZ Coll., Phys. Lett. B **466**, 415 (1999).

[24] A. Piepke *et al.*, Nucl. Phys. Proc. Suppl. **91**, 99 (2001).

[25] K. Nishikawa *et al.*, KEK-PS proposal, Nucl. Phys. Proc. Suppl. **59**, 289 (1997); S.H. Ahn *et al.*, K2K Coll., Phys. Lett. B **511**, 178 (2001).

[26] E. Ables *et al.*, MINOS Coll., FERMILAB-PROPOSAL-P-875 (1995).

[27] H. Shibuya *et al.*, CERN-SPSC-97-24 (1997).

[28] L. Wolfenstein, Phys. Rev. D **17**, 2369 (1978).

[29] A. Halprin, Phys. Rev. D **34**, 3462 (1986).

[30] P.D. Mannheim, Phys. Rev. D **37**, 1935 (1988).

[31] A.J. Baltz, and J. Weneser, Phys. Rev. D **37**, 3364 (1988).

[32] S.P. Mikheyev, and A.Y. Smirnov, Yad. Fiz. **42**, 1441 (1985) [Sov. J. Nucl. Phys. **42**, 913].

[33] A. Friedland, Phys. Rev. D. **64**, 013008 (2001).

[34] E. Lisi, A. Marrone, D. Montanino, A. Palazzo and S.T. Petcov, Phys. Rev. D **63** 093002 (2001).

[35] T.K. Kuo, and J. Pantaleone, Rev. Mod. Phys. **61**, 937 (1989).

[36] L. Landau, Phys. Z. Sov. **2**, 46 (1932); C. Zener, Proc. Roy. Soc. Lon. A **137**, 696 (1932).

[37] S.J. Parke, Phys. Rev. Lett. **57**, 1275 (1986).

[38] W.C. Haxton, Phys. Rev. Lett. **57**, 1271 (1987).

[39] S.T. Petcov, Phys. Lett. B **191**, 299 (1987).

[40] S.T. Petcov, Phys. Lett. B **200**, 373 (1988).

[41] P.I. Krastev, and S. T. Petcov, Phys. Lett. B **207**, 64 (1988).

[42] F. Reines *et al* Phys. Rev. Lett. **15**, 429 (1965).

[43] H. Achar *et al,,* Phys. Lett. **18**, 196 (1965).

[44] K. Daum *et al.*, Frejus Coll., Z. Phys. C **66**, 417 (1995).

[45] M. Aglietta, M. *et al.*, Nusex Coll., Europhys. Lett. **8**, 611 (1989).

[46] R. Becker-Szend *et al.*, IMB Coll., Phys. Rev. D **46**, 3720 (1992).

[47] Y. Fukuda *et al.*, Kamiokande Coll., Phys. Lett. B **335**, 237 (1994).

[48] Y. Fukuda, *et al.*, SuperKamiokande Coll., Phys. Rev. Lett. **81**, 1562 (1998).

[49] W.W.M. Allison *et al.*, Soudan Coll., Phys. Lett. B **449**, 137 (1999).

[50] M. Ambrosio *et al.*, MACRO Coll., Phys. Lett. B **517**, 59 (2001).

[51] J.G. Learned, S. Pakvasa, and T.J. Weiler, Phys. Lett. B **207**, 79 (1988); V. Barger and Whisnant, K., Phys. Lett. B **209**, 365 (1988); K. Hidaka, M. Honda, and S. Midorikawa, Phys. Rev. Lett. **61**, 1537 (1988).

[52] A.M. Dziewonski, D.L. Anderson, Phys. Earth Planet. Inter. **25**, 297 (1981).

[53] J.N. Bahcall, H.M. Pinsonneault, and S. Basu, Astrophys. J. **555**, 990 (2001);

[54] R. Jr. Davis, D.S. Harmer, and K.C. Hoffman, Phys. Rev. Lett. **20**, 1205 (1968); K. Lande *et al.*, Nucl. Phys. Proc. Suppl. **77**, 13 (1999).

[55] W. Hampel *et al.*, GALLEX Coll., Phys. Lett. B **447**, 127 (1999); T. Kirsten, GNO Coll. Talk at ν-2002 conference, Munich, May 2002.

[56] J.N. Abdurashitov *et al.*, SAGE Coll., Phys. Rev. C **60**, 055801 (1999); astro-ph/0204245.

[57] Y. Fukuda, *et al.*, Kamiokande Coll. Phys. Rev. Lett. **77**, 1683 (1996).

[58] M. Smy, 2002, Talk at ν-2002 conference, Munich, May 2002.

[59] R.Q. Ahmad *et al.*, 2001, SNO Coll., Phys. Rev. Lett. **87**, 071301; Ibid **89**, 011301; Ibid **89**, 011302.

[60] J.N. Bahcall, N. A. Bahcall and G. Shaviv, Phys. Rev. Lett. **20**, 1209 (1968); J.N. Bahcall and R. Davis, Science **191**, 264 (1976).

[61] N. Hata, and P.Langacker, Phys. Rev. D **56**, 6107 (1997); G.L. Fogli, E. Lisi, and D. Montanino, Astropart. Phys. **9**, 119 (1998); J.N. Bahcall, P.I. Krastev, and A. Yu Smirnov, Phys. Rev. D **58**, 096016 (1998); M.C. Gonzalez-Garcia, *et. al.* Nucl. Phys. B **573**, 3 (2000).

[62] Barger, V., D. Marfatia, and K. Whisnant, Phys. Rev. Lett. **88**, 011302 (2002); Bahcall, J.N., M.C. Gonzalez-Garcia, and C. Peña-Garay, hep-ph/0204314, To appear in Phys. Rev. C.

[63] Bahcall, J.N., M.C. Gonzalez-Garcia, and C. Peña-Garay, JHEP **0207**, 054 (2002).

THE SUPERSYMMETRIC UNIVERSE

John Ellis

CERN, TH Division

CH 1211 Geneva 23, Switzerland

John.Ellis@cern.ch

Abstract

These lectures provide a phenomenological introduction to supersymmetry, concentrating on the minimal supersymmetric extension of the Standard Model (MSSM). Motivations are provided for thinking that supersymmetry might appear at the TeV scale, including the naturalness of the mass hierarchy, gauge unification and the probable mass of the Higgs boson. Then simple globally supersymmetric field theories are introduced, with the emphasis on features important for model-building. Supersymmetry breaking and local supersymmetry (supergravity) are then introduced, and the structure of sparticle mass matrices and mixing are reviewed. The available experimental and cosmological constraints on MSSM parameters are discussed and combined, and the prospects for discovering supersymmetry in future experiments are previewed. Finally, the observability of leptonic processes violating flavour and CP are discussed, on the basis of the minimal supersymmetric seesaw model of neutrino masses.

1. Getting Motivated

1.1 Defects of the Standard Model

The Standard Model agrees with all confirmed experimental data from accelerators, but is theoretically very unsatisfactory. It does not explain the particle quantum numbers, such as the electric charge Q, weak isospin I, hypercharge Y and colour, and contains at least 19 arbitrary parameters. These include three independent gauge couplings and a possible CP-violating strong-interaction parameter, six quark and three charged-lepton masses, three generalized Cabibbo weak mixing angles and the CP-violating Kobayashi-Maskawa phase, as well as two independent masses for weak bosons.

H.B. Prosper and M. Danilov (eds.), Techniques and Concepts of High-Energy Physics XII, 89–157.

As if 19 parameters were insufficient to appall you, at least nine more parameters must be introduced to accommodate neutrino oscillations: three neutrino masses, three real mixing angles, and three CP-violating phases, of which one is in principle observable in neutrino-oscillation experiments and the other two in neutrinoless double-beta decay experiments. Even more parameters would be needed to generate neutrino masses in a credible way, associated with a heavy-neutrino sector and/or additional Higgs particles.

Eventually, one would like to include gravity in a unified theory along with the other particle interactions, which involves introducing at least two more parameters, Newton's constant $G_N = 1/m_P^2 : m_P \sim 10^{19}$ GeV that characterizes the strength of gravitational interactions, and the cosmological constant Λ or some time-varying form of vacuum energy as seems to be required by recent cosmological data. A complete theory of cosmology will presumably also need parameters to characterize the early inflation of the Universe and to generate its baryon asymmetry, which cannot be explained within the Standard Model.

The Big Issues in physics beyond the Standard Model are conveniently grouped into three categories [1]. These include the problem of **Unification**: is there a simple group framework for unifying all the particle interactions, a so-called Grand Unified Theory (GUT), **Flavour**: why are there so many different types of quarks and leptons and why do their weak interactions mix in the peculiar way observed, and **Mass**: what is the origin of particle masses, are they due to a Higgs boson, why are the masses so small? Solutions to all these problems should eventually be incorporated in a Theory of Everything (TOE) that also includes gravity, reconciles it with quantum mechanics, explains the origin of space-time and why it has four dimensions, etc. String theory, perhaps in its current incarnation of M theory, is the best (only?) candidate we have for such a TOE [2], but we do not yet understand it well enough to make clear experimental predictions.

Supersymmetry is thought to play a rôle in solving many of these problems beyond the Standard Model. As discussed later, GUT predictions for the unification of gauge couplings work best if the effects of relatively light supersymmetric particles are included [3]. Also, the hierarchy of mass scales in physics, and particularly the fact that $m_W \ll m_P$, appears to require relatively light supersymmetric particles: $M \lesssim 1$ TeV for its stabilization [4]. Finally, supersymmetry seems to be essential for the consistency of string theory [5], although this argument does not really restrict the mass scale at which supersymmetric particles should appear.

Thus there are plenty of good reasons to study supersymmetry, and we return later to examine in more detail the motivations provided by unification and the mass hierarchy problem.

1.2   The Electroweak Vacuum

Generating particle masses within the Standard Model requires breaking its gauge symmetry, and the only consistent way to do this is by breaking the symmetry of the electroweak vacuum:

$$m_{W,Z} \neq 0 \ \leftrightarrow \ < 0|X_{I,I_3}|0 > \neq 0 \tag{1}$$

where the symbols I, I_3 denote the weak isospin quantum numbers of whatever object X has a non-zero vacuum expectation value. There are a couple of good reasons to think that X must have (predominantly) isospin $I = 1/2$. One is the ratio of the W and Z boson masses [6]:

$$\rho \equiv \frac{m_W^2}{m_Z^2 \cos^2 \theta_W} \simeq 1, \tag{2}$$

and the other reason is to provide non-zero fermion masses. Since left-handed fermions f_L have $I = 1/2$, right-handed fermions f_R have $I = 0$ and fermion mass terms couple them together: $m_f \bar{f}_L f_R$, we must break isospin symmetry by $1/2$ a unit:

$$m_f \neq 0 \ \leftrightarrow \ < 0|X_{1/2,\pm 1/2}|0 > \neq 0. \tag{3}$$

The next question is, what is the nature of X? Is it elementary or composite? In the initial formulation of the Standard Model, it was assumed that X should be an elementary Higgs-Brout-Englert [7, 8] field H: $< 0|H^0|0 > \neq 0$, which would have a physical excitation that manifested itself as a neutral scalar Higgs boson [7]. However, as discussed in more detail later, an elementary Higgs field has problems with quantum (loop) corrections. Those due to Standard Model particles are quadratically divergent, resulting in a large cutoff-dependent contribution to the physical masses of the Higgs boson, W, Z bosons and other particles:

$$\delta m_H^2 \simeq \mathcal{O}(\frac{\alpha}{\pi})\Lambda^2, \tag{4}$$

where Λ represents the scale at which new physics appears.

The sensitivity (4) disturbs theorists, and lies at the root of the hierarchy problem [4]. One of the suggestions for avoiding it was to postulate replacing an elementary Higgs-Brout-Englert field H by a composite field such as a condensate of fermions: $< 0|\bar{F}F|0 > \neq 0$. This possibility was made more appealing by the fact that fermion condensates are well known in solid-state physics, where Cooper pairs of electrons are responsible for conventional superconductivity, and in strong-interaction physics, where quarks condense in the vacuum: $< 0|\bar{q}q|0 > \neq 0$.

In order to break the electroweak symmetry at a large enough scale, fermions with new interactions that become strong at a higher mass scale would be required. One suggestion was that the Yukawa interactions of the top quark might be strong enough to condense them: $< 0|\bar{t}t|0 > \neq 0$ [9], but this would have required the top quark to weigh more than 200 GeV, in simple models. Alternatively, theorists proposed the existence of new fermions held together by completely new interactions that became strong at a scale ~ 1 TeV, commonly called *Technicolour* models [10].

The simplest version of technicolour predicted three massless technipions that would be eaten by the $W^{\pm}$ and Z^0 to provide their masses via the Higgs-Brout-Englert mechanism, leaving over a single physical scalar state weighing about 1 TeV, that would behave in some ways like a heavy Higgs boson. Unfortunately, this simple technicolour picture must be complicated [11], and runs into various experimental problems, such as the absence of any light technipions [12], the observed suppression of flavour-changing neutral interactions [13], and precision electroweak data, which limit the possible magnitudes of one-loop radiative corrections due to virtual techniparticles [14]. Does this mean that technicolour is dead? Not quite [15], but it has motivated technicolour enthusiasts to pursue epicyclic variations on the original idea, such as walking technicolour [16], in which the technicolour dynamics is not scaled up from QCD in such a naive way.

1.3   Why Supersymmetry?

An alternative approach to the hierarchy problem [4] is offered by supersymmetry, if it appears at an accessible energy scale. One may formulate the hierarchy problem [4] by posing the question why is $m_W \ll m_P$, or, equivalently, why is $G_F \sim 1/m_W^2 \gg G_N = 1/m_P^2$? Another equivalent question is why the Coulomb potential in an atom is so much greater than the Newton potential: $e^2 \gg G_N m^2 = m^2/m_P^2$, where m is a typical particle mass?

Your first thought might simply be to set $m_P \gg m_W$ by hand, and forget about the problem. Life is not so simple, because, as already mentioned, quantum corrections to m_H and hence m_W are quadratically divergent in the Standard Model:

$$\delta m_{H,W}^2 \simeq \mathcal{O}(\frac{\alpha}{\pi})\Lambda^2, \tag{5}$$

which is $\gg m_W^2$ if the cutoff Λ, which represents the scale where new physics beyond the Standard Model appears, is comparable to the GUT or Planck scale. For example, if the Standard Model were to hold unscathed all the way up the Planck mass $m_P \sim 10^{19}$ GeV, the radiative correction (5) would be 36 orders of magnitude than the physical values of $m_{H,W}^2$!

In principle, this is not a problem from the mathematical point of view of renormalization theory. All one has to do is postulate a tree-level value of m_H^2 that is (very nearly) equal and opposite to the 'correction' (5), and the correct physical value may be obtained. However, this strikes many physicists as rather unnatural: they would prefer a mechanism that keeps the 'correction' (5) comparable at most to the physical value.

This is possible in a supersymmetric theory, in which there are equal numbers of bosons and fermions with identical couplings. Since bosonic and fermionic loops have opposite signs, the residual one-loop correction is of the form

$$\delta m_{H,W}^2 \simeq \mathcal{O}(\frac{\alpha}{\pi})(m_B^2 - m_F^2), \tag{6}$$

which is $\lesssim m_{H,W}^2$ and hence naturally small if the supersymmetric partner bosons B and fermions F have similar masses:

$$|m_B^2 - m_F^2| \lesssim 1 \text{ TeV}^2. \tag{7}$$

This is the best motivation we have for finding supersymmetry at relatively low energies [4].

In addition to this first supersymmetric miracle of removing (6) the quadratic divergence (5), many logarithmic divergences are also absent in a supersymmetric theory, as discussed later. This property is logically distinct from the absence of quadratic divergences in a supersymmetric theory, and is also important for supersymmetric model-building.

1.4   What is Supersymmetry?

The basic idea of supersymmetry is the existence of fermionic charges Q_α that relate bosons to fermions. Recall that all previous symmetries, such as flavour $SU(3)$ or electromagnetic $U(1)$, have involved scalar charges Q that link particles with the same spin into multiplets:

$$Q \, |\text{Spin} J > = |\text{Spin} J > . \tag{8}$$

Indeed, Coleman and Mandula [17] proved that it was 'impossible' to mix internal and Lorentz symmetries: $J_1 \leftrightarrow J_2$. However, their 'no-go' theorem assumed implicitly that the prospective charges should have integer spins.

The basic element in their 'proof' was the observation that the only possible conserved tensor charges were those with no Lorentz indices, i.e., scalar charges, and the energy-momentum vector P_μ. To see how their 'proof' worked, consider two-to-two elastic scattering, $1+2 \rightarrow 3+4$, and imagine that there exists a conserved two-index tensor charge, $\Sigma_{\mu\nu}$. By Lorentz invariance, its diagonal matrix elements between single-particle states $|a>$ must take the general form:

$$< a|\Sigma_{\mu\nu}|a > = \alpha P_\mu^{(a)} P_\nu^{(a)} + \beta g_{\mu\nu}, \qquad (9)$$

where α, β are arbitrary reduced matrix elements, and $g_{\mu\nu}$ is the metric tensor. For $\Sigma_{\mu\nu}$ to be conserved in a two-to-two scattering process, one must have

$$P_\mu^{(1)} P_\nu^{(1)} + P_\mu^{(2)} P_\nu^{(2)} = P_\mu^{(3)} P_\nu^{(3)} + P_\mu^{(4)} P_\nu^{(4)}, \qquad (10)$$

where we assume that the symmetry is local, so that two-particle matrix elements of $\Sigma_{\mu\nu}$ play no rôle. Since Lorentz invariance also requires $P_\mu^{(1)} + P_\mu^{(2)} = P_\mu^{(3)} + P_\mu^{(4)}$, the only possible outcomes are $P_\mu^{(1)} = P_\mu^{(3)}$ or $P_\mu^{(4)}$. Thus the only possibilities are completely forward scattering or completely backward scattering. This disagrees with observation, and is in fact theoretically impossible in any local field theory.

This rules out any non-trivial two-index tensor charge, and the argument can clearly be extended to any higher-rank tensor with more Lorentz indices. But what about a spinorial charge Q_α? This can have no diagonal matrix element:

$$< a|Q_\alpha|a > \neq 0, \qquad (11)$$

and hence the Coleman-Mandula argument fails.

So what is the possible form of a 'supersymmetry' algebra that includes such spinorial charges Q_α^i [1]? Since the different Q^i are supposed to generate symmetries, they must commute with the Hamiltonian:

$$[Q^i, H] = 0 : i = 1, 2, , N. \qquad (12)$$

So also must the anticommutator of two spinorial charges:

$$[\{Q^i, Q^j\}, H] = 0 : i, j = 1, 2, , N. \qquad (13)$$

However, the part of the anticommutator $\{Q^i, Q^j\}$ that is symmetric in the internal indices i, j cannot have spin 0. Instead, as we discussed just

---

[1]In what follows, I shall suppress the spinorial subscript α whenever it is not essential. The superscripts $i, j, ..., N$ denote different supersymmetry charges.

above, the only possible non-zero spin choice is $J = 1$, so that

$$\{Q^i, Q^j\} \propto \delta^{ij} P_\mu + \ldots \quad : i, j = 1, 2, , N. \tag{14}$$

In fact, as was proved by Haag, Lopuszanski and Sohnius [18], the only allowed possibility is

$$\{Q^i, Q^j\} = 2\delta^{ij} \gamma^\mu P_\mu \mathcal{C} + \ldots : i, j = 1, 2, , N, \tag{15}$$

where $\mathcal{C}$ is the charge-conjugation matrix discussed in more detail in Lecture 2, and the dots denote a possible 'central charge' that is anti-symmetric in the indices i, j, and hence can only appear when $N > 1$.

According to a basic principle of Swiss law, anything not illegal is compulsory, so there MUST exist physical realizations of the super-symmetry algebra (15). Indeed, non-trivial realizations of the non-relativistic analogue of (15) are known from nuclear physics [19], atomic physics and condensed-matter physics. However, none of these is thought to be fundamental.

In the relativistic limit, supermultiplets consist of massless particles with spins differing by half a unit. In the case of simple $N = 1$ super-symmetry, the basic building blocks are *chiral supermultiplets*:

$$\binom{\frac{1}{2}}{0} \; e.g., \; \left(\begin{array}{c} \ell \; (lepton) \\ \tilde{\ell} \; (slepton) \end{array} \right) or \left(\begin{array}{c} q \; (quark) \\ \tilde{q} \; (squark) \end{array} \right) \tag{16}$$

gauge supermultiplets:

$$\binom{1}{\frac{1}{2}} \; e.g., \; \left(\begin{array}{c} \gamma \; (photon) \\ \tilde{\gamma} \; (photino) \end{array} \right) or \left(\begin{array}{c} g \; (gluon) \\ \tilde{g} \; (gluino) \end{array} \right) \tag{17}$$

and the *graviton supermultiplet* consisting of the spin-2 graviton and the spin-3/2 gravitino.

Could any of the known particles in the Standard Model be linked together in supermultiplets? Unfortunately, none of the known fermions q, ℓ can be paired with any of the known bosons $\gamma, W^\pm Z^0, g, H$, because their internal quantum numbers do not match [20]. For example, quarks q sit in triplet representations of colour, whereas the known bosons are either singlets or octets of colour. Then again, leptons ℓ have non-zero lepton number $L = 1$, whereas the known bosons have $L = 0$. Thus, the only possibility seems to be to introduce new supersymmetric partners (spartners) for all the known particles: quark $\to$ squark, lepton $\to$ slepton, photon $\to$ photino, Z $\to$ Zino, W $\to$ Wino, gluon $\to$ gluino, Higgs $\to$ Higgsino, as suggested in (16, 17) above.

The best that one can say for supersymmetry is that it economizes on principle, not on particles!

1.5   (S)Experimental Hints

By now, you may be wondering whether it makes sense to introduce so many new particles just to deal with a paltry little hierarchy or naturalness problem. But, as they used to say during the First World War, 'if you know a better hole, go to it.' As we learnt above, technicolour no longer seems to be a viable hole, and I am not convinced that theories with large extra dimensions really solve the hierarchy problem, rather than just rewrite it. Fortunately, there are two hints from the high-precision electroweak data that supersymmetry may not be such a bad hole, after all.

One is the fact that there probably exists a Higgs boson weighing less than about 200 GeV [21]. This is perfectly consistent with calculations in the minimal supersymmetric extension of the Standard Model (MSSM), in which the lightest Higgs boson weighs less than about 130 GeV [22], as we discuss later in more detail.

The other hint is provided by the strengths of the different gauge interactions, as measured at LEP [3]. These may be run up to high energy scales using the renormalization-group equations, to see whether they unify as predicted in a GUT. The answer is no, if supersymmetry is not included in the calculations. In that case, GUTs would require

$$\sin^2 \theta_W \ = \ 0.214 \pm 0.004, \tag{18}$$

whereas the experimental value of the effective neutral weak mixing parameter at the Z^0 peak is $\sin^2 \theta = 0.23149 \pm 0.00017$ [21]. On the other hand, minimal supersymmetric GUTs predict

$$\sin^2 \theta_W \ \sim \ 0.232, \tag{19}$$

where the error depends on the assumed sparticle masses, the preferred value being around 1 TeV, as suggested completely independently by the naturalness of the electroweak mass hierarchy. This argument is also discussed later in more detail.

2.   Simple Models

2.1   Deconstructing Dirac

In this Section, we tackle some unavoidable spinorology. The most familiar spinors used in four-dimensional field theories are four-component Dirac spinors ψ. You may recall that it is possible to introduce projection operators

$$P_{L,R} \ \equiv \ \frac{1}{2}(1 \mp \gamma_5), \tag{20}$$

where $\gamma_5 \equiv i\gamma^0\gamma^1\gamma^2\gamma^3$, and the γ_μ can be written in the forms

$$\gamma_\mu = \begin{pmatrix} 0 & \sigma_\mu \\ \bar{\sigma}_\mu & 0 \end{pmatrix}, \tag{21}$$

where $\sigma_\mu \equiv (1, \sigma_i)$, $\bar{\sigma}_\mu \equiv (1, -\sigma_i)$. Then γ_5 can be written in the form $\mathrm{diag}(-1, 1)$, where $-1, 1$ denote 2×2 matrices. Next,we introduce the corresponding left- and right-handed spinors

$$\psi_{L,R} \equiv P_{L,R}\psi, \tag{22}$$

in terms of which one may decompose the four-component spinor into a pair of two-component spinors:

$$\psi = \begin{pmatrix} \psi_L \\ \psi_R \end{pmatrix}. \tag{23}$$

These will serve as our basic fermionic building blocks.

Antifermions can be represented by adjoint spinors

$$\bar{\psi} \equiv \psi^\dagger \gamma^0 = (\bar{\psi}_R, \bar{\psi}_L) \tag{24}$$

where the γ^0 factor has interchanged the left- and right-handed components $\psi_{L,R}$. We can now decompose in terms of these the conventional fermion kinetic term

$$\bar{\psi}\gamma_\mu \partial^\mu \psi = \bar{\psi}_L \gamma_\mu \partial^\mu \psi_L + \bar{\psi}_R \gamma_\mu \partial^\mu \psi_R \tag{25}$$

and the conventional mass term

$$\bar{\psi}\psi = \bar{\psi}_R \psi_L + \bar{\psi}_L \psi_R. \tag{26}$$

We see that the kinetic term keeps separate the left- and right-handed spinors, whereas the mass term mixes them.

The next step is to introduce the charge-conjugation operator $\mathcal{C}$, which changes the overall sign of the vector current $\bar{\psi}\gamma^\mu\psi$. It transforms spinors into their conjugates:

$$\psi^c \equiv \mathcal{C}\,\bar{\psi}^T = \mathcal{C}(\psi^\dagger\gamma^0)^T = \begin{pmatrix} \bar{\psi}_R \\ \bar{\psi}_L \end{pmatrix}, \tag{27}$$

and operates as follows on the γ matrices:

$$\mathcal{C}^{-1}\gamma^\mu\mathcal{C} = -\gamma^{\mu T}. \tag{28}$$

A convenient representation of $\mathcal{C}$ is:

$$\mathcal{C} = i\gamma^0\gamma^2. \tag{29}$$

It is apparent from the above that the conjugate of a left-handed spinor is right-handed:

$$(\psi_L)^c = \begin{pmatrix} 0 \\ \bar{\psi}_L \end{pmatrix}, \tag{30}$$

so that the combination

$$\bar{\psi}_L^c \psi_L = \psi_L \sigma_2 \psi_L \tag{31}$$

mixes left- and right-handed spinors, and has the same form as a mass term (26).

It is apparent from (30) that we can construct four-component Dirac spinors entirely out of two-component left-handed spinors and their conjugates:

$$\psi = \begin{pmatrix} \psi_i \\ \psi_j^c \end{pmatrix}, \tag{32}$$

a trick that will be useful later in our supersymmetric model-building. As examples, instead of working with left- and right-handed quark fields q_L and q_R, or left- and right-handed lepton fields ℓ_L and ℓ_R, we can write the theory in terms of left-handed antiquarks and antileptons: $q_R \to q_L^c$ and $\ell_R \to \ell_L^c$.

2.2 Simplest Supersymmetric Field Theories

Let us now consider a field theory [23] containing just a single left-handed fermion ψ_L and a complex boson ϕ, without any interactions, as described by the Lagrangian

$$L_0 = i\bar{\psi}_L \gamma^\mu \partial_\mu \psi_L + |\partial\phi|^2. \tag{33}$$

We consider the simplest possible non-trivial transformation law for the free theory (33):

$$\phi \to \phi + \delta\phi, \quad \text{where } \delta\phi = \sqrt{2}\bar{E}\psi_L, \tag{34}$$

where E is some constant right-handed spinor. In parallel with (34), we also consider the most general possible transformation law for the fermion ψ:

$$\psi_L \to \psi_L + \delta\psi_L, \quad \text{where } \delta\psi_L = -a\,i\,\sqrt{2}(\gamma_\mu\partial^\mu\phi)E - FE^c, \tag{35}$$

where a and F are constants to be fixed later, and we recall that E^c is a left-handed spinor. We can now consider the resulting transformation of the full Lagrangian (33), which can easily be checked to take the form

$$\delta L_0 = \sqrt{2}\partial_\mu[\bar{\psi}E\partial^\mu\phi + \bar{E}\gamma^\mu\phi^*\gamma_\nu\partial^\nu\psi], \tag{36}$$

if and only if we choose

$$a = 1 \text{ and } F = 0 \tag{37}$$

in this free-field model. With these choices, and the resulting total-derivative transformation law (36) for the free Lagrangian, the free action A_0 is invariant under the transformations (34,35), since

$$\delta A_0 = \delta \int d^4 x L_0 = 0. \tag{38}$$

Fine, you may say, but is this symmetry actually supersymmetry? To convince yourself that it is, consider the sequences of pairs (34,35) of transformations starting from either the boson ϕ or the fermion ψ:

$$\phi \rightarrow \psi \rightarrow \partial\phi, \ \psi \rightarrow \partial\phi \rightarrow \partial\psi. \tag{39}$$

In both cases, the action of two symmetry transformations is equivalent to a derivative, i.e., the momentum operator, corresponding exactly to the supersymmetry algebra. A free boson and a free fermion together realize supersymmetry: like the character in Molière, we have been talking prose all our lives without realizing it!

Now we look at interactions in a supersymmetric field theory [24]. The most general interactions between spin-0 fields ϕ^i and spin-1/2 fields ψ^i that are at most bilinear in the latter, and hence have a chance of being renormalizable in four dimensions, can be written in the form

$$L = L_0 - V(\phi^i, \phi_j^*) - \frac{1}{2} M_{ij}(\phi, \phi^*) \bar{\psi}^{ci} \psi^j \tag{40}$$

where V is a general effective potential, and M_{ij} includes both mass terms and Yukawa interactions for the fermions. Supersymmetry imposes strong constraints on the allowed forms of V and M, as we now see. Suppose that M depended non-trivially on the conjugate fields ϕ^*: then the supersymmetric variation $\delta(M\bar{\psi}^c\psi)$ would contain a term

$$\frac{\partial M}{\partial \phi^*} \psi^* \bar{\psi}^c \psi \tag{41}$$

that could not be compensated by the variation of any other term. We conclude that M must be independent of ϕ^*, and hence $M = M(\phi)$ alone.

Another term in the variation of the last term in (40) is

$$\frac{\partial M_{ij}}{\partial \phi^k} \bar{E} \psi^k \bar{\psi}^{ci} \psi^j. \tag{42}$$

This term cannot be cancelled by the variation of any other term, but can vanish by itself if $\partial M_{ij}/\partial \phi^k$ is completely symmetric in the indices i, j, k. This is possible only if

$$M_{ij} = \frac{\partial W}{\partial \phi^i \partial \phi^j} \tag{43}$$

for some function $W(\phi)$ called the *superpotential*. If the theory is to be renormalizable, W can only be cubic. The trilinear term of W determines the Yukawa couplings, and the bilinear part the mass terms.

We now re-examine the form of the supersymmetric transformation law (35) itself. Yet another term in the variation of the second term in (40) has the form

$$iM_{jk}\bar{\psi}^{cj}\gamma_\mu \partial^\mu \phi^k E + (\text{Herm.Conj.}). \tag{44}$$

This can cancel against an F-dependent term in the variation of the fermion kinetic term

$$-i\bar{\psi}_i \gamma_\mu \partial^\mu F^i E^c + (\text{Herm.Conj.}), \tag{45}$$

if the following relation between F and M holds: $\frac{\partial F_i^*}{\partial \phi^j} = M_{ij}$, which is possible if and only if

$$F_i^* = \frac{\partial W}{\partial \phi^i}. \tag{46}$$

Thus the form of W also determines the required form of the supersymmetry transformation law.

The form of W also determines the effective potential V, as we now see. One of the terms in the variation of V is

$$\frac{\partial V}{\partial \phi^i} \bar{E}\psi^I + (\text{Herm.Conj.}), \tag{47}$$

which can only be cancelled by a term in the variation of $M_{ij}\bar{\psi}^{ci}\psi^j$, which can take the form $M_{ij}\bar{\psi}^{ci}F^j E^c$ if

$$\frac{\partial V}{\partial \phi^i} = M_{ij}F^i, \tag{48}$$

which is in turn possible only if

$$V = |\frac{\partial W}{\partial \phi^i}|^2 = |F^i|^2. \tag{49}$$

We now have the complete supersymmetric field theory for interacting chiral (matter) supermultiplets [24]:

$$L = i\bar{\psi}_i\gamma_\mu\partial^\mu\psi^i + |\partial_\mu\phi^i|^2 - |\frac{\partial W}{\partial\phi^i}|^2 - \frac{1}{2}\frac{\partial^2 W}{\partial\phi^i\partial^j}\bar{\psi}^{ci}\partial\psi^j + (\text{Herm.Conj.}).$$

(50)

This Lagrangian is invariant (up to a total derivative) under the supersymmetry transformations

$$\delta\phi^i = \sqrt{2}\bar{E}\psi^i, \quad \delta\psi^i = -i\sqrt{2}\gamma_\mu\partial^\mu\phi^i E - F^i E^c : \quad F^i = (\frac{\partial W}{\partial\phi^i})^*. \quad (51)$$

The simplest non-trivial superpotential involving a single superfield ϕ is

$$W = \frac{\lambda}{3}\phi^3 + \frac{m}{2}\phi^2.$$

(52)

It is a simple exercise for you to verify using the rules given above that the corresponding Lagrangian is

$$L = i\bar{\psi}\gamma_\mu\partial^\mu\psi + |\partial_\mu\phi|^2 - |m\phi + \lambda\phi^2|^2 - m\bar{\psi}^c\psi - \lambda\phi\bar{\psi}^c\psi. \quad (53)$$

We see explicitly that the bosonic component ϕ of the supermultiplet has the same mass as the fermionic component ψ, and that the Yukawa coupling λ fixes the effective potential.

We now turn to the possible form of a supersymmetric gauge theory [25]. Clearly, it must contain vector fields A_μ^a and fermions χ^a in the same adjoint representation of the gauge group. Once one knows the gauge group and the fermionic matter content, the form of the Lagrangian is completely determined by gauge invariance:

$$L = \frac{i}{2}\bar{\chi}^a\gamma^\mu D_{ab}^\mu\chi^b - \frac{1}{4}F_{\mu\nu}^a F^{a,\mu\nu} \left[-\frac{1}{2}(D^a)^2\right]. \quad (54)$$

Here, the gauge-covariant derivative

$$D_{ab}^\mu \equiv \delta_{ab}\partial^\mu - gf_{abc}A_c^\mu, \quad (55)$$

and the gauge field strength is

$$F_{\mu\nu}^a \equiv \partial_\mu A_\nu^a - \partial_\nu A_\mu^a + gf^{abc}A_\mu^b A_\nu^c, \quad (56)$$

as usual. We return later to the D term at the end of (54). Yet another of the miracles of supersymmetry is that the Lagrangian (54) is automatically supersymmetric, without any further monkeying around. The corresponding supersymmetry transformations may be written as

$$\delta A_\mu^a = -\bar{E}\gamma_\mu\chi^a, \quad (57)$$

$$\delta\chi^a = -\frac{i}{2}F_{\mu\nu}^a\gamma^\mu\gamma^\nu E + D^a E, \quad (58)$$

$$\delta D^a = -i\bar{E}\gamma_5\gamma^\mu D_\mu^{ab}\chi^b. \quad (59)$$

What about the D term in (54)? It is a trivial consequence of equations of motion derived from (54) that $D^a = 0$. However, this is no longer the case if matter is included. Then, it turns out, one must add to (54) the following:

$$\Delta L = -\sqrt{2}g\chi^a\phi_i^*(T^a)_j^i\psi^j + (\text{Herm.Conj.}) + g(\phi_i^*(T^a)_j^i\phi^j)D^a, \quad (60)$$

where T^a is the group representation matrix for the matter fields ϕ^i. With this addition, the equation of motion for D^a tells us that

$$D^a = g\phi_i^*(T^a)_j^i\phi^j, \quad (61)$$

and we find a D term in the full effective potential:

$$V = \Sigma_i|F_i|^2 + \Sigma_a\frac{1}{2}(D^a)^2, \quad (62)$$

where the form of D^a is given in (61).

2.3 Further Aspects of Supersymmetric Field Theories

So far, we have taken a relatively unsophisticated approach to supersymmetry. However, one of the reasons why theorists are so enthusiastic about supersymmetry is because it is not just a new type of symmetry, but extends the concept of space-time itself. Recall the basic form of the supersymmetry algebra:

$$2\delta^{ij}\gamma_\mu P^\mu C = \{Q^i, Q^j\}. \quad (63)$$

The reason this is written backwards here is to emphasize that one can regard supersymmetric charges Q^i as square roots of the translation operator. Recall how the translation operator acts on a bosonic field:

$$\phi(x+a) = e^{ia\cdot P}\phi(x)e^{-ia\cdot P}, \quad (64)$$

where the momentum operator P is the generator of infinitesimal translations:

$$i[P_\mu, \phi(x)] = \partial_\mu\phi(x). \quad (65)$$

Expanding the formula (64), we find the following expression for a small finite translation:

$$\delta_a\phi(x) \equiv \phi(x+a) - \phi(x) \simeq a_\mu\partial^\mu\phi(x) = ia_\mu[P^\mu, \phi(x)]. \quad (66)$$

Following this deconstruction of translations, we now can see better how the supersymmetric charge can be regarded, in some sense, as the square root: '$Q \sim \sqrt{P}$', just as the Dirac equation can be regarded as the square root of the Klein-Gordon equation. There is an exact supersymmetric analogue of (66):

$$\delta_{\bar{E}}\phi(x) = \sqrt{2}\bar{E}\psi(x) = i\sqrt{2}\bar{E}[Q, \phi(x)]. \tag{67}$$

By analogy with (64,66), one may consider the spinor $\bar{E}$ as a sort of 'superspace' coordinate, and one can combine the bosonic field $\phi(x)$ and its fermionic partner $\psi(x)$ into a superfield:

$$\delta_{\bar{E}}\phi(x) = \Phi(x, \bar{E}) - \Phi(x) : \quad \Phi(x, \bar{E}) \equiv \phi(x) + \sqrt{2}\bar{E}\psi(x). \tag{68}$$

At this level, the introduction of superspace and superfields may appear superfluous, but it gives deeper insights into the theory and facilitates the derivation of many important results, such as the non-renormalization theorems of supersymmetry, that we discuss next. In some sense, the next generation of accelerators such as the LHC is 'guaranteed' to discover extra dimensions, either bosonic ones [26], or fermionic.

Many remarkable no-renormalization theorems can be proved in supersymmetric field theories [27]. First and foremost, they have no quadratic divergences. One way to understand this is to compare the renormalizations of bosonic and fermionic mass terms:

$$m_B^2|\phi|^2 \leftrightarrow m_F\bar{\psi}\psi. \tag{69}$$

We know well that fermion masses m_F can only be renormalized logarithmically. Since supersymmetry guarantees that $m_B = m_F$, it follows that there can be no quadratic divergence in m_B. Going further, chiral symmetry guarantees that the one-loop renormalization of a fermion mass has the general multiplicative form:

$$\delta m_F = \mathcal{O}(\frac{\alpha}{\pi}) \, m_F \ln(\frac{\mu_1}{\mu_2}), \tag{70}$$

where $\mu_{1,2}$ are different renormalization scales. This means that if m_F (and hence also m_B) vanish at the tree level in a supersymmetric theory, then both m_F and m_B remain zero after renormalization. This is one example of the reduction in the number of logarithmic divergences in a supersymetric theory.

In general, there is no intrinsic renormalization of any superpotential parameters, including the Yukawa couplings λ, apart from overall multiplicative factors due to wave-function renormalizations:

$$\Phi \to Z\Phi, \tag{71}$$

which are universal for both the bosonic and fermionic components ϕ, ψ in a given superfield Φ. However, gauge couplings *are* renormalized, though the β-function is changed:

$$\beta(g) \neq 0: \quad -11N_c \rightarrow -9N_c \tag{72}$$

at one-loop order in an $SU(N_c)$ supersymmetric gauge theory with no matter, as a result of the extra gaugino contributions.

There are even fewer divergences in theories with more supersymmetries. For example, there is only a finite number of divergent diagrams in a theory with $N = 2$ supersymmetries, which may be cancelled by imposing a few simple relations on the spectrum of supermultiplets. Finally, there are no divergences at all in theories with $N = 4$ supersymmetries, which obey automatically the necessary finiteness conditions.

Many theorists from Dirac onwards have found the idea of a completely finite theory attractive, so it is natural to ask whether theories with $N \geq 2$ supersymmetries could be interesting as realistic field theories. Unfortunately, the answer is 'not immediately', because they do not allow the violation of parity. To see why, consider the simplest possible extended supersymmetric theory containing an $N = 2$ matter multiplet, which contains both left- and right-handed fermions with helicities $\pm 1/2$. Suppose that the left-handed fermion with helicity $+1/2$ sits in a representation R of the gauge group. Now act on it with either of the two supersymmetry charges $Q_{1,2}$: they each yield bosons, that each sit in the same representation R. Now act on either of these with the other supercharge, to obtain a right-handed fermion with helicity $-1/2$: this must also sit in the same representation R of the gauge group. Hence, left- and right-handed fermions have the same interactions, and parity is conserved. There is no way out using gauginos, because they are forced to sit in adjoint representations of the gauge group, and hence also cannot distinguish between right and left.

Thus, if we want to make a supersymmetric extension of the Standard Model, it had better be with just $N = 1$ supersymmetry, and this is what we do in the next Section.

2.4   Building Supersymmetric Models

Any supersymmetric model is based on a Lagrangian that contains a supersymmetric part and a supersymmetry-breaking part:

$$\mathcal{L} = \mathcal{L}_{susy} + \mathcal{L}_{susy\times}. \tag{73}$$

We discuss the supersymmetry-breaking part $\mathcal{L}_{susy\times}$ in the next Lecture: here we concentrate on the supersymmetric part $\mathcal{L}_{susy}$. The minimal supersymmetric extension of the Standard Model (MSSM) has the same gauge interactions as the Standard Model, and Yukawa interactions that are closely related. They are based on a superpotential W that is a cubic function of complex superfields corresponding to left-handed fermion fields. Conventional left-handed lepton and quark doublets are denoted L, Q, and right-handed fermions are introduced via their conjugate fields, which are left-handed, $e_R \to E^c, u_R \to U^c, d_R \to D^c$. In terms of these,

$$W = \Sigma_{L,E^c} \lambda_L L E^c H_1 + \Sigma_{Q,U^c} \lambda_U Q U^c H_2 + \Sigma_{Q,D^c} \lambda_D Q D^c H_1 + \mu H_1 H_2. \tag{74}$$

A few words of explanation are warranted. The first three terms in (74) yield masses for the charged leptons, charge-$(+2/3)$ quarks and charge-$(-1/3)$ quarks respectively. All of the Yukawa couplings $\lambda_{L,U,D}$ are 3×3 matrices in flavour space, whose diagonalizations yield the mass eigenstates and Cabibbo-Kobayashi-Maskawa mixing angles.

Note that two distinct Higgs doublets $H_{1,2}$ have been introduced, for two important reasons. One reason is that the superpotential must be an analytic polynomial: as we saw in (74), it cannot contain both H and H^*, whereas the Standard Model uses both of these to give masses to all the quarks and leptons with just a single Higgs doublet. The other reason is to cancel the triangle anomalies that destroy the renormalizability of a gauge theory. Ordinary Higgs boson doublets do not contribute to these anomalies, but the fermions in Higgs supermultiplets do, and two doublets are required to cancel each others' contributions. Once two Higgs supermultiplets have been introduced, there is the possibility, even the necessity, of a bilinear term $\mu H_1 H_2$ coupling them together.

Once the MSSM superpotential (74) has been specified, the effective potential is also fixed:

$$V = \Sigma_i |F^i|^2 + \frac{1}{2}\Sigma_a(D^a)^2 : \quad F_i^* \equiv \frac{\partial W}{\partial \phi^i}, \quad D^a \equiv g_a \phi_i^*(T^a)_j^i \phi^j, \tag{75}$$

according to the rules explained earlier in this Lecture, where the sums run over the different chiral fields i and the $SU(3), SU(2)$ and $U(1)$ gauge-group factors a.

There are important possible variations on the MSSM superpotential (74), which are impossible in the Standard Model, but are allowed by the gauge symmetries of the MSSM supermultiplets. These are additional superpotential terms that violate the quantity known as R parity:

$$R \equiv (-1)^{3B+L+2S}, \tag{76}$$

where B is baryon number, L is lepton number, and S is spin. It is easy to check that $R = +1$ for all the particles in the Standard Model, and $R = -1$ for all their spartners, which have identical values of B and L, but differ in spin by half a unit. Clearly, R would be conserved if both B and L were conserved, but this is not automatic. Consider the following superpotential terms:

$$\lambda_{ijk}L_iL_jE_k^c + \lambda'_{ijk}L_iQ_jD_k^c + \lambda''_{ijk}U_i^cD_j^cD_k^c + \epsilon_i HL_i, \qquad (77)$$

which are visibly $SU(3) \times SU(2) \times U(1)$ symmetric. The first term in (77) would violate L, causing for example $\tilde{\ell} \to \ell + \ell$, the second would violate both B and L, causing for example $\tilde{q} \to q + \ell$, the third would violate B, causing for example $\tilde{q} \to \bar{q} + \bar{q}$, and the last would violate L by causing $H \leftrightarrow L_i$ mixing. These interactions would provide many exciting signatures for supersymmetry, such as dilepton events, jets plus leptons and multijet events. Such interactions are constrained by direct searches, by the experimental limits on flavour-changing interactions and other rare processes, and by cosmology: they would tend to wipe out the baryon asymmetry of the Universe if they are too strong [28]. They would also cause the lightest supersymmetric particle to be unstable, not necessarily a disaster in itself, but it would remove an excellent candidate for the cold dark matter that apparently abounds throughout the Universe. For simplicity, the conservation of R parity will be assumed in the rest of these Lectures.

We now look briefly at the construction of supersymmetric GUTs, of which the minimal version is based on the group $SU(5)$ [29]. As in the transition from the Standard Model to the MSSM, one simply extends the conventional GUT multiplets to supermultiplets, so that matter particles are assigned to $\bar{5}$ representations $\bar{F}$ and 10 representations T, one doubles the electroweak Higgs fields to include both $H, \bar{H}$ in $5, \bar{5}$ representations, and one postulates a 24 representation Φ to break the $SU(5)$ GUT symmetry down to $SU(3) \times SU(2) \times U(1)$. The superpotential for the Higgs sector takes the general form

$$W_5 = (\mu + \frac{3}{2}\lambda M)H\bar{H} + \lambda H\Phi\bar{H} + f(\Phi). \qquad (78)$$

Here, $f(\Phi)$ is chosen so that the vacuum expectation value of Φ has the form

$$< 0|\Phi|0 > = M \times \mathrm{diag}(1,1,1,-\frac{3}{2},-\frac{3}{2}). \qquad (79)$$

The coefficient of the $H\bar{H}$ term has been chosen so that it almost cancels with the term $\propto H < 0|\Phi|0 > \bar{H}$ coming from the second term in (78), *for the last two components.* In this way, the triplet components of $H, \bar{H}$ acquire large masses $\propto M$, whilst the last two may acquire a vacuum expectation value: $< 0 \mid H \mid 0 > = \text{column}(0,0,0,0,v), < 0|\bar{H}|0 > = \text{column}(0,0,0,0,\bar{v})$, once supersymmetry breaking and radiative corrections are taken into account, as in the next Lecture.

In order that $v, \bar{v} \sim 100$ GeV, it is necessary that the residual $H\bar{H}$ mixing term $\mu \lesssim 1$ TeV. Since, as we recall shortly, $M \sim 10^{16}$ GeV, this means that the parameters of W_5 (78) must be tuned finely to one part in 10^{13}. This fine-tuning may appear very unreasonable, but it is technically natural, in the sense that there are no big radiative corrections. Thanks to the supersymmetric no-renormalization theorem for superpotential parameters, we know that $\delta\lambda, \delta\mu = 0$, apart from wave-function renormalization factors. Thus, if we adjust the input parameters of (78) so that μ is small, it will stay small. However, this begs the more profound question: how did μ get to be so small in the first place?

As already mentioned, a striking piece of circumstantial evidence in favour of the idea of supersymmetric grand unification is provided by the measurements of low-energy gauge couplings at LEP and elsewhere [3]. The three gauge couplings of the Standard Model are renormalized as follows:

$$\frac{dg_a^2}{dt} = b_a \frac{g_a^4}{16\pi^2} + \dots, \tag{80}$$

at one-loop order, and the corresponding value of the electroweak mixing angle $\sin^2 \theta_W (m_Z)$ is given at the one-loop level by:

$$\sin^2 \theta_W (m_Z) = \frac{g'^2}{g_2^2 + g'^2} = \frac{3}{5} \frac{g_1^2(m_Z)}{g_2^2(m_Z) + \frac{3}{5}g_1^2(m_Z)} \tag{81}$$

$$= \frac{1}{1 + 8x} [3x + \frac{\alpha_{em}(m_Z)}{\alpha_3(m_Z)}], \tag{82}$$

where

$$x \equiv \frac{1}{5}(\frac{b_2 - b_3}{b_1 - b_2}). \tag{83}$$

One can distinguish the predictions of different GUTs by their different values of the renormalization coefficients b_i, which are in turn determined by the spectra of light particles around the electroweak scale. In the cases of the Standard Model and the MSSM, these are:

$$\frac{4}{3}N_G - 11 \leftarrow b_3 \rightarrow 2N_G - 9 = -3 \tag{84}$$

$$\frac{1}{6}N_H + \frac{4}{3}N_G - \frac{22}{3} \leftarrow b_2 \rightarrow \frac{1}{2}N_H + 2N_G - 6 = +1 \tag{85}$$

$$\frac{1}{10}N_H + \frac{4}{3}N_G \leftarrow b_1 \rightarrow \frac{3}{10}N_H + 2N_G = \frac{33}{5} \tag{86}$$

$$\frac{23}{218} = 0.1055 \leftarrow \quad x \quad \rightarrow \frac{1}{7} . \tag{87}$$

If we insert the best available values of the gauge couplings:

$$\alpha_{em} = \frac{1}{128}; \ \alpha_3(m_Z) = 0.119 \pm 0.003, \ \sin^2 \theta_W(m_Z) = 0.2315, \tag{88}$$

we find the following value:

$$x = \frac{1}{6.92 \pm 0.07} . \tag{89}$$

We see that experiment strongly favours the inclusion of supersymmetric particles in the renormalization-group equations, as required if the effective low-energy theory is the MSSM (87), as in a simple supersymmetric GUT such as the minimal $SU(5)$ model introduced above.

3. Towards Realistic Models

3.1 Supersymmetry Breaking

This is clearly necessary: $m_e \neq m_{\tilde{e}}, m_\gamma \neq m_{\tilde{\gamma}}$, etc. The Big Issue is whether the breaking of supersymmetry is explicit, i.e., present already in the underlying Lagrangian of the theory, or whether it is spontaneous, i.e., induced by a non-supersymmetric vacuum state. There are in fact several reasons to disfavour explicit supersymmetry breaking. It is ugly, it would be unlike the way in which gauge symmetry is broken, and it would lead to inconsistencies in supergravity theory. For these reasons, theorists have focused on spontaneous supersymmetry breaking.

If the vacuum is not to be supersymmetric, there must be some fermionic state χ that is coupled to the vacuum by the supersymmetry charge Q:

$$< 0|Q|\chi > \equiv f_\chi^2 \neq 0. \tag{90}$$

The fermion χ corresponds to a Goldstone boson in a spontaneously broken bosonic symmetry, and therefore is often termed a Goldstone fermion or a Goldstino.

There is just one small problem in globally supersymmetric models, i.e., those without gravity: spontaneous supersymmetry breaking necessarily entails a positive vacuum energy E_0. To see this, consider the vacuum expectation value of the basic supersymmetry anticommutator:

$$\{Q, Q\} \propto \gamma_\mu P^\mu. \tag{91}$$

According to (90), there is an intermediate state χ, so that

$$< 0|\{Q, Q\}|0 > = |< 0|Q|\chi >|^2 = f_\chi^4 \propto < 0|P_0|0 > = E_0, \tag{92}$$

where we have used Lorentz invariance to set the spatial components $< 0|P_i|0 >= 0$. Spontaneous breaking of global supersymmetry (90) requires

$$E_0 = f_\chi^4 \neq 0. \qquad (93)$$

The next question is how to generate non-zero vacuum energy. Hints are provided by the effective potential in a globally supersymmetric theory:

$$V = \Sigma_i |\frac{\partial W}{\partial \phi^i}|^2 + \frac{1}{2} \Sigma_\alpha g_\alpha^2 |\phi^* T^\alpha \phi|^2. \qquad (94)$$

It is apparent from this expression that either the first 'F term' or the second 'D term' must be positive definite.

The option $D > 0$ requires constructing a model with a $U(1)$ gauge symmetry [30]. The simplest example contains just one chiral (matter) supermultiplet with unit charge, for which the effective potential is:

$$V_D = \frac{1}{2}(\xi + g\phi^*\phi)^2. \qquad (95)$$

the extra constant term ξ is not allowed in a non-Abelian theory, which is why one must use a $U(1)$ theory. We see immediately that the minimum of the effective potential (95) is reached when $< 0|\phi|0 >= 0$, in which case $V_F = 1/2\xi^2 > 0$ and supersymmetry is broken spontaneously. Indeed, it is easy to check that, in this vacuum:

$$m_\phi = g\xi, \, m_\psi = 0, \, m_V = m_{\tilde{V}} = 0, \qquad (96)$$

exhibiting explicitly the boson-fermion mass splitting in the (ϕ, ψ) supermultiplet. Unfortunately, this example cannot be implemented with the $U(1)$ of electromagnetism in the Standard Model, because there are fields with both signs of the hypercharge Y, enabling V_D to vanish. So, one needs a new $U(1)$ gauge group factor, and many new fields in order to cancel triangle anomalies. For these reasons, D-breaking models did not attract much attention for quite some time, though they have had a revival in the context of string theory [31].

The option $F > 0$ also requires additional chiral (matter) fields with somewhat 'artificial' couplings [32]: again, those of the Standard Model do not suffice. The simplest example uses three chiral supermultiplets A, B, C with the superpotential

$$W = \alpha AB^2 + \beta C(B^2 - m^2). \qquad (97)$$

using the rules given in the previous Lecture, it is easy to calculate the corresponding F terms:

$$F_A = \alpha B^2, \, F_B = 2B(\alpha A + \beta C), \, F_C = \beta(B^2 - m^2), \qquad (98)$$

and hence the effective potential

$$V_F = \Sigma_i |F_i|^2 = 4|B(\alpha A + \beta C)|^2 + |\alpha B^2|^2 + |\beta(B^2 - m^2)|^2. \quad (99)$$

Likewise, it is not difficult to check that the three different positive-semidefinite terms in (99) cannot all vanish simultaneously. Hence, necessarily $V_F > 0$, and hence supersymmetry *must* be broken.

The principal outcome of this brief discussion is that there are no satisfactory models of global superysmmetry breaking, which provided some of the motivation for studying local supersymmetry, i.e., supergravity theory.

3.2 Supergravity and Local Supersymmetry Breaking

So far, we have considered global supersymmetry transformations, in which the infinitesimal transformation spinor E is constant throughout space. Now we consider the possibility of a space-time-dependent field $E(x)$. Why?

This step of making symmetries local has become familiar with bosonic symmetries, where it leads to gauge theories, so it is natural to try the analogous step with fermionic symmetries. Moreover, as we see shortly, it leads to an elegant mechanism for spontaneous supersymmetry breaking, again by analogy with gauge theories, the super-Higgs mechanism. Further, as we also see shortly, making supersymmetry local necessarily involves gravity, and even opens the prospect of unifying all the particle interactions and matter fields with extended supersymmetry transformations:

$$G(J = 2) \rightarrow \tilde{G}(J = 3/2) \rightarrow V(J = 1) \rightarrow q, \ell(J = 1/2) \rightarrow H(J = 0)$$
$$(100)$$

in supergravity with $N > 1$ supercharges. In (100), G denotes the graviton, and $\tilde{G}$ the spin-3/2 gravitino, which accompanies it in the graviton supermultiplet:

$$\begin{pmatrix} G \\ \tilde{G} \end{pmatrix} = \begin{pmatrix} 2 \\ \frac{3}{2} \end{pmatrix} \quad (101)$$

Supergravity is in any case an essential ingredient in the discussion of gravitational interactions of supersymmetric particles, needed, for example, for any meaningful discussion of the cosmological constant.

The mechanism for the spontaneous breaking of local supersymmetry is known as the super-Higgs effect [33, 34]. You recall that, in the conventional Higgs effect in spontaneously broken gauge theories, a massless

Goldstone boson is 'eaten' by a gauge boson to provide it with the third polarization state it needs to become massive:

$$(2 \times V_{m=0}) + (1 \times GB) = (3 \times V_{m \neq 0}). \tag{102}$$

In a locally supersymmetric theory, the two polarization states of the massless Goldstone fermion (Goldstino) are 'eaten' by a massless gravitino, giving it the total of four polarization states it needs to become massive:

$$(2 \times \psi^{\mu}_{m=0}) + (2 \times GF) = (4 \times \psi^{\mu}_{m \neq 0}). \tag{103}$$

This process clearly involves the breakdown of local supersymmetry, since the end result is to give the gravitino a different mass from the graviton: $m_G = 0 \neq m_{\tilde{G}} \neq 0$. It is indeed the only known consistent way of breaking local supersymmetry, just as the Higgs mechanism is the only consistent way of breaking gauge symmetry. We shall not go here through all the details of the super-Higgs effect, but there is one noteworthy feature: this local breaking of supersymmetry can be achieved with zero vacuum energy:

$$< 0|V|0 > = 0 \leftrightarrow \Lambda = 0. \tag{104}$$

There is no inconsistency between local supersymmetry breaking and a vanishing cosmological constant Λ, unlike the case of global superymmetry breaking that we discussed earlier.

3.3   Effective Low-Energy Theory

The coupling of matter particles to supergravity is more complicated than the globally supersymmetric case discussed in the previous lecture. Therefore, it is not developed here in detail. Instead, a few key results are presented without proof. The form of the effective low-energy theory suggested by spontaneous supersymmetry breaking in supergravity is:

$$-\frac{1}{2}\Sigma_a m_{1/2_a} \tilde{V}_\alpha \tilde{V}_\alpha - \Sigma_i m^2_{0_i}|\phi^i|^2 - (\Sigma_\lambda A_\lambda \lambda \phi^3 + \Sigma_\mu B_\mu \mu \phi^2 + \text{Herm.Conj.}),$$
$$\tag{105}$$

which contains many free parameters and phases. The breaking of supersymmetry in the effective low-energy theory (105) is explicit but 'soft', in the sense that the renormalization of the parameters $m_{1/2_a}, m_{0_i}, A_\lambda$ and B_μ is logarithmic. Of course, these parameters are not considered to be fundamental, and the underlying mechanism of supersymmetry breaking is thought to be spontaneous, for the reasons described at the beginning of this lecture.

The logarithmic renormalization of the parameters means that one can calculate their low-energy values in terms of high-energy inputs from a supergravity or superstring theory, using standard renormalization-group equations [35]. In the case of the low-energy gaugino masses M_a, the renormalization is multiplicative and identical with that of the corresponding gauge coupling α_a at the one-loop level:

$$\frac{M_a}{m_{1/2_a}} = \frac{\alpha_a}{\alpha_{GUT}} \tag{106}$$

where we assume GUT unification of the gauge couplings at the input supergravity scale. In the case of the scalar masses, there is both multiplicative renormalization and renormalization related to the gaugino masses:

$$\frac{\partial m_{0_i}^2}{\partial t} = \frac{1}{16\pi^2}[\lambda^2(m_0^2 + A_\lambda^2) - g_a^2 M_a^2] \tag{107}$$

at the one-loop level, where $t \equiv \ln(Q^2/m_{GUT}^2)$, and the $\mathcal{O}(1)$ group-theoretical coefficients have been omitted. In the case of the first two generations, the first terms in (107) are negligible, and one may integrate (107) trivially to obtain effective low-energy parameters

$$m_{0_i}^2 = m_0^2 + C_i m_{1/2}^2, \tag{108}$$

where universal inputs are assumed, and the coefficients C_i are calculable in any given model. The first terms in (107) are, however, important for the third generation and for the Higgs bosons of the MSSM, as we now see.

Notice that the signs of the first terms in (107) are positive, and that of the last term negative. This means that the last term tends to *increase* $m_{0_i}^2$ as the renormalization scale Q *decreases*, an effect seen in Fig. 1. The positive signs of the first terms mean that they tend to *decrease* $m_{0_i}^2$ as Q *decreases*, an effect seen for a Higgs squared-mass in Fig. 1. Specifically, the negative effect on H_u seen in Fig. 1 is due to its large Yukawa coupling to the t quark: $\lambda_t \sim g_{2,3}$. The exciting aspect of this observation is that spontaneous electroweak symmetry breaking is possible [35] when $m_H^2(Q) < 0$, as occurs in Fig. 1. Thus the spontaneous breaking of supersymmetry, which normally provides $m_0^2 > 0$, and renormalization, which then drive $m_H^2(Q) < 0$, conspire to make spontaneous electroweak symmetry breaking possible. Typically, this occurs at a renormalization scale that is exponentially smaller than the input supergravity scale:

$$\frac{m_W}{m_P} = exp(\frac{-\mathcal{O}(1)}{\alpha_t}) : \alpha_t \equiv \frac{\lambda_t^2}{4\pi}. \tag{109}$$

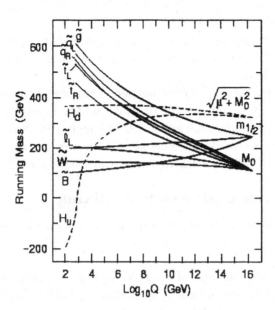

Figure 1. The renormalization-group evolution of the soft supersymmetry-breaking parameters in the MSSM, showing the increase in the squark and slepton masses as the renormalization scale decreases, whilst the Higgs squared-mass may become negative, triggering electroweak symmetry breaking.

Typical dynamical calculations find that $m_W \sim 100$ GeV emerges naturally if $m_t \sim 60$ to 200 GeV, and this was in fact one of the first suggestions that m_t might be as high as was subsequently observed.

To conclude this section, let us briefly review the reasons why soft supersymmetry breaking might be universal, at least in some respects.

There are important constraints on the mass differences of squarks and sleptons with the same internal quantum numbers, coming from flavour-changing neutral interactions [36]. These are suppressed in the Standard Model by the Glashow-Iliopoulos-Maiani mechanism [37], which limits them to magnitudes $\propto \Delta m_q^2 / m_W^2$ for small squared-mass differences Δm_q^2. Depending on the process considered, it is either necessary or desirable that sparticle exchange contributions, which would have expected magnitudes $\sim \Delta m_{\tilde{q}}^2 / m_{\tilde{q}}^2$, be suppressed by a comparable factor. In particular, one would like

$$m_0^2(\text{first generation}) - m_0^2(\text{second generation}) \sim \delta m_q^2 \times \frac{m_{\tilde{q}}^2}{m_W^2}. \quad (110)$$

The limits on third-generation sparticle masses from flavour-changing neutral interactions are less severe, and the first/second-generation degeneracy could be relaxed if $m_{\tilde{q}}^2 \gg m_W^2$, but models with physical values of m_0^2 degenerate to $\mathcal{O}(m_q^2)$ are certainly preferred. However, this restriction is not respected in many low-energy effective theories derived from string models.

The desirability of degeneracy between sparticles of different generations help encourage some people to study models in which this property would emerge naturally, such as models of gauge-mediated supersymmetry breaking or extra dimensions [26]. However, for the rest of these lectures we shall mainly stick to familiar old supergravity.

3.4 Sparticle Masses and Mixing

We now progress to a more complete discussion of sparticle masses and mixing.

Sfermions : Each flavour of charged lepton or quark has both left- and right-handed components $f_{L,R}$, and these have separate spin-0 boson superpartners $\tilde{f}_{L,R}$. These have different isospins $I = \frac{1}{2}$, 0, but may mix as soon as the electroweak gauge symmetry is broken. Thus, for each flavour we should consider a 2×2 mixing matrix for the $\tilde{f}_{L,R}$, which takes the following general form:

$$M_{\tilde{f}}^2 \equiv \begin{pmatrix} m_{\tilde{f}_{LL}}^2 & m_{\tilde{f}_{LR}}^2 \\ m_{\tilde{f}_{LR}}^2 & m_{\tilde{f}_{RR}}^2 \end{pmatrix} \tag{111}$$

The diagonal terms may be written in the form

$$m_{\tilde{f}_{LL,RR}}^2 = \tilde{m}_{\tilde{f}_{L,R}}^2 + m_{\tilde{f}_{L,R}}^{D^2} + m_f^2 \tag{112}$$

where m_f is the mass of the corresponding fermion, $\tilde{m}_{\tilde{f}_{L,R}}^2$ is the soft supersymmetry-breaking mass discussed in the previous section, and $m_{\tilde{f}_{L,R}}^{D^2}$ is a contribution due to the quartic D terms in the effective potential:

$$m_{\tilde{f}_{L,R}}^{D^2} = m_Z^2 \cos 2\beta \ (I_3 + \sin^2 \theta_W Q_{em}) \tag{113}$$

where the term $\propto I_3$ is non-zero only for the $\tilde{f}_L$. Finally, the off-diagonal mixing term takes the general form

$$m_{\tilde{f}_{L,R}}^2 = m_f \left(A_f + \mu_{\cot \beta}^{\tan \beta} \right) \quad \text{for} \quad f = _{u,c,t}^{e,\mu,\tau,d,s,b} \tag{114}$$

It is clear that $\tilde{f}_{L,R}$ mixing is likely to be important for the $\tilde{t}$, and it may also be important for the $\tilde{b}_{L,R}$ and $\tilde{\tau}_{L,R}$ if $\tan \beta$ is large.

We also see from (112) that the diagonal entries for the $\tilde{t}_{L,R}$ would be different from those of the $\tilde{u}_{L,R}$ and $\tilde{c}_{L,R}$, even if their soft supersymmetry-breaking masses were universal, because of the m_f^2 contribution. In fact, we also expect non-universal renormalization of $m_{\tilde{t}_{LL,RR}}^2$ (and also $m_{\tilde{b}_{LL,RR}}^2$ and $m_{\tilde{\tau}_{LL,RR}}^2$ if $\tan\beta$ is large), because of Yukawa effects analogous to those discussed in the previous section for the renormalization of the soft Higgs masses. For these reasons, the $\tilde{t}_{L,R}$ are not usually assumed to be degenerate with the other squark flavours. Indeed, one of the $\tilde{t}$ could well be the lightest squark, perhaps even lighter than the t quark itself [38].

Charginos: These are the supersymmetric partners of the $W^\pm$ and $H^\pm$, which mix through a 2×2 matrix

$$-\frac{1}{2}\,(\tilde{W}^-,\tilde{H}^-)\;M_C\;\begin{pmatrix}\tilde{W}^+ \\ \tilde{H}^+\end{pmatrix} \;+\; \text{herm.conj.} \qquad (115)$$

where

$$M_C \equiv \begin{pmatrix} M_2 & \sqrt{2}m_W\sin\beta \\ \sqrt{2}m_W\cos\beta & \mu \end{pmatrix} \qquad (116)$$

Here M_2 is the unmixed $SU(2)$ gaugino mass and μ is the Higgs mixing parameter introduced in (74). Fig. 2 displays (among other lines to be discussed later) the contour $m_{\chi^\pm} = 91$ GeV for the lighter of the two chargino mass eigenstates [39].

Neutralinos: These are characterized by a 4×4 mass mixing matrix [40], which takes the following form in the $(\tilde{W}^3, \tilde{B}, \tilde{H}_2^0, \tilde{H}_1^0)$ basis:

$$m_N = \begin{pmatrix} M_2 & 0 & \frac{-g_2 v_2}{\sqrt{2}} & \frac{g_2 v_1}{\sqrt{2}} \\ 0 & M_1 & \frac{g'v_2}{\sqrt{2}} & \frac{-g'v_1}{\sqrt{2}} \\ \frac{-g_2 v_2}{\sqrt{2}} & \frac{g'v_2}{\sqrt{2}} & 0 & \mu \\ \frac{g_2 v_1}{\sqrt{2}} & \frac{-g'v_1}{\sqrt{2}} & \mu & 0 \end{pmatrix} \qquad (117)$$

Note that this has a structure similar to M_C (116), but with its entries replaced by 2×2 submatrices. As has already been mentioned, one conventionally assumes that the $SU(2)$ and $U(1)$ gaugino masses $M_{1,2}$ are universal at the GUT or supergravity scale, so that

$$M_1 \simeq M_2\,\frac{\alpha_1}{\alpha_2} \qquad (118)$$

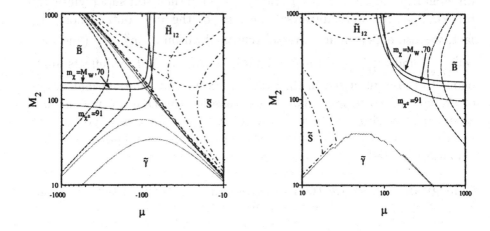

Figure 2. The (μ, M_2) plane characterizing charginos and neutralinos, for (a) $\mu < 0$ and (b) $\mu > 0$, including contours of m_χ and $m_{\chi^\pm}$, and of neutralino purity [39].

so the relevant parameters of (117) are generally taken to be $M_2 = (\alpha_2/\alpha_{GUT})m_{1/2}$, μ and $\tan\beta$.

Fig. 2 also displays contours of the mass of the lightest neutralino χ, as well as contours of its gaugino and Higgsino contents [39]. In the limit $M_2 \to 0$, χ would be approximately a photino and it would be approximately a Higgsino in the limit $\mu \to 0$. Unfortunately, these idealized limits are excluded by unsuccessful LEP and other searches for neutralinos and charginos, as discussed in more detail below.

3.5 The Lightest Supersymmetric Particle

This is expected to be stable in the MSSM, and hence should be present in the Universe today as a cosmological relic from the Big Bang [41, 40]. Its stability arises because there is a multiplicatively-conserved quantum number called R parity, that takes the values $+1$ for all conventional particles and -1 for all sparticles [20]. The conservation of R parity can be related to that of baryon number B and lepton number L, since

$$R = (-1)^{3B+L+2S} \tag{119}$$

where S is the spin. Note that R parity could be violated either spontaneously if $< 0|\tilde{\nu}|0 > \neq 0$ or explicitly if one of the supplementary couplings (77) is present. There could also be a coupling HL, but this can be defined away be choosing a field basis such that $\bar{H}$ is defined as the superfield with a bilinear coupling to H. Note that R parity is not violated by the simplest models for neutrino masses, which have $\Delta L = 0, \pm 2$, nor by simple GUTs, which violate combinations of B and L that leave R invariant. There are three important consequences of R conservation:

1 sparticles are always produced in pairs, e.g., $\bar{p}p \to \tilde{q}\tilde{g}X$, $e^+e^- \to \tilde{\mu}^+ + \tilde{\mu}^-$,

2 heavier sparticles decay to lighter ones, e.g., $\tilde{q} \to q\tilde{g}, \tilde{\mu} \to \mu\tilde{\gamma}$, and

3 the lightest sparticles is stable,

because it has no legal decay mode.

This feature constrains strongly the possible nature of the lightest supersymmetric sparticle. If it had either electric charge or strong interactions, it would surely have dissipated its energy and condensed into galactic disks along with conventional matter. There it would surely have bound electromagnetically or via the strong interactions to conventional nuclei, forming anomalous heavy isotopes that should have been detected. There are upper limits on the possible abundances of such bound relics, as compared to conventional nucleons:

$$\frac{n(\text{relic})}{n(p)} \lesssim 10^{-15} \text{ to } 10^{-29} \tag{120}$$

for 1 GeV $\lesssim m_{\text{relic}} \lesssim$ 1 TeV. These are far below the calculated abundances of such stable relics:

$$\frac{n(\text{relic})}{n(p)} \gtrsim 10^{-6} \ (10^{-10}) \tag{121}$$

for relic particles with electromagnetic (strong) interactions. We may conclude [40] that any supersymmetric relic is probably electromagnetically neutral with only weak interactions, and could in particular not be a gluino. Whether the lightest hadron containing a gluino is charged or neutral, it would surely bind to some nuclei. Even if one pleads for some level of fractionation, it is difficult to see how such gluino nuclei could avoid the stringent bounds established for anomalous isotopes of many species.

Plausible scandidates of different spins are the sneutrinos $\tilde{\nu}$ of spin 0, the lightest neutralino χ of spin 1/2, and the gravitino $\tilde{G}$ of spin 3/2. The sneutrinos have been ruled out by the combination of LEP experiments and direct searches for cosmological relics. The gravitino cannot be ruled out, but we concentrate on the neutralino possibility for the rest of these Lectures.

A very attractive feature of the neutralino candidature for the lightest supersymmetric particle is that it has a relic density of interest to astrophysicists and cosmologists: $\Omega_\chi h^2 = \mathcal{O}(0.1)$ over generic domains of the MSSM parameter space [40], as discussed in the next Lecture. In these domains, the lightest neutralino χ could constitute the cold dark matter favoured by theories of cosmological structure formation.

3.6 Supersymmetric Higgs Bosons

As was discussed in Lecture 2, one expects two complex Higgs doublets $H_2 \equiv (H_2^+, H_2^0)$, $H_1 \equiv (H_1^+, H_1^0)$ in the MSSM, with a total of 8 real degrees of freedom. Of these, 3 are eaten via the Higgs mechanism to become the longitudinal polarization states of the $W^\pm$ and Z^0, leaving 5 physical Higgs bosons to be discovered by experiment. Three of these are neutral: the lighter CP-even neutral h, the heavier CP-even neutral H, the CP-odd neutral A, and charged bosons $H^\pm$. The quartic potential is completely determined by the D terms

$$V_4 = \frac{g^2 + g'^2}{8} \left(|H_1^0|^2 - |H_2^0|^2 \right) \tag{122}$$

for the neutral components, whilst the quadratic terms may be parametrized at the tree level by

$$m_{H_1}^2 \, |H_1|^2 + m_{H_2}^2 \, |H_2|^2 + (m_3^2 \, H_1 H_2 + \text{herm.conj.}) \tag{123}$$

where $m_3^2 = B_\mu \mu$. One combination of the three parameters $(m_{H_1}^2, m_{H_2}^2, m_3^2)$ is fixed by the Higgs vacuum expectation $v = \sqrt{v_1^2 + v_2^2} = 246$ GeV, and the other two combinations may be rephrased as $(m_A, \tan\beta)$. These characterize all Higgs masses and couplings in the MSSM at the tree level. Looking back at (122), we see that the gauge coupling strength of the quartic interactions suggests a relatively low mass for at least the lightest MSSM Higgs boson h, and this is indeed the case, with $m_h \leq m_Z$ at the tree level:

$$m_h^2 = m_Z^2 \, \cos^2 2\beta \tag{124}$$

This raised considerable hope that the lightest MSSM Higgs boson could be discovered at LEP, with its prospective reach to $m_H \sim 100$ GeV.

However, radiative corrections to the Higgs masses are calculable in a supersymmetric model (this was, in some sense, the whole point of introducing supersymmetry!), and they turn out to be non-negligible for $m_t \sim 175$ GeV [22]. Indeed, the leading one-loop corrections to m_h^2

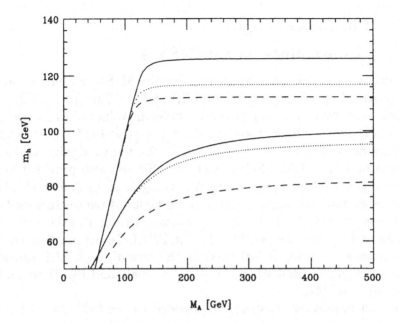

Figure 3. The lightest Higgs boson mass in the MSSM, for different values of $\tan\beta$ and the CP-odd Higgs boson mass M_A.

depend quartically on m_t:

$$\Delta m_h^2 = \frac{3m_t^4}{4\pi^2 v^2} \ \ln\ \left(\frac{m_{\tilde{t}_1} m_{\tilde{t}_2}}{m_t^2}\right) + \frac{3m_t^4 \hat{A}_t^2}{8\pi^2 v^2} \left[2h(m_{\tilde{t}_1}^2, m_{\tilde{t}_2}^2) + \hat{A}_t^2 \ f(m_{\tilde{t}_1}^2, m_{\tilde{t}_2}^2)\right] + \ldots \tag{125}$$

where $m_{\tilde{t}_{1,2}}$ are the physical masses of the two stop squarks $\tilde{t}_{1,2}$ to be discussed in more detail shortly, $\hat{A}_t \equiv A_t - \mu \cot\beta$, and

$$h(a,b) \equiv \frac{1}{a-b} \ \ln\left(\frac{a}{b}\right) \ , \ \ f(a,b) = \frac{1}{(a-b)^2} \left[2 - \frac{a+b}{a-b} \ \ln\left(\frac{a}{b}\right)\right] \tag{126}$$

Non-leading one-loop corrections to the MSSM Higgs masses are also known, as are corrections to coupling vertices, two-loop corrections and renormalization-group resummations [42]. For $m_{\tilde{t}_{1,2}} \lesssim 1$ TeV and a plausible range of A_t, one finds

$$m_h \lesssim 130 \text{ GeV} \tag{127}$$

as seen in Fig. 14. There we see the sensitivity of m_h to $(m_A, \tan\beta)$, and we also see how m_A, m_H and $m_{H^\pm}$ approach each other for large m_A.

4.   Phenomenology

4.1   Constraints on the MSSM

Important experimental constraints on the MSSM parameter space are provided by direct searches at LEP and the Tevatron collider, as compiled in the $(m_{1/2}, m_0)$ planes for different values of $\tan\beta$ and the sign of μ in Fig. 4. One of these is the limit $m_{\chi^\pm} \gtrsim 103.5$ GeV provided by chargino searches at LEP [43], where the fourth significant figure depends on other CMSSM parameters. LEP has also provided lower limits on slepton masses, of which the strongest is $m_{\tilde{e}} \gtrsim 99$ GeV [44], again depending only sightly on the other CMSSM parameters, as long as $m_{\tilde{e}} - m_\chi \gtrsim 10$ GeV. The most important constraints on the u, d, s, c, b squarks and gluinos are provided by the FNAL Tevatron collider: for equal masses $m_{\tilde{q}} = m_{\tilde{g}} \gtrsim 300$ GeV. In the case of the $\tilde{t}$, LEP provides the most stringent limit when $m_{\tilde{t}} - m_\chi$ is small, and the Tevatron for larger $m_{\tilde{t}} - m_\chi$ [43].

Another important constraint is provided by the LEP lower limit on the Higgs mass: $m_H > 114.4$ GeV [46]. This holds in the Standard Model, for the lightest Higgs boson h in the general MSSM for $\tan\beta \lesssim 8$, and almost always in the CMSSM for all $\tan\beta$, at least as long as CP is conserved [2]. Since m_h is sensitive to sparticle masses, particularly $m_{\tilde{t}}$, via loop corrections:

$$\delta m_h^2 \propto \frac{m_t^4}{m_W^2} \ln\left(\frac{m_{\tilde{t}}^2}{m_t^2}\right) + \ldots \tag{128}$$

the Higgs limit also imposes important constraints on the soft supersymmetry-breaking CMSSM parameters, principally $m_{1/2}$ [50] as seen in Fig. 4. The constraints are here evaluated using FeynHiggs [42], which is estimated to have a residual uncertainty of a couple of GeV in m_h.

Also shown in Fig. 4 is the constraint imposed by measurements of $b \to s\gamma$ [47]. These agree with the Standard Model, and therefore provide bounds on MSSM particles, such as the chargino and charged Higgs masses, in particular. Typically, the $b \to s\gamma$ constraint is more important for $\mu < 0$, as seen in Fig. 4a and c, but it is also relevant for $\mu > 0$, particularly when $\tan\beta$ is large as seen in Fig. 4d.

The final experimental constraint we consider is that due to the measurement of the anomolous magnetic moment of the muon. Following its first result last year [51], the BNL E821 experiment has recently reported a new measurement [52] of $a_\mu \equiv \frac{1}{2}(g_\mu - 2)$, which deviates by 3.0

---

[2]The lower bound on the lightest MSSM Higgs boson may be relaxed significantly if CP violation feeds into the MSSM Higgs sector [49].

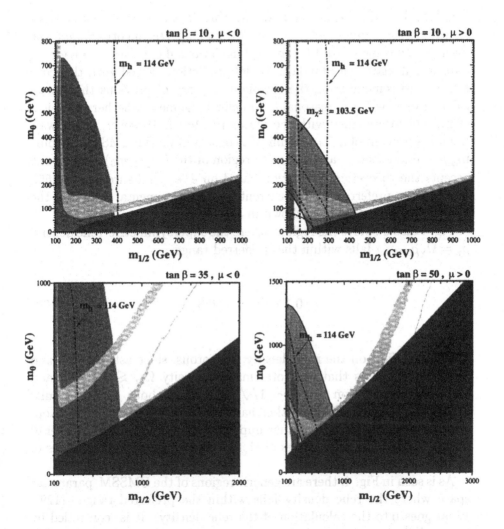

Figure 4. Compilations of phenomenological constraints on the CMSSM for (a) $\tan\beta = 10, \mu < 0$, (b) $\tan\beta = 10, \mu > 0$, (c) $\tan\beta = 35, \mu < 0$ and (d) $\tan\beta = 50, \mu > 0$, assuming $A_0 = 0, m_t = 175$ GeV and $m_b(m_b)^{\overline{MS}}_{SM} = 4.25$ GeV [45]. The near-vertical lines are the LEP limits $m_{\chi^\pm} = 103.5$ GeV (dashed and black) [43], shown in (b) only, and $m_h = 114$ GeV (dotted and red) [46]. Also, in the lower left corner of (b), we show the $m_{\tilde{e}} = 99$ GeV contour [44]. In the dark (brick red) shaded regions, the LSP is the charged $\tilde{\tau}_1$, so this region is excluded. The light (turquoise) shaded areas are the cosmologically preferred regions with $0.1 \leq \Omega_\chi h^2 \leq 0.3$ [45]. The medium (dark green) shaded regions that are most prominent in panels (a) and (c) are excluded by $b \to s\gamma$ [47]. The shaded (pink) regions in the upper right regions show the $\pm 2\sigma$ ranges of $g_\mu - 2$. For $\mu > 0$, the $\pm 2(1)\sigma$ contours are also shown as solid (dashed) black lines [48].

standard deviations from the best available Standard Model predictions based on low-energy $e^+e^- \to$ hadrons data [53]. On the other hand, the discrepancy is more like 1.6 standard deviations if one uses $\tau \to$ hadrons data to calculate the Standard Model prediction. Faced with this confusion, and remembering the chequered history of previous theoretical calculations [54], it is reasonable to defer judgement whether there is a significant discrepancy with the Standard Model. However, either way, the measurement of a_μ is a significant constraint on the CMSSM, favouring $\mu > 0$ in general, and a specific region of the $(m_{1/2}, m_0)$ plane if one accepts the theoretical prediction based on $e^+e^- \to$ hadrons data [55]. The regions preferred by the current $g - 2$ experimental data and the $e^+e^- \to$ hadrons data are shown in Fig. 4.

Fig. 4 also displays the regions where the supersymmetric relic density $\rho_\chi = \Omega_\chi \rho_{critical}$ falls within the preferred range

$$0.1 < \Omega_\chi h^2 < 0.3 \qquad (129)$$

The upper limit on the relic density is rigorous, since astrophysics and cosmology tell us that the total matter density $\Omega_m \lesssim 0.4$ [56], and the Hubble expansion rate $h \sim 1/\sqrt{2}$ to within about 10 % (in units of 100 km/s/Mpc). On the other hand, the lower limit in (129) is optional, since there could be other important contributions to the overall matter density. Smaller values of $\Omega_\chi h^2$ correspond to smaller values of $(m_{1/2}, m_0)$, in general.

As is seen in Fig. 4, there are generic regions of the CMSSM parameter space where the relic density falls within the preferred range (129). What goes into the calculation of the relic density? It is controlled by the annihilation cross section [40]:

$$\rho_\chi = m_\chi n_\chi, \quad n_\chi \sim \frac{1}{\sigma_{ann}(\chi\chi \to \ldots)}, \qquad (130)$$

where the typical annihilation cross section $\sigma_{ann} \sim 1/m_\chi^2$. For this reason, the relic density typically increases with the relic mass, and this combined with the upper bound in (129) then leads to the common expectation that $m_\chi \lesssim \mathcal{O}(1)$ GeV.

However, there are various ways in which the generic upper bound on m_χ can be increased along filaments in the $(m_{1/2}, m_0)$ plane. For example, if the next-to-lightest sparticle (NLSP) is not much heavier than χ: $\Delta m/m_\chi \lesssim 0.1$, the relic density may be suppressed by coannihilation: $\sigma(\chi+\text{NLSP}\to \ldots)$ [57]. In this way, the allowed CMSSM region may acquire a 'tail' extending to larger sparticle masses. An example of this possibility is the case where the NLSP is the lighter stau: $\tilde{\tau}_1$ and $m_{\tilde{\tau}_1} \sim m_\chi$, as seen in Fig. 4 [58].

Another mechanism for extending the allowed CMSSM region to large m_χ is rapid annihilation via a direct-channel pole when $m_\chi \sim \frac{1}{2}m_{Higgs,Z}$ [59, 45]. This may yield a 'funnel' extending to large $m_{1/2}$ and m_0 at large $\tan\beta$, as seen in panels (c) and (d) of Fig. 4 [45]. Yet another allowed region at large $m_{1/2}$ and m_0 is the 'focus-point' region [60], which is adjacent to the boundary of the region where electroweak symmetry breaking is possible. The lightest supersymmetric particle is relatively light in this region.

4.2 Benchmark Supersymmetric Scenarios

As seen in Fig. 4, all the experimental, cosmological and theoretical constraints on the MSSM are mutually compatible. As an aid to understanding better the physics capabilities of the LHC, various e^+e^- linear collider designs and non-accelerator experiments, a set of benchmark supersymmetric scenarios have been proposed [62]. Their distribution in the $(m_{1/2}, m_0)$ plane is sketched in Fig. 5. These benchmark scenarios are compatible with all the accelerator constraints mentioned above, including the LEP searches and $b \to s\gamma$, and yield relic densities of LSPs in the range suggested by cosmology and astrophysics. The benchmarks are not intended to sample 'fairly' the allowed parameter space, but rather to illustrate the range of possibilities currently allowed.

In addition to a number of benchmark points falling in the 'bulk' region of parameter space at relatively low values of the supersymmetric particle masses, as see in Fig. 5, we also proposed [62] some points out along the 'tails' of parameter space extending out to larger masses. These clearly require some degree of fine-tuning to obtain the required relic density and/or the correct $W^\pm$ mass, and some are also disfavoured by the supersymmetric interpretation of the $g_\mu - 2$ anomaly, but all are logically consistent possibilities.

4.3 Prospects for Discovering Supersymmetry

In the CMSSM discussed here, there are just a few prospects for discovering supersymmetry at the FNAL *Tevatron collider* [62], but these could be increased in other supersymmetric models [63]. Fig. 6 shows the physics reach for observing pairs of supersymmetric particles at the

124

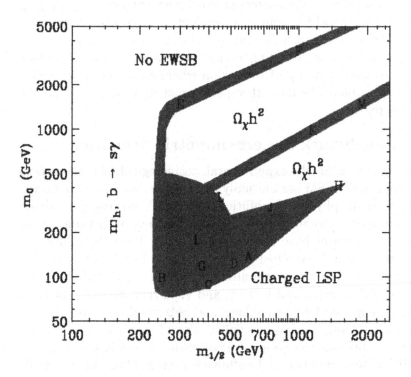

Figure 5. The locations of the benchmark points proposed in [62] in the region of the $(m_{1/2}, m_0)$ plane where $\Omega_\chi h^2$ falls within the range preferred by cosmology (shaded blue). Note that the filaments of the allowed parameter space extending to large $m_{1/2}$ and/or m_0 are sampled.

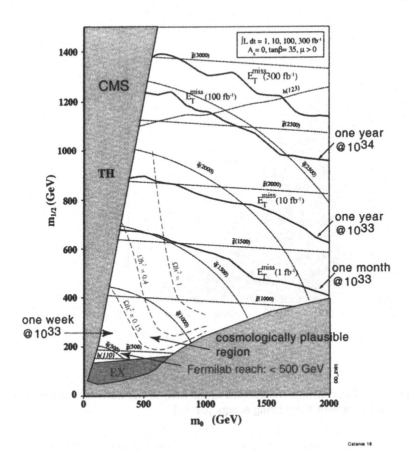

Figure 6. The regions of the $(m_0, m_{1/2})$ plane that can be explored by the LHC with various integrated luminosities [66], using the missing energy + jets signature [65].

LHC. The signature for supersymmetry - multiple jets (and/or leptons) with a large amount of missing energy - is quite distinctive, as seen in Fig. 7 [64, 65]. Therefore, the detection of the supersymmetric partners of quarks and gluons at the LHC is expected to be quite easy if they weigh less than about 2.5 TeV [66]. Moreover, in many scenarios one should be able to observe their cascade decays into lighter supersymmetric particles, as seen in Fig. 8 [67]. As seen in Fig. 9, large fractions of the supersymmetric spectrum should be seen in most of the benchmark scenarios, although there are a couple where only the lightest supersymmetric Higgs boson would be seen [62], as seen in Fig. 9.

Electron-positron colliders provide very clean experimental environments, with egalitarian production of all the new particles that are kinematically accessible, including those that have only weak interactions.

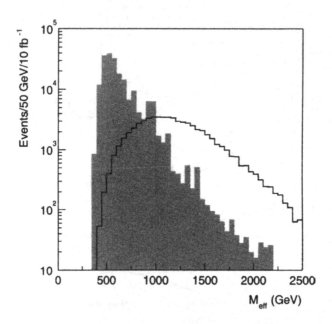

Figure 7. The distribution expected at the LHC in the variable M_{eff} that combines the jet energies with the missing energy [64, 65].

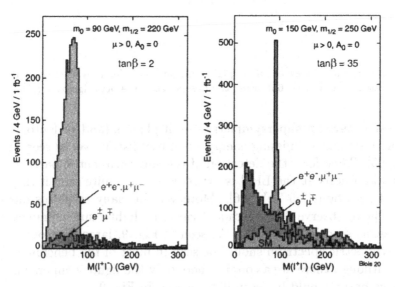

Figure 8. The dilepton mass distributions expected at the LHC due to sparticle decays in two different supersymmetric scenarios [66, 65].

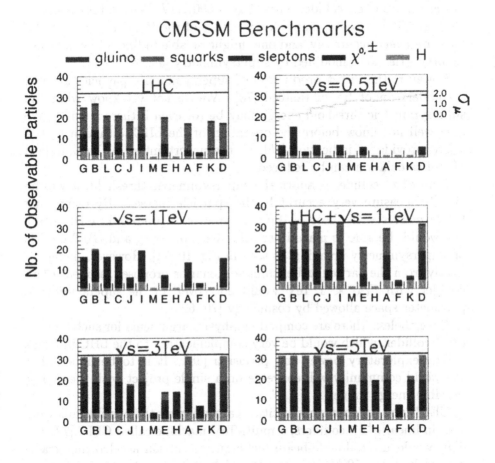

Figure 9. The numbers of different sparticles expected to be observable at the LHC and/or linear e^+e^- colliders with various energies, in each of the proposed benchmark scenarios [62], ordered by their difference from the present central experimental value of $g_\mu - 2$.

Moreover, polarized beams provide a useful analysis tool, and $e\gamma$, $\gamma\gamma$ and e^-e^- colliders are readily available at relatively low marginal costs.

The $e^+e^- \to \bar{t}t$ threshold is known to be at $E_{\mathrm{CM}} \sim 350$ GeV. Moreover, if the Higgs boson indeed weighs less than 200 GeV, as suggested by the precision electroweak data, its production and study would also be easy at an e^+e^- collider with $E_{\mathrm{CM}} \sim 500$ GeV. With a luminosity of 10^{34} cm^{-2}s^{-1} or more, many decay modes of the Higgs boson could be measured very accurately, and one might be able to find a hint whether its properties were modified by supersymmetry [68, 69].

However, the direct production of supersymmetric particles at such a collider cannot be guaranteed [70]. We do not yet know what the supersymmetric threshold energy may be (or even if there is one!). We may well not know before the operation of the LHC, although $g_\mu - 2$ might provide an indication [55], if the uncertainties in the Standard Model calculation can be reduced.

If an e^+e^- collider is above the supersymmetric threshold, it will be able to measure very accurately the sparticle masses. By comparing their masses with those of different sparticles produced at the LHC, one would be able to make interesting tests of string and GUT models of supersymmetry breaking, as seen in Fig. 10 [71]. However, independently from the particular benchmark scenarios proposed, a linear e^+e^- collider with $E_{\mathrm{CM}} < 1$ TeV would not cover all the supersymmetric parameter space allowed by cosmology [70, 62].

Nevertheless, there are compelling physics arguments for such a linear e^+e^- collider, which would be very complementary to the LHC in terms of its exploratory power and precision [68]. It is to be hoped that the world community will converge on a single project with the widest possible energy range.

CERN and collaborating institutes are studying the possible following step in linear e^+e^- colliders, a multi-TeV machine called CLIC [72, 73]. This would use a double-beam technique to attain accelerating gradients as high as 150 MV/m, and the viability of accelerating structures capable of achieving this field has been demonstrated in the CLIC test facility [74]. Parameter sets have been calculated for CLIC designs with $E_{\mathrm{CM}} = 3, 5$ TeV and luminosities of 10^{35} cm^{-2}s^{-1} or more [72].

In many of the proposed benchmark supersymmetric scenarios, CLIC would be able to complete the supersymmetric spectrum and/or measure in much more detail heavy sparticles found previously at the LHC, as seen in Fig. 9 [62]. CLIC produces more beamstrahlung than lower-energy linear e^+e^- colliders, but the supersymmetric missing-energy signature would still be easy to distinguish, and accurate measurements of masses and decay modes could still be made, as seen in Fig. 11 [75].

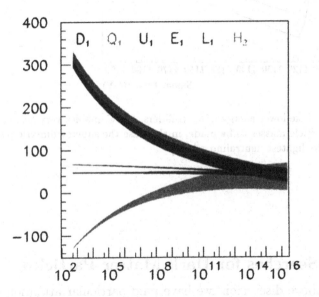

Figure 10. Measurements of sparticle masses at the LHC and a linear e^+e^- linear collider will enable one to check their universality at some input GUT scale, and check possible models of supersymmetry breaking [71]. Both axes are labelled in GeV units.

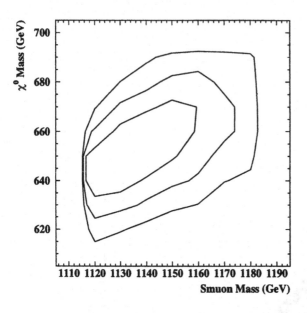

Figure 11. Like lower-energy e^+e^- colliders, CLIC enables very accurate measurements of sparticle masses to be made, in this case the supersymmetric partner of the muon and the lightest neutralino χ^0 [75].

4.4   Searches for Dark Matter Particles

In the above discussion, we have paid particular attention to the region of parameter space where the lightest supersymmetric particle could constitute the cold dark matter in the Universe [40]. How easy would this be to detect? Fig. 12 shows rates for the elastic spin-independent scattering of supersymmetric relics [76], including the projected sensitivities for CDMS II [77] and CRESST [78] (solid) and GENIUS [79] (dashed). Also shown are the cross sections calculated in the proposed benchmark scenarios discussed in the previous section, which are considerably below the DAMA [80] range ($10^{-5} - 10^{-6}$ pb), but may be within reach of future projects. The prospects for detecting elastic spin-independent scattering are less bright, as also shown in Fig. 12. Indirect searches for supersymmetric dark matter via the products of annihilations in the galactic halo or inside the Sun also have prospects in some of the benchmark scenarios [76], as seen in Fig. 13.

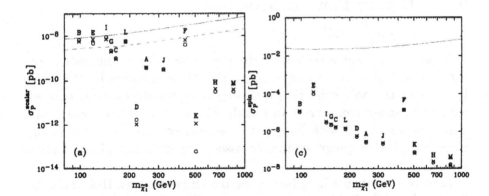

Figure 12. Left panel: elastic spin-independent scattering of supersymmetric relics on protons calculated in benchmark scenarios [76], compared with the projected sensitivities for CDMS II [77] and CRESST [78] (solid) and GENIUS [79] (dashed). The predictions of the SSARD code (blue crosses) and Neutdriver[81] (red circles) for neutralino-nucleon scattering are compared. The labels A, B, ...,L correspond to the benchmark points as shown in Fig. 5. Right panel: prospects for detecting elastic spin-independent scattering in the benchmark scenarios, which are less bright.

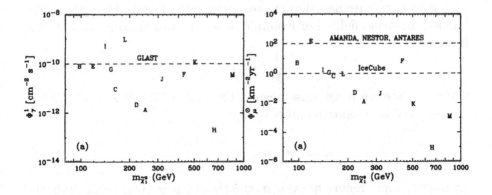

Figure 13. Left panel: prospects for detecting photons with energies above 1 GeV from annihilations in the centre of the galaxy, assuming a moderate enhancement there of the overall halo density, and right panel: prospects for detecting muons from energetic solar neutrinos produced by relic annihilations in the Sun, as calculated [76] in the benchmark scenarios using Neutdriver[81].

5.   Lepton Flavour Violation

5.1   Why Not?

There is no good reason why either the total lepton number L or the   individual lepton flavours $L_{e,\mu,\tau}$ should be conserved. We have learnt that the only conserved quantum numbers are those associated with exact gauge symmetries, just as the conservation of electromagnetic charge is associated with $U(1)$ gauge invariance. On the other hand, there is no exact gauge symmetry associated with any of the lepton numbers.

Moreover, neutrinos have been seen to oscillate between their different flavours [82, 83], showing that the separate lepton flavours $L_{e,\mu,\tau}$ are indeed not conserved, though the conservation of total lepton number L is still an open question. The observation of such oscillations strongly suggests that the neutrinos have different masses. Again, massless particles are generally associated with exact gauge symmetries, e.g., the photon with the $U(1)$ symmetry of the Standard Model, and the gluons with its $SU(3)$ symmetry. In the absence of any leptonic gauge symmetry, non-zero lepton masses are to be expected, in general.

The conservation of lepton number is an accidental symmetry of the renormalizable terms in the Standard Model lagrangian. However, one could easily add to the Standard Model non-renormalizable terms that would generate neutrino masses, even without introducing a 'right-handed' neutrino field. For example, a non-renormalizable term of the form [84]

$$\frac{1}{M}\nu H \cdot \nu H, \tag{131}$$

where M is some large mass beyond the scale of the Standard Model, would generate a neutrino mass term:

$$m_\nu \nu \cdot \nu : \; m_\nu = \frac{\langle 0|H|0\rangle^2}{M}. \tag{132}$$

Of course, a non-renormalizable interaction such as (131) seems unlikely to be fundamental, and one should like to understand the origin of the large mass scale M.

The minimal renormalizable model of neutrino masses requires the introduction of weak-singlet 'right-handed' neutrinos N. These will in general couple to the conventional weak-doublet left-handed neutrinos via Yukawa couplings Y_ν that yield Dirac masses $m_D \sim m_W$. In addition, these 'right-handed' neutrinos N can couple to themselves via Majorana masses M that may be $\gg m_W$, since they do not require electroweak summetry breaking. Combining the two types of mass term, one obtains the seesaw mass matrix [85]:

$$(\nu_L, N) \begin{pmatrix} 0 & M_D \\ M_D^T & M \end{pmatrix} \begin{pmatrix} \nu_L \\ N \end{pmatrix}, \tag{133}$$

where each of the entries should be understood as a matrix in generation space.

In order to provide the two measured differences in neutrino masses-squared, there must be at least two non-zero masses, and hence at least two heavy singlet neutrinos N_i [86, 87]. Presumably, all three light neutrino masses are non-zero, in which case there must be at least three N_i. This is indeed what happens in simple GUT models such as SO(10), but some models [88] have more singlet neutrinos [89]. In this Lecture, for simplicity we consider just three N_i.

As we discuss in the next Section, this seesaw model can accommodate the neutrino mixing seen experimentally, and naturally explains the small differences in the masses-squared of the light neutrinos. By itself, it would lead to unobservably small transitions between the different charged-lepton flavours. However, supersymmetry may enhance greatly the rates for processes violating the different charged-lepton flavours, rendering them potentially observable, as we discuss in subsequent Sections.

5.2 Neutrino Masses and Mixing in the Seesaw Model

The effective mass matrix for light neutrinos in the seesaw model may be written as:

$$\mathcal{M}_\nu = Y_\nu^T \frac{1}{M} Y_\nu v^2 \left[\sin^2 \beta\right] \tag{134}$$

where we have used the relation $m_D = Y_\nu v \left[\sin \beta\right]$ with $v \equiv \langle 0|H|0\rangle$, and the factors of $\sin \beta$ appear in the supersymmetric version of the seesaw model. It is convenient to work in the field basis where the charged-lepton masses $m_{\ell^\pm}$ and the heavy singlet-neutrino mases M are real and diagonal. The seesaw neutrino mass matrix $\mathcal{M}_\nu$ (134) may then be diagonalized by a unitary transformation U:

$$U^T \mathcal{M}_\nu U = \mathcal{M}_\nu^d. \tag{135}$$

This diagonalization is reminiscent of that required for the quark mass matrices in the Standard Model. In that case, it is well known that one can redefine the phases of the quark fields [90] so that the mixing matrix U_{CKM} has just one CP-violating phase [91]. However, in the neutrino case, there are fewer independent field phases, and one is left with three physical CP-violating parameters:

$$U = \tilde{P}_2 V P_0 : P_0 \equiv \mathrm{Diag}\left(e^{i\phi_1}, e^{i\phi_2}, 1\right). \tag{136}$$

Here $\tilde{P}_2 = \mathrm{Diag}\left(e^{i\alpha_1}, e^{i\alpha_2}, e^{i\alpha_3}\right)$ contains three phases that can be removed by phase rotations and are unobservable in light-neutrino physics, V is the light-neutrino mixing matrix first considered by Maki, Nakagawa and Sakata (MNS) [92], and P_0 contains 2 observable CP-violating phases $\phi_{1,2}$. The MNS matrix describes neutrino oscillations

$$V = \begin{pmatrix} c_{12} & s_{12} & 0 \\ -s_{12} & c_{12} & 0 \\ 0 & 0 & 1 \end{pmatrix} \begin{pmatrix} 1 & 0 & 0 \\ 0 & c_{23} & s_{23} \\ 0 & -s_{23} & c_{23} \end{pmatrix} \begin{pmatrix} c_{13} & 0 & s_{13} \\ 0 & 1 & 0 \\ -s_{13}e^{-i\delta} & 0 & c_{13}e^{-i\delta} \end{pmatrix} \tag{137}$$

The Majorana phases $\phi_{1,2}$ are in principle observable in neutrinoless double-β decay, whose matrix element is proportional to

$$\langle m_\nu \rangle_{ee} \equiv \Sigma_i U_{ei}^* m_{\nu_i} U_{ie}^\dagger. \tag{138}$$

Later we discuss how other observable quantities might be sensitive indirectly to the Majorana phases.

The first matrix factor in (137) is measurable in solar neutrino experiments. As seen in Fig. 14, the recent data from SNO [83] and Super-Kamiokande [93] prefer quite strongly the large-mixing-angle (LMA) solution to the solar neutrino problem with $\Delta m_{12}^2 \sim 6 \times 10^{-5}$ eV^2, though the LOW solution with lower δm^2 cannot yet be ruled out. The data favour large but non-maximal mixing: $\theta_{12} \sim 30^o$. The second matrix factor in (137) is measurable in atmospheric neutrino experiments. As seen in Fig. 15, the data from Super-Kamiokande in particular [82] favour maximal mixing of atmospheric neutrinos: $\theta_{23} \sim 45^o$ and $\Delta m_{23}^2 \sim 2.5 \times 10^{-3}$ eV^2. The third matrix factor in (137) is basically unknown, with experiments such as Chooz [94] and Super-Kamiokande only establishing upper limits on θ_{13}, and a fortiori no information on the CP-violating phase δ.

The phase δ could in principle be measured by comparing the oscillation probabilities for neutrinos and antineutrinos and computing the CP-violating asymmetry [95]:

$$P\left(\nu_e \to \nu_\mu\right) - P\left(\bar{\nu}_e \to \bar{\nu}_\mu\right) = 16s_{12}c_{12}s_{13}c_{13}^2 s_{23}c_{23}\sin\delta \tag{139}$$

$$\sin\left(\frac{\Delta m_{12}^2}{4E}L\right)\sin\left(\frac{\Delta m_{13}^2}{4E}L\right)\sin\left(\frac{\Delta m_{23}^2}{4E}L\right),$$

as seen in Fig. 16 [96, 97]. This is possible only if Δm_{12}^2 and s_{12} are large enough - as now suggested by the success of the LMA solution to

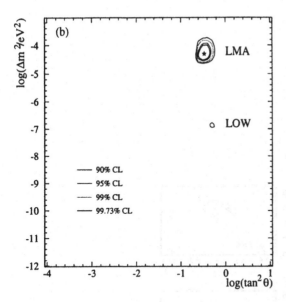

Figure 14. A global fit to solar neutrino data, following the SNO measurements of the total neutral-current reaction rate, the energy spectrum and the day-night asymmetry, favours large mixing and $\Delta m^2 \sim 6 \times 10^{-5}$ eV^2 [83].

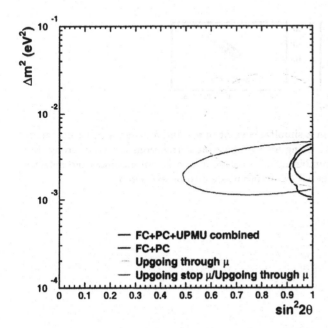

Figure 15. A fit to the Super-Kamiokande data on atmospheric neutrinos [82] indicates near-maximal $\nu_\mu - \nu_\tau$ mixing with $\Delta m^2 \sim 2.5 \times 10^{-3}$ eV^2.

136

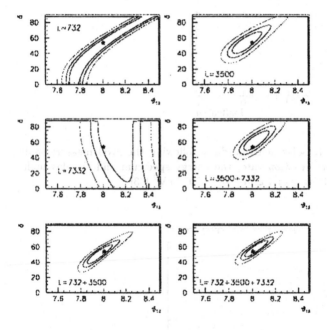

Figure 16. Correlations in a simultaneous fit of θ_{13} and δ, using a neutrino energy threshold of about 10 GeV. Using a single baseline correlations are very strong, but can be largely reduced by combining information from different baselines and detector techniques [96], enabling the CP-violating phase δ to be extracted.

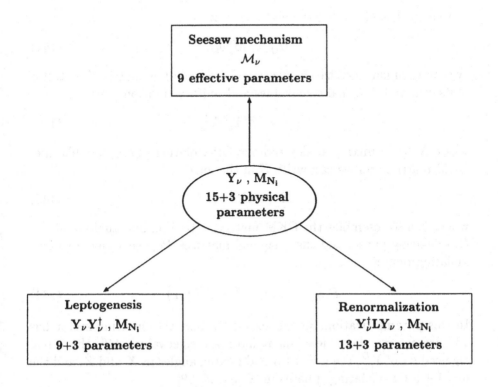

Figure 17. Roadmap for the physical observables derived from Y_ν and N_i [104].

the solar neutrino problem, and if s_{13} is large enough - which remains an open question.

We have seen above that the effective low-energy mass matrix for the light neutrinos contains 9 parameters, 3 mass eigenvalues, 3 real mixing angles and 3 CP-violating phases. However, these are not all the parameters in the minimal seesaw model. As shown in Fig. 17, this model has a total of 18 parameters [98, 99]. Most of the rest of this Lecture is devoted to understanding better the origins and possible manifestations of the remaining parameters, many of which may have controlled the generation of matter in the Universe via leptogenesis [100] and may be observable via renormalization in supersymmetric models [101, 99, 102, 103].

To see how the extra 9 parameters appear [99], we reconsider the full lepton sector, assuming that we have diagonalized the charged-lepton mass matrix:

$$(Y_\ell)_{ij} = Y^d_{\ell_i}\delta_{ij}, \tag{140}$$

as well as that of the heavy singlet neutrinos:

$$M_{ij} = M^d_i\delta_{ij}. \tag{141}$$

We can then parametrize the neutrino Dirac coupling matrix Y_ν in terms of its real and diagonal eigenvalues and unitary rotation matrices:

$$Y_\nu = Z^* Y^d_{\nu_k} X^\dagger, \tag{142}$$

where X has 3 mixing angles and one CP-violating phase, just like the CKM matrix, and we can write Z in the form

$$Z = P_1\bar{Z}P_2, \tag{143}$$

where $\bar{Z}$ also resembles the CKM matrix, with 3 mixing angles and one CP-violating phase, and the diagonal matrices $P_{1,2}$ each have two CP-violating phases:

$$P_{1,2} = \mathrm{Diag}\left(e^{i\theta_{1,3}}, e^{i\theta_{2,4}}, 1\right). \tag{144}$$

In this parametrization, we see explicitly that the neutrino sector has 18 parameters: the 3 heavy-neutrino mass eigenvalues M^d_i, the 3 real eigenvalues of $Y^D_{\nu_i}$, the $6 = 3 + 3$ real mixing angles in X and $\bar{Z}$, and the $6 = 1 + 5$ CP-violating phases in X and $\bar{Z}$ [99].

As we discuss later in more detail, leptogenesis [100] is proportional to the product

$$Y_\nu Y^\dagger_\nu = P^*_1\bar{Z}^*\left(Y^d_\nu\right)^2\bar{Z}^T P_1, \tag{145}$$

which depends on 13 of the real parameters and 3 CP-violating phases, whilst the leading renormalization of soft supersymmetry-breaking masses depends on the combination

$$Y^\dagger_\nu Y_\nu = X\left(Y^d_\nu\right)^2 X^\dagger, \tag{146}$$

which depends on just 1 CP-violating phase, with two more phases appearing in higher orders, when one allows the heavy singlet neutrinos to be non-degenerate [102].

In order to see how the low-energy sector is embedded in this full parametrization, we first recall that the 3 phases in $\tilde{P}_2$ (136) become observable when one also considers high-energy quantities. Next, we introduce a complex orthogonal matrix

$$R \equiv \sqrt{M^d}^{-1} Y_\nu U \sqrt{M^d}^{-1} \left[v\sin\beta\right], \tag{147}$$

which has 3 real mixing angles and 3 phases: $R^T R = 1$. These 6 additional parameters may be used to characterize Y_ν, by inverting (147):

$$Y_\nu = \frac{\sqrt{M^d} R \sqrt{M^d} U^\dagger}{[v \sin \beta]},\qquad(148)$$

giving us the same grand total of $18 = 9 + 3 + 6$ parameters [99]. The leptogenesis observable (145) may now be written in the form

$$Y_\nu Y_\nu^\dagger = \frac{\sqrt{M^d} R \mathcal{M}_\nu^d R^\dagger \sqrt{M^d}}{[v^2 \sin^2 \beta]},\qquad(149)$$

which depends on the 3 phases in R, but *not* the 3 low-energy phases $\delta, \phi_{1,2}$, nor the 3 real MNS mixing angles [99]! Conversely, the leading renormalization observable (146) may be written in the form

$$Y_\nu^\dagger Y_\nu = U \frac{\sqrt{\mathcal{M}_\nu^d R^\dagger M^d R \sqrt{\mathcal{M}_\nu^d}}}{[v^2 \sin^2 \beta]} U^\dagger,\qquad(150)$$

which depends explicitly on the MNS matrix, including the CP-violating phases δ and $\phi_{1,2}$, but only one of the three phases in $\tilde{P}_2$ [99].

5.3 Renormalization of Soft Supersymmetry-Breaking Parameters

Let us now discuss the renormalization of soft supersymmetry-breaking parameters m_0^2 and A in more detail, assuming that the input values at the GUT scale are flavour-independent. If they are not, there will be additional sources of flavour-changing processes, beyond those discussed in this and subsequent sections [36, 105]. In the leading-logarithmic approximation, and assuming degenerate heavy singlet neutrinos, one finds the following radiative corrections to the soft supersymmetry-breaking terms for sleptons:

$$\left(\delta m_{\tilde{L}}^2\right)_{ij} = -\frac{1}{8\pi^2}\left(3m_0^2 + A_0^2\right)\left(Y_\nu^\dagger Y_\nu\right)_{ij} \mathrm{Ln}\left(\frac{M_{GUT}}{M}\right),$$

$$(\delta A_\ell)_{ij} = -\frac{1}{8\pi^2} A_0 Y_{\ell_i} \left(Y_\nu^\dagger Y_\nu\right)_{ij} \mathrm{Ln}\left(\frac{M_{GUT}}{M}\right),\qquad(151)$$

where we have intially assumed that the heavy singlet neutrinos are approximately degenerate with $M \ll M_{GUT}$. In this case, there is a single analogue of the Jarlskog invariant of the Standard Model [106]:

$$J_{\tilde{L}} \equiv \mathrm{Im}\left[\left(m_{\tilde{L}}^2\right)_{12}\left(m_{\tilde{L}}^2\right)_{23}\left(m_{\tilde{L}}^2\right)_{31}\right],\qquad(152)$$

which depends on the single phase that is observable in this approximation. There are other Jarlskog invariants defined analogously in terms of various combinations with the A_ℓ, but these are all proportional [99].

There are additional contributions if the heavy singlet neutrinos are not degenerate:

$$\left(\tilde{\delta}m_{\tilde{L}}^2\right)_{ij} = -\frac{1}{8\pi^2}\left(3m_0^2 + A_0^2\right)\left(Y_\nu^\dagger LY_\nu\right)_{ij} : L \equiv \mathrm{Ln}\left(\frac{\bar{M}}{M_i}\right)\delta_{ij}, \quad (153)$$

where $\bar{M} \equiv \sqrt[3]{M_1 M_2 M_3}$, with $\left(\tilde{\delta}A_\ell\right)_{ij}$ being defined analogously. These new contributions contain the matrix factor

$$Y^\dagger LY = XY^d P_2 \bar{Z}^T L\bar{Z}^* P_2^* y^d X^\dagger, \quad (154)$$

which introduces dependences on the phases in $\bar{Z}P_2$, though not P_1. In this way, the renormalization of the soft supersymmetry-breaking parameters becomes sensitive to a total of 3 CP-violating phases [102].

5.4   Exploration of Parameter Space

Now that we have seen how the 18 parameters in the minimal supersymmetric seesaw model might in principle be observable, we would like to explore the range of possibilities in this parameter space. This requires confronting two issues: the unwieldy large dimensionality of the parameter space, and the inclusion of the experimental information already obtained (or obtainable) from low-energy studies of neutrinos. Of the 9 parameters accessible to these experiments: $m_{\nu_1}, m_{\nu_2}, m_{\nu_3}, \theta_{12}, \theta_{23}, \theta_{31}, \delta, \phi_1$ and ϕ_2, we have measurements of 4 combinations: $\Delta m_{12}^2, \Delta m_{23}^2, \theta_{12}$ and θ_{23}, and upper limits on the overall light-neutrino mass scale, θ_{13} and the double-β decay observable (138).

The remaining 9 parameters not measurable in low-energy neutrino physics may be characterized by an auxiliary Hermitean matrix of the following form [101, 103]:

$$H \equiv Y_\nu^\dagger DY_\nu, \quad (155)$$

where D is an arbitrary real and diagonal matrix. Possible choices for D include $\mathrm{Diag}(\pm1, \pm1, \pm1)$ and the logarithmic matrix L defined in (153). Once one specifies the 9 parameters in H, either in a statistical survey or in some definite model, one can calculate

$$H' \equiv \sqrt{\mathcal{M}_\nu^d}U^\dagger HU\sqrt{\mathcal{M}_\nu^d}, \quad (156)$$

which can then be diagonalized by a complex orthogonal matrix R':

$$H' = R'^\dagger \mathcal{M}'^d R' : R'^T R' = 1. \quad (157)$$

In this way, we can calculate all the remaining physical parameters:

$$(\mathcal{M}_\nu, H) \to (\mathcal{M}_\nu, \mathcal{M}'^d, R') \to (Y_\nu, M_i) \tag{158}$$

and then go on to calculate leptogenesis, charged-lepton violation, etc [101, 103].

A freely chosen model will in general violate the experimental upper limit on $\mu \to e\gamma$ [107]. It is easy to avoid this problem using the parametrization (155) [103]. If one chooses $D = L$ and requires the entry $H_{12} = 0$, the leading contribution to $\mu \to e\gamma$ from renormalization of the soft supersymmetry-breaking masses will be suppressed. To suppress $\mu \to e\gamma$ still further, one may impose the constraint $H_{13}H_{23} = 0$. This condition evidently has two solutions: either $H_{13} = 0$, in which case $\tau \to e\gamma$ is suppressed but not $\tau \to \mu\gamma$, or alternatively $H_{23} = 0$, which favours $\tau \to e\gamma$ over $\tau \to \mu\gamma$. Thus we may define two generic textures H^1 and H^2:

$$H^1 \equiv \begin{pmatrix} a & 0 & 0 \\ 0 & b & d \\ 0 & d^\dagger & c \end{pmatrix}, \quad H^2 \equiv \begin{pmatrix} a & 0 & d \\ 0 & b & 0 \\ d^\dagger & 0 & c \end{pmatrix}. \tag{159}$$

We use these as guides in the following, whilst recalling that they represent extremes, and the truth may not favour one $\tau \to \ell\gamma$ decay mode so strongly over the other.

5.5 Leptogenesis

In addition to the low-energy neutrino constraints, we frequently employ the constraint that the model parameters be compatible with the leptogenesis scenario for creating the baryon asymmetry of the Universe [100]. We recall that the baryon-to-entropy ratio Y_B in the Universe today is found to be in the range $10^{-11} < Y_B < 3 \times 10^{-10}$. This is believed to have evolved from a similar asymmetry in the relative abundances of quarks and antiquarks before they became confined inside hadrons when the temperature of the Universe was about 100 MeV. In the leptogenesis scenario [100], non-perturbative electroweak interactions caused this small asymmetry to evolve out of a similar small asymmetry in the relative abundances of leptons and antileptons that had been generated by CP violation in the decays of heavy singlet neutrinos.

The total decay rate of such a heavy neutrino N_i may be written in the form

$$\Gamma_i = \frac{1}{8\pi} \left(Y_\nu Y_\nu^\dagger \right)_{ii} M_i. \tag{160}$$

One-loop CP-violating diagrams involving the exchange of heavy neutrino N_j would generate an asymmetry in N_i decay of the form:

$$\epsilon_{ij} = \frac{1}{8\pi} \frac{1}{\left(Y_\nu Y_\nu^\dagger\right)_{ii}} \text{Im}\left(\left(Y_\nu Y_\nu^\dagger\right)_{ij}\right)^2 f\left(\frac{M_j}{M_i}\right), \qquad (161)$$

where $f(M_j/M_i)$ is a known kinematic function.

As already remarked, the relevant combination

$$\left(Y_\nu Y_\nu^\dagger\right) = \sqrt{M^d} R \mathcal{M}^d R^\dagger \sqrt{M^d} \cdot \qquad (162)$$

is independent of U and hence of the light neutrino mixing angles and CP-violating phases. The basic reason for this is that one makes a unitary sum over all the light lepton species in evaluating the asymmetry ϵ_{ij}. It is easy to derive a compact expression for ϵ_{ij} in terms of the heavy neutrino masses and the complex orthogonal matrix R:

$$\epsilon_{ij} = \frac{1}{8\pi} M_j f\left(\frac{M_j}{M_i}\right) \frac{\text{Im}\left(\left(R\mathcal{M}_\nu^d R^\dagger\right)_{ij}\right)^2}{(R\mathcal{M}_\nu^d R^\dagger)_{ii}}. \qquad (163)$$

This depends explicitly on the extra phases in R: how can we measure them?

The basic principle of a strategy to do this is the following [99, 102, 103]. The renormalization of soft supersymmetry-breaking parameters, and hence flavour-changing interactions and CP violation in the lepton sector, depend on the leptogenesis parameters as well as the low-energy neutrino parameters $\delta, \phi_{1,2}$. If one measures the latter in neutrino experiments, and the discrepancy in the soft supersymmetry-breaking determines the leptogenesis parameters.

An example how this could work is provided by the two-generation version of the supersymmetric seesaw model [99]. In this case, we have $\mathcal{M}_\nu^d = \text{Diag}(m_{\nu_1}, m_{\nu_1})$ and $M^d = \text{Diag}(M_1, M_2)$, and we may parameterize

$$R = \begin{pmatrix} \cos(\theta_r + i\theta_i) & \sin(\theta_r + i\theta_i) \\ -\sin(\theta_r + i\theta_i) & \cos(\theta_r + i\theta_i) \end{pmatrix}. \qquad (164)$$

In this case, the leptogenesis decay asymmetry is proportional to

$$\text{Im}\left(\left(Y_\nu Y_\nu^\dagger\right)^{21}\right)^2 = \frac{(m_{\nu_1}^2 - m_{\nu_2}^2) M_1 M_2}{2v^4 \sin^4 \beta} \sinh 2\theta_i \sin 2\theta_r. \qquad (165)$$

We see that this is related explicitly to the CP-violating phase and mixing angle in R (164), and is independent of the low-energy neutrino parameters. Turning now to the renormalization of the soft supersymmetry-breaking parameters, assuming for simplicity maximal mixing in the MNS matrix V and setting the diagonal Majorana phase matrix $P_0 = \text{Diag}(e^{-i\phi}, 1)$, we find that

$$\text{Re}\left[\left(Y_\nu^\dagger Y_\nu\right)^{12}\right] = -\frac{(m_{\nu_2} - m_{\nu_1})}{4v^2 \sin^2 \beta}(M_1 + M_2)\cosh 2\theta_i + \cdots,$$

$$\text{Im}\left[\left(Y_\nu^\dagger Y_\nu\right)^{12}\right] = \frac{\sqrt{m_{\nu_2} m_{\nu_1}}}{2v^2 \sin^2 \beta}(M_1 + M_2)\sinh 2\theta_i \cos \phi + \cdots (166)$$

In this case, the strategy for relating leptogenesis to low-energy observables would be: (i) use double-β decay to determine ϕ, (ii) use low-energy observables sensitive to $\text{Re}, \text{Im}\left[\left(Y_\nu^\dagger Y_\nu\right)^{12}\right]$ to determine θ_r and θ_i (166), which then (iii) determine the leptogenesis asymmetry (165) in this two-generation model.

In general, one may formulate the following strategy for calculating leptogenesis in terms of laboratory observables:

- Measure the neutrino oscillation phase δ and the Majorana phases $\phi_{1,2}$,

- Measure observables related to the renormalization of soft supersymmetry-breaking parameters, that are functions of $\delta, \phi_{1,2}$ and the leptogenesis phases,

- Extract the effects of the known values of δ and $\phi_{1,2}$, and isolate the leptogenesis parameters.

In the absence of complete information on the first two steps above, we are currently at the stage of preliminary explorations of the multi-dimensional parameter space. As seen in Fig. 18, the amount of the leptogenesis asymmetry is explicitly independent of δ [104]. An important observation is that there is a non-trivial lower bound on the mass of the lightest heavy singlet neutrino N:

$$M_{N_1} \gtrsim 10^{10} \text{ GeV} \qquad (167)$$

if the light neutrinos have the conventional hierarchy of masses, and

$$M_{N_1} \gtrsim 10^{11} \text{ GeV} \qquad (168)$$

if they have an inverted hierarchy of masses [104]. This observation is potentially important for the cosmological abundance of gravitinos, which would be problematic if the cosmological temperature was once high enough for leptogenesis by thermally-produced singlet neutrinos weighing as much as (167, 168) [108]. However, these bounds could be relaxed if the two lightest N_i were near-degenerate, as seen in Fig. 19 [109].

144

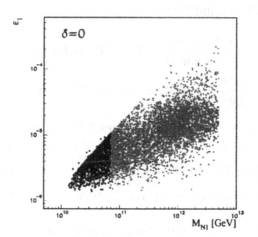

Figure 18.   Heavy singlet neutrino decay may exhibit a CP-violating asymmetry, leading to leptogenesis and hence baryogenesis, even if the neutrino oscillation phase δ vanishes [104].

Striking aspects of this scenario include the suppression of $\mu \to e\gamma$, the relatively large value of $\tau \to \mu\gamma$, and a preferred value for the neutrino-less double-β decay observable:

$$\langle m \rangle_{ee} \sim \sqrt{\Delta m^2_{solar}} \sin^2 \theta_{12}. \tag{169}$$

5.6   Flavour-Violating Decays of Charged Leptons

Several such decays can be studied within this framework, including $\mu \to e\gamma, \tau \to e\gamma, \tau \to \mu\gamma, \mu \to 3e$, and $\tau \to 3\mu/e$ [110].

The effective Lagrangian for $\mu \to e\gamma$ and $\mu \to 3e$ can be written in the form [111, 99]:

$$
\begin{aligned}
\mathcal{L} = -\frac{4G_F}{\sqrt{2}} \{ & m_\mu A_R \overline{\mu_R} \sigma^{\mu\nu} e_L F_{\mu\nu} + m_\mu A_L \overline{\mu_L} \sigma^{\mu\nu} e_R F_{\mu\nu} \\
& + g_1 (\overline{\mu_R} e_L)(\overline{e_R} e_L) + g_2 (\overline{\mu_L} e_R)(\overline{e_L} e_R) \\
& + g_3 (\overline{\mu_R} \gamma^\mu e_R)(\overline{e_R} \gamma_\mu e_R) + g_4 (\overline{\mu_L} \gamma^\mu e_L)(\overline{e_L} \gamma_\mu e_L) \\
& + g_5 (\overline{\mu_R} \gamma^\mu e_R)(\overline{e_L} \gamma_\mu e_L) + g_6 (\overline{\mu_L} \gamma^\mu e_L)(\overline{e_R} \gamma_\mu e_R) + h.c. \}. \tag{170}
\end{aligned}
$$

The decay $\mu \to e\gamma$ is related directly to the coefficients $A_{L,R}$:

$$Br(\mu^+ \to e^+ \gamma) = 384\pi^2 \left(|A_L|^2 + |A_R|^2 \right), \tag{171}$$

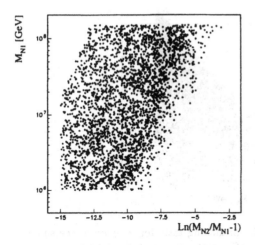

Figure 19. *Figure 19.* The lower limit on the mass of the lightest heavy singlet neutrino may be significantly reduced if the two lightest singlet neutrinos are almost degenerate [109].

and the branching ratio for $\mu \to 3e$ is given by

$$
\begin{aligned}
B(\mu \to e\gamma) &= 2(C_1 + C_2) + C_3 + C_4 + 32 \left(\ln \frac{m_\mu^2}{m_e^2} - \frac{11}{4} \right)(C_5 + C_6) \\
&+ 16(C_7 + C_8) + 8(C_9 + C_{10}),
\end{aligned}
\tag{172}
$$

where

$$
\begin{aligned}
C_1 &= \frac{|g_1|^2}{16} + |g_3|^2, \quad C_2 = \frac{|g_2|^2}{16} + |g_4|^2, \\
C_3 &= |g_5|^2, \quad C_4 = |g_6|^2, C_5 = |eA_R|^2, \quad C_6 = |eA_L|^2, \quad C_7 = \text{Re}(eA_R g_4^*), \\
C_8 &= \text{Re}(eA_L g_3^*), \quad C_9 = \text{Re}(eA_R g_6^*), \quad C_{10} = \text{Re}(eA_L g_5^*) \,..
\end{aligned}
\tag{173}
$$

These coefficients may easily be calculated using the renormalization-group equations for soft supersymmetry-breaking parameters [99, 103].

Fig. 20 displays a scatter plot of $B(\mu \to e\gamma)$ in the texture H^1 mentioned earlier, as a function of the singlet neutrino mass M_{N_3}. We see that $\mu \to e\gamma$ may well have a branching ratio close to the present experimental upper limit, particularly for larger M_{N_3}. Predictions for $\tau \to \mu\gamma$ and $\tau \to e\gamma$ decays are shown in Figs. 21 and 22 for the textures H^1 and H^2, respectively. As advertised earlier, the H^1 texture favours $\tau \to \mu\gamma$ and the H^2 texture favours $\tau \to e\gamma$. We see that the branching ratios decrease with increasing sparticle masses, but that the range due to variations in the neutrino parameters is considerably larger than that

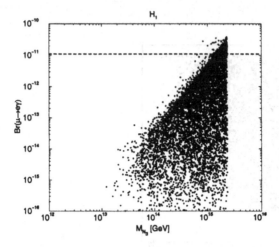

Figure 20. Scatter plot of the branching ratio for $\mu \to e\gamma$ in the supersymmetric seesaw model for various values of its unknown parameters [103].

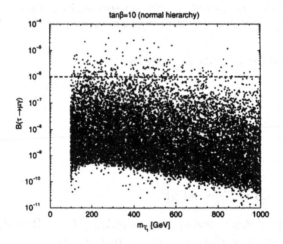

Figure 21. Scatter plot of the branching ratio for $\tau \to \mu\gamma$ in one variant of the supersymmetric seesaw model for various values of its unknown parameters [103].

due to the sparticle masses. The present experimental upper limits on $\tau \to \mu\gamma$, in particular, already exclude significant numbers of parameter choices.

The branching ratio for $\mu \to 3e$ is usually dominated by the photonic penguin diagram, which contributes the $C_{5,6}$ terms in (172), yielding an essentially constant ratio for $B(\mu \to 3e)/B(\mu \to e\gamma)$. However, if $\mu \to e\gamma$ decay is parametrically suppressed, as it may have to be in order to respect the experimental upper bound on this decay, then other

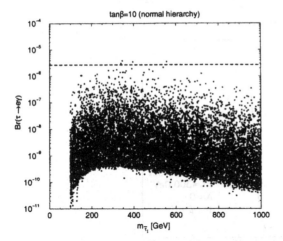

Figure 22. Scatter plot of the branching ratio for $\tau \to e\gamma$ in a variant the supersymmetric seesaw model for various values of its unknown parameters [103].

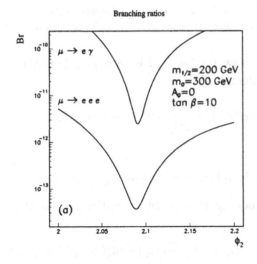

Figure 23. The branching ratio for $\mu \to e\gamma$ may be suppressed for some particular values of the model parameters, in which case the branching ratio for $\mu \to 3e$ gets significant contributions form other diagrams besides the photonic penguin diagram [99].

diagrams may become important in $\mu \to 3e$ decay. In this case, the ratio $B(\mu \to 3e)/B(\mu \to e\gamma)$ may be enhanced, as seen in Fig. 23.

As a result, interference between the photonic penguin diagram and the other diagrams may in principle generate a measurable T-odd asymmetry in $\mu \to 3e$ decay. This is sensitive to the CP-violating parameters

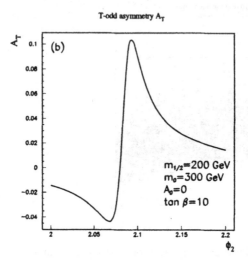

Figure 24.   The T-violating asymmetry A_T in $\mu \to 3e$ decay is enhanced in the regions of parameter space shown in Fig. 23 where the branching ratio for $\mu \to e\gamma$ is suppressed, and different diagrams may interfere in the $\mu \to 3e$ decay amplitude [99].

in the supersymmetric seesaw model, and is in principle observable in polarized $\mu^+ \to e^+ e^- e^+$ decay:

$$A_T(\mu^+ \to e^+ e^- e^+) = \frac{3}{2B}(2.0C_{11} - 1.6C_{12}), \qquad (174)$$

where

$$C_{11} = \mathrm{Im}\,(eA_R g_4^* + eA_L g_3^*), C_{12} = \mathrm{Im}\,(eA_R g_6^* + eA_L g_5^*), \qquad (175)$$

and B is the $\mu \to 3e$ branching ratio with an optimized cutoff for the more energetic positron:

$$B = 1.8(C_1+C_2)+0.96(C_3+C_4)+88(C_5+C_6)+14(C_7+C_8)+8(C_9+C_{10}). \qquad (176)$$

As seen in Fig. 24, the T-odd asymmetry is enhanced in regions of parameter space where $B(\mu \to e\gamma)$ is suppressed [99]. If/when $\mu \to e\gamma$ and/or $\mu \to 3e$ decays are observed, measuring A_T (174) may provide an interesting window on CP violation in the seesaw model.

5.7   Lepton Electric Dipole Moments

This CP violation may also be visible in electric dipole moments for the electron and muon d_e and d_μ [112]. It is usually thought that these are unobservably small in the minimal supersymmetric seesaw model, and that $|d_e/d_\mu| = m_e/m_\mu$. However, d_e and d_μ may be strongly enhanced if the heavy singlet neutrinos are not degenerate [102], and depend on new phases that contribute to leptogenesis [3]. The leading contributions to d_e and d_μ in the presence of non-degenerate heavy-singlet neutrinos are produced by the following terms in the renormalization of soft supersymmetry-breaking parameters:

$$\left(\tilde{\delta} m_{\tilde{L}}^2\right)_{ij} = \frac{18}{(4\pi)^4} \left(m_0^2 + A_e^2\right) \{Y_\nu^\dagger L Y_\nu, Y_\nu^\dagger Y_\nu\}_{ij} \ln\left(\frac{M_{GUT}}{\bar{M}}\right), \quad (177)$$

and similarly for the trilinear terms, where the mean heavy-neutrino mass $\bar{M} \equiv \sqrt[3]{M_1 M_2 M_3}$ and the matrix $L \equiv \ln(\bar{M}/M_i)\delta_{ij}$ were introduced in (153).

It should be emphasized that non-degenerate heavy-singlet neutrinos are actually expected in most models of neutrino masses. Typical examples are texture models of the form

$$Y_\nu \sim Y_0 \begin{pmatrix} 0 & c\epsilon_\nu^3 & d\epsilon_\nu^3 \\ c\epsilon_\nu^3 & a\epsilon_\nu^2 & b\epsilon_\nu^2 \\ d\epsilon_\nu^3 & b\epsilon_\nu^2 & e^{i\psi} \end{pmatrix},$$

where Y_0 is an overall scale, ϵ_ν characterizes the hierarchy, a, b, c and d are $\mathcal{O}(1)$ complex numbers, and ψ is an arbitrary phase. For example, there is an SO(10) GUT model of this form with $d = 0$ and a flavour SU(3) model with $a = b$ and $c = d$. The hierarchy of heavy-neutrino masses in such a model is

$$M_1 : M_2 : M_3 = \epsilon_N^6 : \epsilon_N^4 : 1, \quad (178)$$

and indicative ranges of the hierarchy parameters are

$$\epsilon_\nu \sim \sqrt{\frac{\Delta m_{solar}^2}{\Delta m_{atmo}^2}}, \quad \epsilon_N \sim 0.1 \text{ to } 0.2. \quad (179)$$

Fig. 25 shows how much d_e and d_μ may be increased as soon as the degeneracy between the heavy neutrinos is broken: $\epsilon \neq 1$. We also see that $|d_\mu/d_e| \gg m_\mu/m_e$ when $\epsilon_N \sim 0.1$ to 0.2. Scatter plots of d_e and d_μ are shown in Fig. 26, where we see that values as large as $d_\mu \sim 10^{-27}$ e.cm and $d_e \sim 3 \times 10^{-30}$ e.cm are possible. For comparison, the present experimental upper limits are $d_e < 1.6 \times 10^{-27}$ e.cm [113] and $d_\mu < 10^{-18}$ e.cm [51]. An ongoing series of experiments might be able to reach $d_e < 3 \times 10^{-30}$ e.cm, and a type of solid-state experiment

---

[3]This effect makes lepton electric dipole moments possible even in a two-generation model.

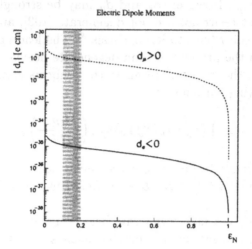

Electric Dipole Moments

Figure 25. The electric dipole moments of the electron and muon, d_e and d_μ, may be enhanced if the heavy singlet neutrinos are non-degenerate. The horizontal axis parameterizes the breaking of their degeneracy, and the vertical strip indicates a range favoured in certain models [102].

that might be sensitive to $d_e \sim 10^{-33}$ e.cm has been proposed [114]. Also, $d_\mu \sim 10^{-24}$ e.cm might be accessible with the PRISM experiment proposed for the JHF [115], and $d_\mu \sim 5\times 10^{-26}$ e.cm might be attainable at the front end of a neutrino factory [116]. It therefore seems that d_e might be measurable with foreseeable experiments, whilst d_μ would present more of a challenge.

6.   Concluding Remarks

The title assigned to my lectures was a tad grandiose. Nevertheless, I hope I have convinced you that supersymmetry is among the more plausible extensions of the Standard Model. As such, it may indeed have played a key rôle in the history of our Universe, as well as in future experiments both with and without accelerators. In particular, the latter might even reveal that most of the matter in the Universe is supersymmetric, in which case the title may not seem too immodest!

Acknowledgments

It is a pleasure to thank Harrison Prosper for organizing such an interesting school in such enjoyable surroundings.

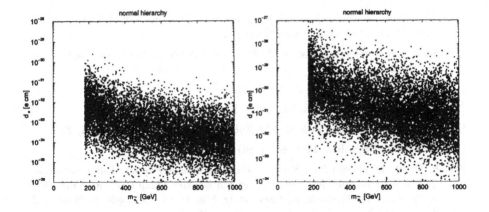

Figure 26. Scatter plots of d_e and d_μ in variants of the supersymmetric seesaw model, for different values of the unknown parameters [103].

References

[1] J. R. Ellis, Lectures at 1998 CERN Summer School, St. Andrews, *Beyond the Standard Model for Hillwalkers*, arXiv:hep-ph/9812235.

[2] J. R. Ellis, *The Superstring: Theory Of Everything, Or Of Nothing?*, Nature **323** (1986) 595.

[3] J. Ellis, S. Kelley and D. V. Nanopoulos, Phys. Lett. B **260** (1991) 131; U. Amaldi, W. de Boer and H. Furstenau, Phys. Lett. B **260** (1991) 447; P. Langacker and M. x. Luo, Phys. Rev. D **44** (1991) 817; C. Giunti, C. W. Kim and U. W. Lee, Mod. Phys. Lett. A **6** (1991) 1745.

[4] L. Maiani, *Proceedings of the 1979 Gif-sur-Yvette Summer School On Particle Physics*, 1; G. 't Hooft, in *Recent Developments in Gauge Theories, Proceedings of the Nato Advanced Study Institute, Cargese, 1979*, eds. G. 't Hooft *et al.*, (Plenum Press, NY, 1980); E. Witten, Phys. Lett. B **105** (1981) 267.

[5] M. B. Green, J. H. Schwarz and E. Witten, *Superstring Theory*, (Cambridge Univ. Press, 1987).

[6] D. A. Ross and M. J. Veltman, Nucl. Phys. B **95** (1975) 135.

[7] P. W. Higgs, Phys. Lett. **12** (1964) 132; Phys. Rev. Lett. **13** (1964) 508.

[8] F. Englert and R. Brout, Phys. Rev. Lett. **13** (1964) 321.

[9] C. T. Hill, Phys. Lett. B **266** (1991) 419; for a recent review, see: C. T. Hill and E. H. Simmons, arXiv:hep-ph/0203079.

[10] For a historical reference, see: E. Farhi and L. Susskind, Phys. Rept. **74** (1981) 277.

[11] S. Dimopoulos and L. Susskind, Nucl. Phys. B **155** (1979) 237; E. Eichten and K. Lane, Phys. Lett. B **90** (1980) 125.

[12] J. R. Ellis, M. K. Gaillard, D. V. Nanopoulos and P. Sikivie, Nucl. Phys. B **182** (1981) 529.

[13] S. Dimopoulos and J. R. Ellis, Nucl. Phys. B **182** (1982) 505.

[14] G. Altarelli, F. Caravaglios, G. F. Giudice, P. Gambino and G. Ridolfi, JHEP **0106** (2001) 018 [arXiv:hep-ph/0106029].

[15] For a recent reference, see: K. Lane, *Two lectures on technicolor*, arXiv:hep-ph/0202255.

[16] B. Holdom, Phys. Rev. D **24** (1981) 1441.

[17] S. R. Coleman and J. Mandula, Phys. Rev. **159** (1967) 1251.

[18] R. Haag, J. T. Lopuszanski and M. Sohnius, Nucl. Phys. B **88** (1975) 257.

[19] F. Iachello, Phys. Rev. Lett. **44** (1980) 772.

[20] P. Fayet, as reviewed in *Supersymmetry, Particle Physics And Gravitation*, CERN-TH-2864, published in *Proc. of Europhysics Study Conf. on Unification of Fundamental Interactions*, Erice, Italy, Mar 17-24, 1980, eds. S. Ferrara, J. Ellis, P. van Nieuwenhuizen (Plenum Press, 1980).

[21] LEP Electroweak Working Group,
http://lepewwg.web.cern.ch/LEPEWWG/Welcome.html.

[22] Y. Okada, M. Yamaguchi and T. Yanagida, Prog. Theor. Phys. **85** (1991) 1; J. R. Ellis, G. Ridolfi and F. Zwirner, Phys. Lett. B **257** (1991) 83; H. E. Haber and R. Hempfling, Phys. Rev. Lett. **66** (1991) 1815.

[23] For an early review, see: P. Fayet and S. Ferrara, Phys. Rept. **32** (1977) 249; see also: H. P. Nilles, Phys. Rept. **110** (1984) 1; H. E. Haber and G. L. Kane, Phys. Rept. **117** (1985) 75.

[24] J. Wess and B. Zumino, Phys. Lett. B **49** (1974) 52; Nucl. Phys. B **70** (1974) 39.

[25] J. Wess and B. Zumino, Nucl. Phys. B **78** (1974) 1.

[26] I. Antoniadis and K. Benakli, Int. J. Mod. Phys. A **15** (2000) 4237 [arXiv:hep-ph/0007226].

[27] S. Ferrara, J. Wess and B. Zumino, Phys. Lett. B **51** (1974) 239; S. Ferrara, J. Iliopoulos and B. Zumino, Nucl. Phys. B **77** (1974) 413.

[28] B. A. Campbell, S. Davidson, J. R. Ellis and K. A. Olive, Phys. Lett. B **256** (1991) 457; W. Fischler, G. F. Giudice, R. G. Leigh and S. Paban, Phys. Lett. B **258** (1991) 45.

[29] S. Dimopoulos and H. Georgi, Nucl. Phys. B **193** (1981) 150; N. Sakai, Z. Phys. C **11** (1981) 153.

[30] P. Fayet and J. Iliopoulos, Phys. Lett. B **51** (1974) 461.

[31] M. Dine, N. Seiberg and E. Witten, Nucl. Phys. B **289** (1987) 589.

[32] L. O'Raifeartaigh, Nucl. Phys. B **96** (1975) 331; P. Fayet, Phys. Lett. B **58** (1975) 67.

[33] J. Polonyi, Hungary Central Inst. Res. preprint KFKI-77-93 (1977).

[34] E. Cremmer, B. Julia, J. Scherk, S. Ferrara, L. Girardello and P. van Nieuwenhuizen, Nucl. Phys. B **147** (1979) 105.

[35] K. Inoue, A. Kakuto, H. Komatsu and S. Takeshita, Prog. Theor. Phys. **68** (1982) 927 [Erratum-ibid. **70** (1982) 330]; L.E. Ibáñez and G.G. Ross, Phys. Lett. B **110** (1982) 215; L.E. Ibáñez, Phys. Lett. B **118** (1982) 73; J. Ellis, D.V. Nanopoulos and K. Tamvakis, Phys. Lett. B **121** (1983) 123; J. Ellis, J.

Hagelin, D.V. Nanopoulos and K. Tamvakis, Phys. Lett. B **125** (1983) 275; L. Alvarez-Gaumé, J. Polchinski, and M. Wise, Nucl. Phys. B **221** (1983) 495.

[36] J. R. Ellis and D. V. Nanopoulos, Phys. Lett. B **110** (1982) 44; R. Barbieri and R. Gatto, Phys. Lett. B **110** (1982) 211.

[37] S. L. Glashow, J. Iliopoulos and L. Maiani, Phys. Rev. D **2** (1970) 1285.

[38] J. R. Ellis and S. Rudaz, Phys. Lett. B **128** (1983) 248.

[39] J. R. Ellis, T. Falk, G. Ganis, K. A. Olive and M. Schmitt, Phys. Rev. D **58** (1998) 095002 [arXiv:hep-ph/9801445].

[40] J. Ellis, J.S. Hagelin, D.V. Nanopoulos, K.A. Olive and M. Srednicki, Nucl. Phys. B **238** (1984) 453; see also H. Goldberg, Phys. Rev. Lett. **50** (1983) 1419.

[41] H. Goldberg, Phys. Rev. Lett. **50** (1983) 1419.

[42] S. Heinemeyer, W. Hollik and G. Weiglein, Comput. Phys. Commun. **124**, 76 (2000) [arXiv:hep-ph/9812320]; S. Heinemeyer, W. Hollik and G. Weiglein, Eur. Phys. J. C **9** (1999) 343 [arXiv:hep-ph/9812472].

[43] Joint LEP 2 Supersymmetry Working Group, *Combined LEP Chargino Results, up to 208 GeV*,
http://lepsusy.web.cern.ch/lepsusy/www/inos_moriond01/charginos_pub.html.

[44] Joint LEP 2 Supersymmetry Working Group, *Combined LEP Selectron/Smuon/Stau Results, 183-208 GeV*,
http://alephwww.cern.ch/~ganis/SUSYWG/SLEP/sleptons_2k01.html.

[45] J. R. Ellis, T. Falk, G. Ganis, K. A. Olive and M. Srednicki, Phys. Lett. B **510** (2001) 236 [arXiv:hep-ph/0102098].

[46] LEP Higgs Working Group for Higgs boson searches, OPAL Collaboration, ALEPH Collaboration, DELPHI Collaboration and L3 Collaboration, *Search for the Standard Model Higgs Boson at LEP*, ALEPH-2001-066, DELPHI-2001-113, CERN-L3-NOTE-2699, OPAL-PN-479, LHWG-NOTE-2001-03, CERN-EP/2001-055, arXiv:hep-ex/0107029; *Searches for the neutral Higgs bosons of the MSSM: Preliminary combined results using LEP data collected at energies up to 209 GeV*, LHWG-NOTE-2001-04, ALEPH-2001-057, DELPHI-2001-114, L3-NOTE-2700, OPAL-TN-699, arXiv:hep-ex/0107030.

[47] M.S. Alam et al., [CLEO Collaboration], Phys. Rev. Lett. **74** (1995) 2885 as updated in S. Ahmed et al., CLEO CONF 99-10; BELLE Collaboration, BELLE-CONF-0003, contribution to the 30th International conference on High-Energy Physics, Osaka, 2000. See also K. Abe *et al.*, [Belle Collaboration], [arXiv:hep-ex/0107065]; L. Lista [BaBar Collaboration], [arXiv:hep-ex/0110010]; C. Degrassi, P. Gambino and G. F. Giudice, JHEP **0012** (2000) 009 [arXiv:hep-ph/0009337]; M. Carena, D. Garcia, U. Nierste and C. E. Wagner, Phys. Lett. B **499** (2001) 141 [arXiv:hep-ph/0010003].

[48] J. R. Ellis, K. A. Olive and Y. Santoso, New J. Phys. **4** (2002) 32 [arXiv:hep-ph/0202110].

[49] M. Carena, J. R. Ellis, A. Pilaftsis and C. E. Wagner, Nucl. Phys. B **586** (2000) 92 [arXiv:hep-ph/0003180], Phys. Lett. B **495** (2000) 155 [arXiv:hep-ph/0009212]; and references therein.

[50] J. R. Ellis, G. Ganis, D. V. Nanopoulos and K. A. Olive, Phys. Lett. B **502** (2001) 171 [arXiv:hep-ph/0009355].

[51] H. N. Brown *et al.* [Muon g-2 Collaboration], Phys. Rev. Lett. **86**, 2227 (2001) [arXiv:hep-ex/0102017].

[52] G. W. Bennett *et al.* [Muon g-2 Collaboration], ppm," Phys. Rev. Lett. **89** (2002) 101804 [Erratum-ibid. **89** (2002) 129903] [arXiv:hep-ex/0208001].

[53] M. Davier, S. Eidelman, A. Hocker and Z. Zhang, arXiv:hep-ph/0208177; see also K. Hagiwara, A. D. Martin, D. Nomura and T. Teubner, arXiv:hep-ph/0209187; F. Jegerlehner, unpublished, as reported in M. Krawczyk, arXiv:hep-ph/0208076.

[54] M. Knecht and A. Nyffeler, arXiv:hep-ph/0111058; M. Knecht, A. Nyffeler, M. Perrottet and E. De Rafael, arXiv:hep-ph/0111059; M. Hayakawa and T. Kinoshita, arXiv:hep-ph/0112102; I. Blokland, A. Czarnecki and K. Melnikov, arXiv:hep-ph/0112117; J. Bijnens, E. Pallante and J. Prades, arXiv:hep-ph/0112255.

[55] L. L. Everett, G. L. Kane, S. Rigolin and L. Wang, Phys. Rev. Lett. **86**, 3484 (2001) [arXiv:hep-ph/0102145]; J. L. Feng and K. T. Matchev, Phys. Rev. Lett. **86**, 3480 (2001) [arXiv:hep-ph/0102146]; E. A. Baltz and P. Gondolo, Phys. Rev. Lett. **86**, 5004 (2001) [arXiv:hep-ph/0102147]; U. Chattopadhyay and P. Nath, Phys. Rev. Lett. **86**, 5854 (2001) [arXiv:hep-ph/0102157]; S. Komine, T. Moroi and M. Yamaguchi, Phys. Lett. B **506**, 93 (2001) [arXiv:hep-ph/0102204]; J. Ellis, D. V. Nanopoulos and K. A. Olive, Phys. Lett. B **508** (2001) 65 [arXiv:hep-ph/0102331]; R. Arnowitt, B. Dutta, B. Hu and Y. Santoso, Phys. Lett. B **505** (2001) 177 [arXiv:hep-ph/0102344] S. P. Martin and J. D. Wells, Phys. Rev. D **64**, 035003 (2001) [arXiv:hep-ph/0103067]; H. Baer, C. Balazs, J. Ferrandis and X. Tata, Phys. Rev. D **64**, 035004 (2001) [arXiv:hep-ph/0103280].

[56] N. A. Bahcall, J. P. Ostriker, S. Perlmutter and P. J. Steinhardt, Science **284** (1999) 1481 [arXiv:astro-ph/9906463].

[57] S. Mizuta and M. Yamaguchi, Phys. Lett. B **298** (1993) 120 [arXiv:hep-ph/9208251]; J. Edsjo and P. Gondolo, Phys. Rev. D **56** (1997) 1879 [arXiv:hep-ph/9704361].

[58] J. Ellis, T. Falk and K. A. Olive, Phys. Lett. B **444** (1998) 367 [arXiv:hep-ph/9810360]; J. Ellis, T. Falk, K. A. Olive and M. Srednicki, Astropart. Phys. **13** (2000) 181 [arXiv:hep-ph/9905481]; M. E. Gómez, G. Lazarides and C. Pallis, Phys. Rev. D **61** (2000) 123512 [arXiv:hep-ph/9907261] and Phys. Lett. B **487** (2000) 313 [arXiv:hep-ph/0004028]; R. Arnowitt, B. Dutta and Y. Santoso, Nucl. Phys. B **606** (2001) 59 [arXiv:hep-ph/0102181].

[59] M. Drees and M. M. Nojiri, Phys. Rev. D **47** (1993) 376 [arXiv:hep-ph/9207234]; H. Baer and M. Brhlik, Phys. Rev. D **53** (1996) 597 [arXiv:hep-ph/9508321] and Phys. Rev. D **57** (1998) 567 [arXiv:hep-ph/9706509]; H. Baer, M. Brhlik, M. A. Diaz, J. Ferrandis, P. Mercadante, P. Quintana and X. Tata, Phys. Rev. D **63** (2001) 015007 [arXiv:hep-ph/0005027]; A. B. Lahanas, D. V. Nanopoulos and V. C. Spanos, Mod. Phys. Lett. A **16** (2001) 1229 [arXiv:hep-ph/0009065].

[60] J. L. Feng, K. T. Matchev and T. Moroi, Phys. Rev. Lett. **84**, 2322 (2000) [arXiv:hep-ph/9908309]; J. L. Feng, K. T. Matchev and T. Moroi, Phys. Rev. D **61**, 075005 (2000) [arXiv:hep-ph/9909334]; J. L. Feng, K. T. Matchev and F. Wilczek, Phys. Lett. B **482**, 388 (2000) [arXiv:hep-ph/0004043].

[61] See, for example: I. Hinchliffe, F. E. Paige, M. D. Shapiro, J. Soderqvist and W. Yao, Phys. Rev. D **55** (1997) 5520; TESLA Technical Design Report, DESY-01-011, Part III, *Physics at an e^+e^- Linear Collider* (March 2001).

[62] M. Battaglia *et al.*, Eur. Phys. J. C **22** (2001) 535 [arXiv:hep-ph/0106204].

[63] G. L. Kane, J. Lykken, S. Mrenna, B. D. Nelson, L. T. Wang and T. T. Wang, arXiv:hep-ph/0209061.

[64] D. R. Tovey, Phys. Lett. B **498** (2001) 1 [arXiv:hep-ph/0006276].

[65] F. E. Paige, hep-ph/0211017.

[66] ATLAS Collaboration, *ATLAS detector and physics performance Technical Design Report*, CERN/LHCC 99-14/15 (1999); S. Abdullin *et al.* [CMS Collaboration], arXiv:hep-ph/9806366; S. Abdullin and F. Charles, Nucl. Phys. B **547** (1999) 60 [arXiv:hep-ph/9811402]; CMS Collaboration, Technical Proposal, CERN/LHCC 94-38 (1994).

[67] D. Denegri, W. Majerotto and L. Rurua, Phys. Rev. D **60** (1999) 035008.

[68] TESLA Technical Design Report, DESY-01-011, Part III, *Physics at an e^+e^- Linear Collider* (March 2001).

[69] J. Ellis, S. Heinemeyer, K. A. Olive and G. Weiglein, hep-ph/0211206.

[70] J. R. Ellis, G. Ganis and K. A. Olive, Phys. Lett. B **474** (2000) 314 [arXiv:hep-ph/9912324].

[71] G. A. Blair, W. Porod and P. M. Zerwas, Phys. Rev. **D63** (2001) 017703 [arXiv:hep-ph/0007107]; arXiv:hep-ph/0210058.

[72] R. W. Assmann *et al.* [CLIC Study Team], *A 3-TeV e^+e^- Linear Collider Based on CLIC Technology*, ed. G. Guignard, CERN 2000-08; for more information about this project, see:
http://ps-div.web.cern.ch/ps-div/CLIC/Welcome.html.

[73] CLIC Physics Study Group,
http://clicphysics.web.cern.ch/CLICphysics/.

[74] For more information about this project, see:
http://ctf3.home.cern.ch/ctf3/CTFindex.htm.

[75] M. Battaglia, private communication.

[76] J. Ellis, J. L. Feng, A. Ferstl, K. T. Matchev and K. A. Olive, arXiv:astro-ph/0110225.

[77] CDMS Collaboration, R. W. Schnee *et al.*, Phys. Rept. **307**, 283 (1998).

[78] CRESST Collaboration, M. Bravin *et al.*, Astropart. Phys. **12**, 107 (1999) [arXiv:hep-ex/9904005].

[79] H. V. Klapdor-Kleingrothaus, arXiv:hep-ph/0104028.

[80] DAMA Collaboration, R. Bernabei *et al.*, Phys. Lett. B **436** (1998) 379.

[81] G. Jungman, M. Kamionkowski and K. Griest, Phys. Rept. **267**, 195 (1996) [arXiv:hep-ph/9506380];
http://t8web.lanl.gov/people/jungman/neut-package.html.

[82] Y. Fukuda *et al.* [Super-Kamiokande Collaboration], Phys. Rev. Lett. **81**, 1562 (1998) [arXiv:hep-ex/9807003].

[83] Q. R. Ahmad *et al.* [SNO Collaboration], Phys. Rev. Lett. **89** (2002) 011301 [arXiv:nucl-ex/0204008]; Phys. Rev. Lett. **89** (2002) 011302 [arXiv:nucl-ex/0204009].

[84] R. Barbieri, J. R. Ellis and M. K. Gaillard, Phys. Lett. B **90** (1980) 249.

[85] M. Gell-Mann, P. Ramond and R. Slansky, Proceedings of the Supergravity Stony Brook Workshop, New York, 1979, eds. P. Van Nieuwenhuizen and D. Freedman (North-Holland, Amsterdam); T. Yanagida, Proceedings of the Workshop on Unified Theories and Baryon Number in the Universe, Tsukuba, Japan 1979 (edited by A. Sawada and A. Sugamoto, KEK Report No. 79-18, Tsukuba); R. Mohapatra and G. Senjanovic, Phys. Rev. Lett. **44** (1980) 912.

[86] P. H. Frampton, S. L. Glashow and T. Yanagida, arXiv:hep-ph/0208157.

[87] T. Endoh, S. Kaneko, S. K. Kang, T. Morozumi and M. Tanimoto, arXiv:hep-ph/0209020.

[88] J. R. Ellis, J. S. Hagelin, S. Kelley and D. V. Nanopoulos, Nucl. Phys. B **311** (1988) 1.

[89] J. R. Ellis, M. E. Gómez, G. K. Leontaris, S. Lola and D. V. Nanopoulos, Eur. Phys. J. C **14** (2000) 319.

[90] J. R. Ellis, M. K. Gaillard and D. V. Nanopoulos, Nucl. Phys. B **109** (1976) 213.

[91] M. Kobayashi and T. Maskawa, Prog. Theor. Phys. **49** (1973) 652.

[92] Z. Maki, M. Nakagawa and S. Sakata, Prog. Theor. Phys. **28** (1962) 870.

[93] S. Fukuda et al. [Super-Kamiokande Collaboration], Phys. Lett. B **539** (2002) 179 [arXiv:hep-ex/0205075].

[94] Chooz Collaboration, Phys. Lett. B **420** (1998) 397.

[95] A. De Rújula, M.B. Gavela and P. Hernández, Nucl. Phys. **B547** (1999) 21, hep-ph/9811390.

[96] A. Cervera et al., Nucl. Phys. B **579**, 17 (2000) [Erratum-ibid. B **593**, 731 (2001)].

[97] M. Apollonio et al., *Oscillation physics with a neutrino factory*, arXiv:hep-ph/0210192; and references therein.

[98] J. A. Casas and A. Ibarra, Nucl. Phys. B **618** (2001) 171 [arXiv:hep-ph/0103065].

[99] J. R. Ellis, J. Hisano, S. Lola and M. Raidal, Nucl. Phys. B **621**, 208 (2002) [arXiv:hep-ph/0109125].

[100] M. Fukugita and T. Yanagida, Phys. Lett. **B174**, 45 (1986).

[101] S. Davidson and A. Ibarra, JHEP **0109** (2001) 013.

[102] J. R. Ellis, J. Hisano, M. Raidal and Y. Shimizu, Phys. Lett. B **528**, 86 (2002) [arXiv:hep-ph/0111324].

[103] J. R. Ellis, J. Hisano, M. Raidal and Y. Shimizu, arXiv:hep-ph/0206110.

[104] J. R. Ellis and M. Raidal, Nucl. Phys. B **643** (2002) 229 [arXiv:hep-ph/0206174].

[105] A. Masiero and O. Vives, New J. Phys. **4** (2002) 4.

[106] C. Jarlskog, Phys. Rev. Lett. **55** (1985) 1039; Z. Phys. C **29** (1985) 491.

[107] K. Hagiwara et al. [Particle Data Group Collaboration], Phys. Rev. D **66** (2002) 010001.

[108] R. Cyburt, J. R. Ellis, B. Fields and K. A. Olive, astro-ph/0211258, and references therein.

[109] J. R. Ellis, M. Raidal and T. Yanagida, arXiv:hep-ph/0206300.

[110] Y. Kuno and Y. Okada, Rev. Mod. Phys. **73** (2001) 151; J. Hisano, T. Moroi, K. Tobe and M. Yamaguchi, Phys. Rev. D **53** (1996) 2442; J. Hisano, D. Nomura and T. Yanagida, Phys. Lett. B **437** (1998) 351; J. Hisano and D. Nomura, Phys. Rev. D **59** (1999) 116005; W. Buchmüller, D. Delepine and F. Vissani, Phys. Lett. B **459** (1999) 171; M. E. Gómez, G. K. Leontaris, S. Lola and J. D. Vergados, Phys. Rev. D **59** (1999) 116009; W. Buchmüller, D. Delepine and L. T. Handoko, Nucl. Phys. B **576** (2000) 445; J. L. Feng, Y. Nir and Y. Shadmi, Phys. Rev. D **61** (2000) 113005; J. Sato and K. Tobe, Phys. Rev. D **63** (2001) 116010; J. Hisano and K. Tobe, Phys. Lett. B **510** (2001) 197; D. Carvalho, J. Ellis, M. Gómez and S. Lola, Phys. Lett. B **515** (2001) 323; S. Baek, T. Goto, Y. Okada and K. Okumura, hep-ph/0104146; S. Lavignac, I. Masina and C.A. Savoy, hep-ph/0106245.

[111] Y. Okada, K. Okumura and Y. Shimizu, Phys. Rev. D **58** (1998) 051901; Phys. Rev. D **61** (2000) 094001.

[112] T. Ibrahim and P. Nath, Phys. Rev. D **57** (1998) 478 [Erratum - *ibid.* **58** (1998) 019901]; S. Abel, S. Khalil and O. Lebedev, Nucl. Phys. B **606** (2001) 151; S. Abel, D. Bailin, S. Khalil and O. Lebedev, Phys. Lett. B **504** (2001) 241.

[113] B. C. Regan, E. D. Commins, C. J. Schmidt and D. DeMille, Phys. Rev. Lett. **88** (2002) 071805.

[114] S. K. Lamoreaux, arXiv:nucl-ex/0109014.

[115] M. Furusaka *et al.*, JAERI/KEK Joint Project Proposal *The Joint Project for High-Intensity Proton Accelerators*, KEK-REPORT-99-4, JAERI-TECH-99-056.

[116] J. Äystö *et al.*, *Physics with Low-Energy Muons at a Neutrino Factory Complex*, CERN-TH/2001-231, hep-ph/0109217; and references therein.

WEIGHING THE UNIVERSE

James Rich
SPP-Dapnia, CEA, Saclay
91191 Gif-sur-Yvette, France
rich@hep.saclay.cea.fr

Abstract

These lectures present an overview of attempts to measure the cosmological density parameters and to explain their values in term of fundamental physics.

1.   Introduction

The enormous progress in observational cosmology over the last 10 years has led to the tentative conclusion that the Universe is made up of the curious mixture of substances shown in Table 1. It is believed that most particles in the Universe are cold photons or neutrinos. On the other hand, most of the energy is in the form of ordinary "baryonic" matter (protons and nuclei plus electrons), and not-so-ordinary "cold dark matter" and an effective vacuum energy or cosmological constant.

The densities in table 1 are normalized to the "critical density"

$$\rho_c = \frac{3H_0^2}{8\pi G},$$ (1)

where G is the gravitational constant and the Hubble constant is the factor of proportionality between the distance of a galaxy and its recession velocity (Hubble's Law)

$$\frac{dR}{dt} = H_0 R.$$ (2)

Most recent measurements [1] give values consistent with

$$H_0 = 70 \pm 10 \ \mathrm{km \, s^{-1} Mpc^{-1}}.$$ (3)

H.B. Prosper and M. Danilov (eds.), Techniques and Concepts of High-Energy Physics XII, 159–188.

This gives a critical density of

$$\rho_c = \frac{3H_0^2}{8\pi G} = 0.92\, h_{70}^2 \times 10^{-26}\,\mathrm{kg\, m^{-3}} \tag{4}$$

$$= 1.4\, h_{70}^2 \times 10^{11}\, M_\odot\, \mathrm{Mpc^{-3}} = 0.51\, h_{70}^2 \times 10^{10}\, \mathrm{eV\, m^{-3}}\,, \tag{5}$$

where

$$h_{70} = \frac{H_0}{70\,\mathrm{km\, s^{-1} Mpc^{-1}}} \tag{6}$$

The critical density corresponds to about one galaxy/Mpc^3 or about 5 protons/m^3. In a Universe where most of the energy is in the form of non-relativistic objects, the critical density corresponds to the dividing line between "open" universes ($\Omega_T < 1$) that expand eternally and "closed" universes ($\Omega_T > 1$) that eventually stop expanding and start to contract. This connection no longer holds when the Universe is dominated by vacuum energy where the expansion can be eternal even if $\Omega_T > 1$.

In table 1 we see that the total density of the Universe is nearly critical

$$\Omega_T = 1.00 \pm 0.03\,. \tag{7}$$

Since vacuum energy is dominant, we might conclude that the expansion will be eternal, though it is dangerous to extrapolate so far into the future.

The job of cosmologists is, first, to determine the cosmological densities in Table 1 and, second, to explain how these densities developed from "primordial" initial conditions. Traditionally, the primordial conditions were generally taken to be a plasma of elementary particles whose temperature and density rise to arbitrarily high values as one goes back in time towards a "big bang." Modern cosmology generally avoids the question of whether or not such a bang really occurred by saying that the present epoch started with a temperature of order the grand-unification scale ($kT \sim 10^{16}$ GeV) that was the result of a phase transition that ended en epoch of "inflation" when the energy of the Universe was contained in a simple scalar field. What went on before inflation is a purely speculative question.

In the following subsections, we will describe the different components of Table 1 and give the methods used to estimate their densities. In Section 2 we will describe current attempts to determine the nature of the Cold Dark Matter. More details can be found in the author's recent textbook [8].

Table 1. The known and suspected occupants of the Universe. For each species, i, the table gives estimated number density of particles, n_i, and the estimated mass or energy density, $\Omega_i = \rho_i/\rho_c$, normalized to the " critical density," $\rho_c = 0.92\, h_{70}^2 \times 10^{-26}\,\mathrm{kg\,m^{-3}}$. Some of the estimated densities depend of the numerical value of the Hubble constant H_0 resulting in the factors of $h_{70} = H_0/(70\,\mathrm{km\ s^{-1}Mpc^{-1}}) \sim 1$. Other than the density of the directly observed photons, all numbers in this table are estimated by methods that are more or less indirect and should be considered as provisional. For this reason, the error bars should not be taken too seriously.

species	n_i $(\mathrm{m^{-3}})$	$\Omega_i = \rho_i/\rho_c$	Reference
CBR photons	$n_\gamma = (4.11 \pm 0.02) \times 10^8$	$\Omega_\gamma = 5.06\, h_{70}^{-2} \times 10^{-5}$	[1, 2]
ν_e, ν_μ, ν_τ	$n_\nu = (3/11) n_\gamma$ (per species)	$\Omega_\nu > 4\, h_{70}^{-2} \times 10^{-4}$	[3]
baryons (+electrons)	$n_b \sim 0.2 \pm 0.05$ $(n_{He}/n_H \sim 0.08)$	$\Omega_b \sim (0.04 \pm 0.01)\, h_{70}^{-2}$	[4, 5]
cold dark matter	?	$\Omega_{CDM} \sim 0.3 \pm 0.1$	[1]
"vacuum"	0	$\Omega_\Lambda \sim 0.7 \pm 0.1$	[6, 7, 5]
total		$\Omega_T \sim 1.00 \pm 0.03$	[5]

1.1   Baryons

The total amount of "visible matter" (stars in galaxies) can be estimated from the amount of visible light generated by galaxies. To do this, one uses the fact that typical stars have masses of order one solar mass $M_\odot = 1.988 \times 10^{30}$ kg, and generate energy at a rate of order one solar luminosity, $L_\odot = 2.4 \times 10^{45}$ eV s^{-1}. Quantitatively, the mass-to-light ratio used for visible mass estimates is about $3\,(M_\odot/L_\odot)$. The total light output in the local universe is of order

$$J \sim 10^8\, L_\odot\, \text{Mpc}^{-3} . \tag{8}$$

Multiplying by $3\,(M_\odot/L_\odot)$ and dividing by the critical density gives a mean visible density of

$$\Omega_{\text{vis}} = \frac{\rho_{\text{vis}}}{\rho_{\text{c}}} \sim 0.002 \qquad (\text{vis} = \text{visible stars}) . \tag{9}$$

The total density of baryonic matter (protons, nuclei and electrons) is estimated to be an order of magnitude greater than that of visible baryons (9):

$$\Omega_{\text{b}} = (0.04 \pm 0.01)\, h_{70}^{-2} . \tag{10}$$

This estimate comes from the theory of the nucleosynthesis of the light elements [4] which correctly predicts the *relative* abundances of the light elements only if Ω_{b} is near this value. To briefly summarize the argument, the nuclear composition of the Universe changes with time as the stars transform their hydrogen into helium and then into heavier elements. Nevertheless, there appears to be a "primordial" mix of nuclei consisting of about 75% hydrogen (by mass) and 25% ^{4}He along with traces of ^{2}H, ^{3}He, and ^{7}Li. This mixture is approximately the observed mixture in certain locations unpolluted by stellar nucleosynthesis. The primordial abundances were determined by nuclear reactions that took place when the universal temperature was ~ 60 keV. Primordial nucleosynthesis calculations predict the abundances as a function of the total baryon density at $T \sim 60$ keV, since it is this density that determines the nuclear reaction rates. The predicted nuclear abundances match the observed abundances if the current baryon density has the value given by (10).

Of particular importance for this estimation of Ω_{b} is the observed ration of deuterium to ordinary hydrogen. In the early Universe, deuterium was an intermediate step in the reactions that collected the available neutrons into ^{4}He nuclei. For high baryon densities, the reactions proceed easily and the efficiency for transformation to ^{4}He is high. The amount of surviving deuterium is therefore a rapidly decreasing function of the total baryon density, as can be seen in Fig. 1.

Since $\Omega_b > \Omega_{vis}$ one can wonder where the missing "dark" baryons are. Most of them are thought to be in the intergalactic medium in the form of an ionized gas [12]. Some fraction of them may be in dark compact objects such as dead stars (neutron stars or white dwarfs) or stars too light to burn hydrogen (brown dwarfs). It has also been suggested [13] that a significant fraction of the baryons are contained in cold molecular clouds.

Finally, we mention that there are apparently very few antibaryons in the visible universe [14]. Any antimatter consisting of antibaryons and positrons would quickly annihilate in collisions with ordinary matter. Even if the antimatter where somehow separated from the matter, annihilations in intergalactic space at the boundaries between matter and antimatter domains would lead to a flux of high-energy annihilation photons higher than the observed flux from other sources. It thus seems probable that the density of antimatter is exponentially small within our horizon.

The lack of antibaryons and the small baryon to photon ratio is believed to have developed from a plasma a relativistic quarks and gluons where to good approximation the numbers of photons, quarks and anti quarks where nearly equal, but with a small excess, of order 10^{-9} of quarks over antiquarks. As the universe cooled, the quarks and antiquarks combined to form pions but the small excess of quarks resulted in the nucleons that exist to this day. The origin of the excess is not known but is believed to have been generated by baryon-number and CP violating processes at high temperatures.

1.2 Cold Dark Matter

Galaxies and galaxy clusters were formed by gravitational collapse of non-relativistic matter. One of the main results of the theory of this process is that it is difficult to understand how baryons of an amount given by (10) could have created the observed structures. The basic problem stems from the fact that the photons of cosmological origin (Sect. 1.3) are observed to have an energy spectrum that is nearly independent of the direction of observation. These photons where thermally coupled to the baryons until the Universe became transparent when atoms formed at a temperature of $kT \sim 0.25\,\text{eV}$. Because of this coupling, the homogeneity of the photons implies that the baryons were extremely homogeneous at early times. Going from this homogeneous state to the present inhomo-

164

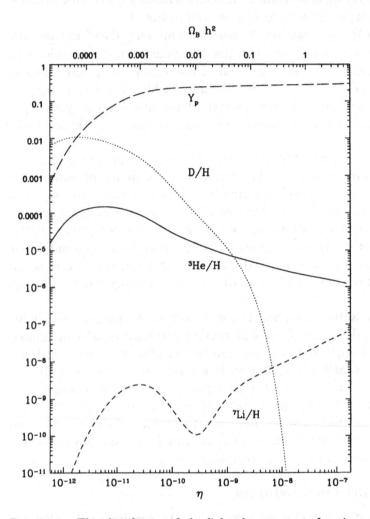

Figure 1. The abundances of the light elements as a function of the assumed baryon-photon ratio η (Bottom Horizontal Axis) or of the assumed value of $\Omega_b h^2$ (Top Horizontal Axis), as calculated in [9]. For ^{4}He, the abundance is given as the fraction $Y_p = \rho_{He}/\rho_b$ of the total baryonic mass that is in the form of ^{4}He, while the other elements are reported as number densities normalized to ^{1}H. The abundance by mass of ^{4}He is a slowly increasing function of η. The abundances of the loosely bound intermediate nuclei ^{2}H and ^{3}He are decreasing functions of η. The form of the curve for ^{7}Li is due to the fact that the production is mostly direct for $\eta < 3 \times 10^{-10}$ and mostly indirect via production and subsequent decay of ^{7}Be for $\eta > 3 \times 10^{-10}$. Observations [10] indicate that ^{2}H$/^1$H $\sim 3.4 \times 10^{-5}$ in high-redshift intergalactic clouds. If this figure reflects the primordial abundance, we can conclude that $\eta \sim 5 \times 10^{-10}$ corresponding to $\Omega_b h^2 \sim 0.02$ ($\Omega_b h^2_{70} \sim 0.04$). Figure courtesy of Elisabeth Vangioni-Flam.

geneous state is "difficult" if there are only baryons because they suffer from their tight coupling with the homogeneous photons.

Popular models of structure formation generally assume the existence of dark matter in some form that has been only weakly interacting and non-relativistic since the epoch of $T \sim$ MeV. This matter is generically called "cold dark matter" (CDM). Often, it is assumed to be in the form of nonbaryonic weakly interacting massive particles called generically "wimps". The fact that they are only weakly interacting allows them to gravitate freely without being inhibited by interactions with other particles (baryons and photons).

The present density of CDM is estimated to be an order of magnitude greater than that of baryons but still less than critical

$$\Omega_{\text{CDM}} \sim 0.3 . \tag{11}$$

This dark matter is believed to make up most of the mass of galactic halos and galaxy clusters.

The value of Ω_{CDM} can be most easily estimated by measuring the ratio of baryonic to total mass in galaxy clusters. The largest clusters contain thousands of galaxies and up to $10^{15} \, h_{70}^{-1} M_\odot$ in volumes of a few Mpc^3. This mass corresponds to the mean mass contained in a sphere of radius $\sim 10 \, h_{70}^{-1}$ Mpc so rich clusters have over-densities of order 1000. The nearest rich cluster is the Virgo Cluster at a distance of ~ 20 Mpc.

Clusters are rather ill-defined when viewed as simple collections of galaxies. They become much more distinct when observed through their X-ray emission. The X-rays are produced through bremsstrahlung by electrons in the ionized intergalactic gas. It is believed that this gas constitutes the majority of the baryonic mass of clusters.

Cluster masses can be most easily estimated by measuring the velocity dispersion of the member galaxies and then applying the virial theorem. More modern techniques use the X-ray temperature or the shapes of background galaxies that are deformed by the lensing action of the cluster. The most massive clusters have mass-to-light ratios of order 200 [15]:

$$\frac{M}{L} \sim 200 \, h_{70} \, M_\odot / L_\odot . \tag{12}$$

The mass of the cluster gas can be estimated from the observed X-ray luminosity. Observations indicate that it is an order of magnitude less than the total mass [16]:

$$\frac{M_{\text{gas}}}{M_{\text{total}}} \sim \frac{M_{\text{baryons}}}{M_{\text{total}}} \sim 0.12 \, h_{70}^{-3/2} . \tag{13}$$

This factor of 10 is perhaps the best evidence for nonbaryonic dark matter since it would be difficult to hide 90% of the cluster baryons.

We can now estimate Ω_M/Ω_b by assuming that the baryon-total ratio in clusters is equal to the universal value:

$$\Omega_M = \frac{\Omega_b}{0.12\,h_{70}^{-3/2}} = 0.33\,h_{70}^{-1/2}\,, \tag{14}$$

where we have used the value of Ω_b derived from the nucleosynthesis value of the baryon-photon ratio.

Ω_M can also be estimated through its influence on structure formation. The simplest method of this type is the evolution of the number of galaxy clusters with time or, equivalently, with redshift [24]. Structures can form from small density perturbations only if the universe is mass-dominated with near critical density. If this condition is satisfied, a region of a small over-density acts as mini-closed universe that will eventually detach from the expansion to form a virialized structure. If $\Omega_M \sim 1$ structures are still forming today, whereas if $\Omega_M \ll 1$ structure formation took place only in the past when the mass density was near critical. This effect should be visible in the number of clusters as a function of redshift, a large dependence signaling $\Omega_M \sim 1$. A non-zero vacuum energy density complicates the analysis but the cluster evolution still gives a relatively robust value for Ω_M. One recent study [25] gives $\Omega_M = 0.49 \pm 0.12$ for $\Omega_\Lambda = 0$ and $\Omega_M = 0.44 \pm 0.12$ for $\Omega_T = 1$.

At super-cluster scales, Ω_M is one of the parameters that determine the shape of the spectrum of density fluctuations. By this we mean the variations of density observed on a "scale" R, i.e. when the density is measured by measuring the mass in spheres of radium R and dividing by the volume of the sphere. At "large" scales, the density fluctuations are observed to decrease with scale:

$$\frac{\Delta\rho}{\rho}(R) \propto R^{-2}\,. \tag{15}$$

This is a "Harrison-Zeldovich" spectrum of fluctuations that is believed to have been generated by quantum fluctuations of the scalar field driving the inflation epoch. At "small" scales, the spectrum flattens and rises less rapidly than R^{-2} because the modes associated with small scales oscillated as acoustic waves during the epoch when relativistic matter dominated the Universe. The dividing line between large and small scales depends on the amount of non-relativistic matter since it is this quantity that determines when the radiation-dominated epoch ends. The observed dividing line (~ 20Mpc) is consistent with what is expected for $\Omega_M \sim 0.3$.

In summary, it appears that most authors agree that a critical universe containing only matter is disfavored

$$0.2 < \Omega_M < 0.5 . \tag{16}$$

It can be hoped that better redshift surveys and studies of gravitational weak-lensing by large-scale structure will clarify the situation.

While the existence of some sort of cold dark matter is generally supposed, there is at present no indication of what it is actually made of. It is often supposed that they are made of wimps but, unfortunately, there are no known wimps in the zoo of elementary particles and their existence is a bold prediction of cosmology. Some extensions of the standard model of particle physics predict the existence of wimps that are sufficiently heavy that they would not yet have been produced at accelerators. Originally, it was proposed that a new heavy neutrino with $m_\nu > 1\,\mathrm{GeV}$ could be the CDM but this possibility was excluded [17] by a combination of accelerator results and direct searches (Sect. 2).

A more speculative class of models that predict the existence of wimps are "supersymmetric" models. In these models, each of the known fermions (bosons) is paired with a heavy supersymmetric partner that is a boson (fermion). The lightest of the supersymmetric partners (LSP) is expected to be stable and to have only weak interactions, making it an ideal wimp candidate. The parameters of the supersymmetric model can be chosen so that the wimp has the required present-day density, $\Omega_{CDM} \sim 0.3$. The mass would be expected to be between 10 GeV and 10 TeV. Efforts are underway to detect supersymmetric particles at accelerators and in the Galaxy (Sect. 2).

Other candidates for nonbaryonic dark matter are light neutrinos and primordial black holes. We will see that neutrinos with the density $\Omega_\nu \sim 0.3$ would have masses near 10 eV. Such light particles have difficulties in forming the observed structures because they were relativistic when $T \sim \mathrm{MeV}$ and constitute " hot dark matter." This type of dark matter leads to a spectrum of density fluctuations that is strongly suppressed for scales less than $\sim 70\mathrm{Mpc}$. This is because neutrinos remain relativistic until late times and freely move over distances that correspond to this scale. This would wash out any initial fluctuation on scales smaller than this.

Primordial black holes work well in structure formation but cosmologists lack a convincing scenario for their production in the early Universe

Finally, we mention that estimates of the dark-matter content of the Universe suppose that normal Newtonian/Einsteinian gravity can be used at super-kiloparsec distances. There is no direct test of this theory at super-solar-system scales. In fact it has long been emphasized by

Milgrom [19] that it is possible to modify Newton's gravitational law in way to obviate the need for any dark matter. This very speculative "MOND" (MOdified Newtonian Dynamics) hypothesis has proved to be very difficult do eliminate.

1.3 Photons

The most abundant particles in the Universe are the photons of the " cosmic background radiation" (CBR) (also referred to in the literature as " CMB"). These photons have a nearly perfect thermal spectrum as shown in Fig. 2. The photon temperature is $T_\gamma = 2.725\,\mathrm{K} = 2.35 \times 10^{-4}\,\mathrm{eV}$ corresponding to a number density of $n_\gamma = 411\,\mathrm{cm}^{-3}$. This is considerably greater than the number of photons that have been generated by stars. Despite their great abundance, the low temperature of the CBR results in a small photon energy density:

$$\Omega_\gamma = 5.06\, h_{70}^{-2} \times 10^{-5} \ . \tag{17}$$

The present-day Universe is nearly transparent to photons. The CBR photons were thermalized in the early Universe when the temperature was $> 0.26\,\mathrm{eV}$ and baryonic matter was completely ionized. At $T \sim 0.26\,\mathrm{eV}$, baryonic matter "recombined" to form atoms and the resulting decrease in the photon-matter cross-section made the Universe transparent.

The CBR temperature is not completely isotropic but is observed to vary by factors of order 10^{-5} according to the direction of observation. These small variations are believed to be due to the density inhomogeneities present at the moment of recombination. As such, the temperature anisotropies provide information about the "initial conditions" for structure formation. The spectrum of anisotropies when interpreted within the framework of popular models provides information that constrains Ω_T to be very near unity.

1.4 Neutrinos

In addition to thermal photons, it is believed that the Universe is filled with neutrinos, ν_e, ν_μ and ν_τ and the corresponding antineutrinos. The three neutrino " flavors" should be identified with the neutrino created in interactions with the associated charged leptons e, μ and τ. For instance, the ν_e is created in β decay in association with an electron or positron:

$$(A, Z) \ \rightarrow \ (A, Z \mp 1)\, e^\pm \ \nu_e\,(\bar\nu_e)\,, \tag{18}$$

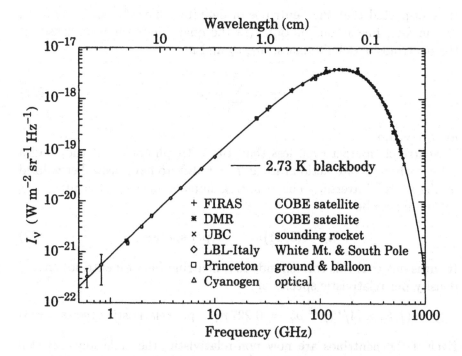

Figure 2. The observed spectrum of the cosmic (microwave) background radiation (CBR) [1]. The points at wavelengths < 1 cm come from ground-based experiments. At shorter wavelengths the Earth's atmosphere is opaque and measurements must be made from balloons, rockets or satellites. The high precision points around the peak of the spectrum were made by the FIRAS instrument of the COBE satellite which observed from 1989 to 1995 [2]. Compilation courtesy of the Particle Data Group.

where A is the number of nucleons and Z is the number of protons in the nucleus. The ν_μ is created in pion decay in association with a muon:

$$\pi^\pm \to \mu^\pm \, \nu_\mu \, (\bar{\nu}_\mu) \,. \tag{19}$$

The ν_τ is created in τ decay, e.g.:

$$\tau^\pm \to \pi^\pm \, \bar{\nu}_\tau \, (\nu_\tau) \,. \tag{20}$$

It is suspected that the neutrinos of definite " flavor", ν_e, ν_μ and ν_τ, are, in fact, linear combinations (in the quantum-mechanical sense) of the neutrinos of definite mass, ν_1, ν_2 and ν_3:

$$\nu_f = \sum_{i=1}^{3} \alpha_{fi} \nu_i \tag{21}$$

for $f = e, \mu, \tau$.

Neutrinos interact even less than the CBR photons but they had a sufficiently high interaction rate at $T > 1\,\text{MeV}$ to have been thermalized in the early Universe. The resulting number density of neutrinos is calculated to be

$$n_\nu = (3/11) n_\gamma \qquad \text{per species.} \tag{22}$$

It turns out that this corresponds to a neutrino ($+$ antineutrino) energy density per relativistic species of

$$\rho_\nu = (7/8) \times (4/11)^{4/3} \rho_\gamma = 0.227 \rho_\gamma \qquad \text{per relativistic species .} \tag{23}$$

Even if the neutrinos are now non-relativistic, the relation (22) still holds so we expect $n_\nu = 112\,\text{cm}^{-3}$ per species today.

For an effectively massless neutrino species, i.e. $m_\nu \ll T_\nu$, the summed neutrino and antineutrino contribution of that species to the energy density is even less than that of CBR photons:

$$\Omega_\nu = 1.15 \, h_{70}^{-2} \times 10^{-5} \qquad \text{if } m_\nu \ll 10^{-4}\,\text{eV} \,. \tag{24}$$

For a species of mass greater than the temperature, the neutrinos are currently non-relativistic and the summed neutrino and antineutrino mass density is

$$\Omega_\nu = \frac{m_\nu n_\nu}{\rho_c} = 0.2 \, h_{70}^{-2} \, \frac{m_\nu}{10\,\text{eV}} \qquad \text{if } m_\nu \gg 10^{-4}\,\text{eV} \,. \tag{25}$$

If one of the neutrinos species has a mass in the eV range it would contribute significantly to the universal energy density. We can also conclude that any neutrino species with a number density given by (22) must have $m_\nu < 30\,\mathrm{eV}$ to avoid giving a cosmological density greater than the observed density.

Limits on the neutrino masses come from kinematic studies of the decays (18), (19) and (20). The present limits [1] are not very constraining: $m_1 < 10\,\mathrm{eV}$, $m_2 < 170\,\mathrm{keV}$ and, $m_3 < 18.2\,\mathrm{MeV}$, where $i = 1, 2, 3$ correspond to the neutrinos of definite mass most closely aligned with ν_e, ν_μ and ν_τ.

Evidence for non-zero neutrino masses comes from searches for " neutrino oscillations", i.e. the transformation of given flavor neutrino into a different flavor, e.g.

$$\pi^+ \rightarrow \mu^+ \nu_\mu , \tag{26}$$

followed at some distance by the neutrino acting as different flavor

$$\nu\,\mathrm{p} \rightarrow \tau^- (!)\,\mathrm{p}\pi^+ . \tag{27}$$

These experiments are, unfortunately, only sensitive to *differences* in the squares of neutrino masses.[1] Recent observation of oscillations of neutrinos produced in the atmospheric interactions of cosmic rays [3] have given results that are most easily interpreted as

$$m_3^2 - m_2^2 \sim 10^{-3}\,\mathrm{eV}^2 . \tag{28}$$

Anomalies in the spectrum of solar neutrinos [20] can be explained by

$$m_2^2 - m_1^2 \sim 10^{-5}\,\mathrm{eV}^2 . \tag{29}$$

If these results are confirmed there are at least two species of non-relativistic neutrinos.   The mass differences only give lower limits on the masses themselves obtained by supposing that they are widely separated, i.e. $m_3 \gg m_2 \gg m_1$. In this case it follows that $m_3 > 0.02\,\mathrm{eV}$ and $m_2 > 0.002\,\mathrm{eV}$. This implies:

$$\Omega_\nu > 0.0004\,h_{70}^{-2} . \tag{30}$$

It is often supposed that neutrinos have a " hierarchical" mass pattern like that of the charged leptons, i.e. $m_3 \gg m_2 \gg m_1$. If this is the case, the above inequalities become approximate equalities. It is not possible, for the moment, to directly verify this hypothesis, so a cosmologically important neutrino is possible if the neutrino species have similar masses.

Finally, we note that, because of their extremely weak interactions, there is little hope of directly detecting the cosmic neutrino background [21].

---

[1]The differences determine the " oscillation length", i.e. the characteristic distance traveled before a neutrino starts to interact like a neutrino of a different flavor.

1.5   The Vacuum

Perhaps the most surprising recent discovery is that the Universe appears to be dominated by an apparent "vacuum energy" or "cosmological constant" Λ:

$$\Omega_\Lambda \sim \frac{\Lambda}{3H_0^2} \sim 0.7 . \tag{31}$$

Vacuum energy is, by definition, energy that is not associated with particles and is therefore not diluted by the expansion of the Universe. Unless the present vacuum is only metastable, this implies the the vacuum energy density is independent of time. The value implied by $\Omega_\Lambda = 0.7$ is

$$\rho_\Lambda(t) \sim 3\, h_{70}^2 \times 10^9 \, \text{eV}\, \text{m}^{-3} . \tag{32}$$

The observational evidence for the existence of such an energy involves the apparent luminosity of high redshift supernovae which can provide information on whether the universal expansion is accelerating or decelerating. One would expect that the gravitational attraction of ordinary matter would cause the expansion to decelerate. On the other hand, it can be shown in general relativity that a positive vacuum energy causes the expansion to accelerate.

The acceleration or deceleration impacts on the flux of supernovae because a given redshift, z, defined by the the emitted (λ_0) and observed (λ_1) wavelengths of photons

$$1 + z \equiv \frac{\lambda_1}{\lambda_0} . \tag{33}$$

For nearby objects, this redshift is most simply interpreted as a Doppler shift, but for cosmological distances, it corresponds to a given universal expansion factor, $1 + z$ between the time of the explosion and the time of the observation. If the expansion is accelerating (decelerating), the expansion rate in the past was relatively slower (faster) implying a longer (shorter) time between the explosion and observation. This time is just the time-of-flight for the observed photons and the large (small) time-of-flight corresponds to a large (small) distance to the supernova. Since the observed flux is a decreasing function of the distance to the supernova, we see by this chain of reasoning that an accelerating (decelerating) expansion yields a relatively small (large) photon flux from a supernova of a given redshift.

Two teams [6, 7] have observed that the fluxes of supernova at $z \sim 0.5$ are about 20% smaller than that expected for a marginally decelerated universe with $\Omega_M = 0.3$ (Fig. 3). The effect can be explained by a vacuum energy (assuming $\Omega_M \sim 0.3$) of

$$\Omega_\Lambda \sim 0.7 \qquad \text{if } \Omega_M = 0.3 \ . \qquad (34)$$

The low supernovae fluxes could, of course, be explained by supposing that typical supernovae were less luminous in the past than at present, or that 20% of the light is absorbed somewhere between the supernova and us. The evolution hypothesis can be tested locally by observing supernovae in galaxies of varying "effective age" as given for instance by their metalicity. Because any absorbing dust would be heated in the process, the absorption hypothesis can be ruled out if the diffuse infrared background can be ascribed to discrete sources [27].

Fundamental physics cannot currently be used to calculate the value of the vacuum energy even though it is a concept used throughout modern gauge theories of particle physics. It is expected to be a temperature-dependent quantity which changes in a calculable manner during phase transitions, e.g. the electroweak transition at $T \sim 300\,\text{GeV}$ when the intermediate vector bosons, $W^\pm$ and Z^0, became massive. While the vacuum energy does not change in particle collisions, so its existence can usually be ignored, it does lead to certain observable effects like the Casimir force between uncharged conductors. Unfortunately, all calculable quantities involving vacuum energy concern differences in energy densities and there are no good ideas on how to calculate the absolute value.

Despite the lack of ideas, the existence of a vacuum energy density of the magnitude given by (32) is especially surprising. In natural units, an energy density has the dimension of the fourth power of mass, so a vacuum energy density can be associated with a mass scale M

$$\rho_\Lambda \sim \frac{M^4}{(\hbar c)^3} \ . \qquad (35)$$

Particle physicists are tempted to choose the Planck mass

$$m_{\text{pl}} = (\hbar c^5 / G)^{1/2} \sim 10^{19}\,\text{GeV} \qquad (36)$$

as the most fundamental scale giving $\rho_\Lambda \sim 3 \times 10^{132}\,\text{eV}\,\text{m}^{-3}$. This is 123 orders of magnitude too large making it perhaps the worst guess in the history of physics. In fact, the density (32) implies a scale of $M \sim 10^{-3}\,\text{eV}$ which is not obviously associated with any other fundamental scale in particle physics, though it is near the estimated masses of the neutrinos.

A second problem with an energy density (31) is that it is comparable to the matter density $\Omega_M \sim 0.3$. Since the matter density changes

174

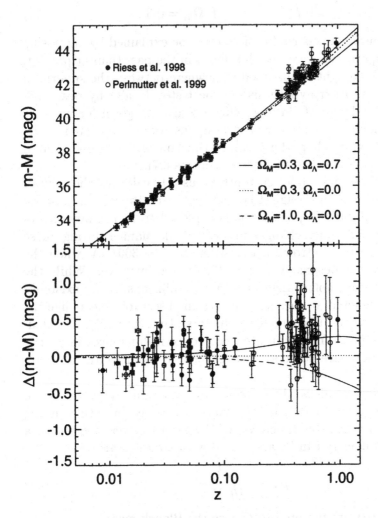

Figure 3. Measurements of the apparent luminosities of type Ia supernovae by the Supernova Cosmology Project [6] and the High-Z Supernova Search [7]. (compilation courtesy of A. Riess [31]). The upper panel shows the Hubble diagram. The lower panel shows the difference between the observed apparent magnitudes and the magnitude expected for a universe with $(\Omega_M = 0.3, \Omega_\Lambda = 0.0)$. At $z \sim 0.5$, supernovae are on average $\sim 0.2\,\mathrm{mag}$ (20%) less luminous than expected for $(\Omega_M = 0.3, \Omega_\Lambda = 0.0)$. This effect can be explained by supposing $(\Omega_M \sim 0.3, \Omega_\Lambda \sim 0.7)$.

with the expansion of the Universe while the vacuum energy does not, it appears that we live in a special epoch when the two energies are comparable.

Finally, we note that the observational evidence for a strictly constant vacuum energy density is also consistent with a component that slowly varies with time. Such a component could be the energy associated with a hypothetical scalar field. "Quintessence" models involving such fields are currently widely discussed in the literature [22]. They are motivated by the'aforementioned problems with a pure vacuum energy but a completely satisfactory model has not yet been formulated.

1.6 The total density Ω_T

The sum of the individually measured Ωs gives $\Omega_T \sim \Omega_\Lambda + \Omega_M \sim 1$. It is now possible to measure Ω_T directly via the spectrum of temperature fluctuations of the cosmic background radiation. The CBR has a spectrum of anisotropies that depends on the angular scale, i.e. the pixel size used to measure the temperature variations. To be more precise, we denote by $T(\vec{\theta}, \Delta\theta)$ the temperature observed in the direction $\vec{\theta}$ and averaged over a disk on the sky of radius $\Delta\theta$. The variance of this quantity over all directions defines $\Delta T(\Delta\theta)$. Observations show a peak in this quantity for $\Delta\theta \sim 1\,\mathrm{deg}$ (Fig. 4).

According to a popular class of models of structure formation (CDM models with adiabatic scale-invariant primordial fluctuations), the scale size expected to show the greatest fluctuations is that corresponding to the Hubble distance (i.e. the inverse of the expansion rate) at recombination. The corresponding angular size depends on Ω_M and Ω_Λ with $\Omega_T = 1$ corresponding to $\Delta\theta \sim 1\,\mathrm{deg}$. This is in good agreement with the value recently observed by the Archeops experiment [5]:

$$\Omega_T = 1.00 \pm 0.03 . \tag{37}$$

Future satellite experiments [30] on CBR anisotropies should measure this most important Ω to a very high precision.

2. Dark Matter

It is clear that there will always be doubts about the standard cosmological model as long the dark matter has not been identified. Here, we review efforts to detect the two favored nonbaryonic candidates, wimps and axions. We also review searches for dark astrophysical objects in the galactic halo using the gravitational lensing effect.

176

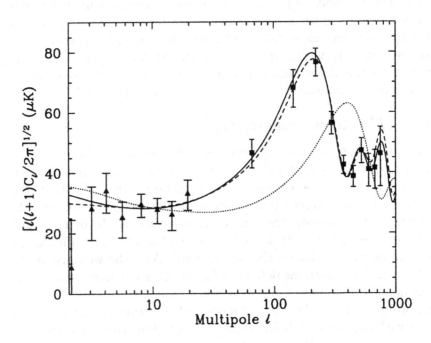

Figure 4. Measurements of the spectrum of CBR anisotropies as a function of angular scale. The large angular scale (small l) Triangles are from the DMR instrument of the COBE satellite [32]. The small angular scale (large l) Squares are the combined Boomerang [28] and Maxima [29] data. The temperature fluctuations at large angular scales are roughly independent of l at a value of 10^{-5} of the CBR temperature. The fact that the large angle fluctuations are independent of scale is evidence for scale-invariant primordial density fluctuations. The peak at $l \sim 200$ is generally interpreted as the first acoustic peak. Its position indicates that $\Omega_T \sim 1$. The curves correspond to calculations [33] assuming primordial adiabatic fluctuations. The Solid Line is the best fit yielding the parameters ($\Omega_T = 1.2, \Omega_\Lambda = 0.5, \Omega_b h^2 = 0.03$). The Dashed Line is the best Ω_T fit yielding the parameters ($\Omega_T = 1, \Omega_\Lambda = 0.7, \Omega_b h^2 = 0.03$). The Dotted Line is for an open universe with ($\Omega_T = 0.3, \Omega_\Lambda = 0.0, \Omega_b h^2 = 0.03$). For this model, the position of the peak is shifted to $200/\sqrt{\Omega_T} \sim 400$. The most recent results [5] give $\Omega_T = 1.00 \pm 0.03$ and $\Omega_b h^2 = 0.02$ (in agreement with the nucleosynthesis result). Figure courtesy of K. Ganga.

2.1 Wimps

Since the 1980s, wimps (weakly interacting massive particles) have been the standard cold dark matter (CDM) candidate. The primary motivation for introducing such objects comes from the theory of structure formation that suggests the existence of CDM. Supersymmetric extensions of the standard model of particle physics (invented to solve problems unrelated to dark matter) predict the existence of such particles. The fact that they have not yet been seen at accelerators suggests that they must have a mass $m_\chi > 30\,\mathrm{GeV}$ [34]. The particles would have been thermally produced in the early Universe yielding a cosmological abundance inversely proportional to their annihilation cross-section. Supersymmetric models contain many free parameters yielding relic densities within a few orders of magnitude on either side of the critical density.

Today, wimps would be expected to inhabit the halos of spiral galaxies like our own. From the galactic rotation velocity, one can estimate the local density to be about $0.3\ \mathrm{GeV\,cm^{-3}}$ [21]. The orbital velocities of objects trapped in the Galaxy are of order $250\,\mathrm{km\,s^{-1}}$ so the local wimp flux is of order $10^7 \times (1\ \mathrm{GeV}/m_\chi)\,\mathrm{cm^{-2}\,s^{-1}}$.

Goodman and Witten [35] suggested that these wimps could be detected via the observation of nuclei recoiling from wimp-nucleus elastic scatters. Galactic wimps with masses in the GeV range have kinetic energies in the keV range so we can also expect nuclear recoils in the keV range. The rate is proportional to the elastic wimp-nucleus scattering cross-section which depends on the parameters of the particle physics model. Typical values of the wimp-nucleon cross-section are of order $10^{-44}\,\mathrm{cm^2}$, corresponding to a very weak interaction.

Wimp scatters can be observed with "calorimetric" techniques (Fig. 5). Unfortunately, it is difficult to distinguish wimp events from events due to beta or gamma radioactivity (also shown in the figure). Statistically, a signal from wimps can be isolated through the expected $\sim 5\%$ seasonal modulation of the event rate [21, 36]. This modulation is due to the fact that while the Solar System moves through the (isotropic) wimp gas, the Earth's motion around the Sun alternately adds or subtracts from the wimp-detector velocity. Alternatively, certain detectors can distinguish nuclear recoils from the Compton-electron background [37, 38], e.g. hybrid cryogenic detectors that detect two types of excitations, phonons and ionization or phonons and scintillation. The ratio of the two signals is different for nuclear recoils and Compton electrons, allowing background rejection.

The present generation of experiments using NaI scintillators [39], germanium diodes [40], or germanium/silicon hybrid cryogenic calorimeters [37, 38] have backgrounds that make them insensitive to most supersymmetric dark matter. The supersymmetric models with the highest elastic scattering cross-sections give rates somewhat below the current background levels but most give rates at least two orders of magnitude below the present limits.

In spite of the lack of theoretical encouragement, one experiment [39] using NaI scintillators has reported a counting rate with the annual modulation expected from wimps of mass

$$m_\chi \sim 50 \pm 10 \, \text{GeV} \, . \tag{38}$$

The derived wimp-nucleon cross-section is

$$\sigma \sim 6 \pm 2 \times 10^{-42} \, \text{cm}^2 \, . \tag{39}$$

This signal has not been confirmed by competing experiments [38, 37] using hybrid germanium and silicon cryogenic detectors. They report an upper limit for $m_\chi \sim 50$ of

$$\sigma < 2 \times 10^{-42} \, \text{cm}^2 \, . \tag{40}$$

Experiments are rapidly improving their sensitivity so the conflict should be soon completely resolved.

Besides direct detection, it is possible to detect wimps "indirectly" through the detection of particles produced in present-day wimp-antiwimp annihilation. While wimp annihilation ceased in the early Universe because of the universal expansion, it started up again once the wimps became gravitationally bound in galactic halos. The annihilation rate is further enhanced inside material objects like the Sun or Earth. This is because it is possible for wimps to be trapped in such objects if, while traversing the object, the wimp suffers an elastic collision with a nucleus (Fig. 6). If the scatter results in a wimp velocity below the object's escape velocity, the wimp will find itself in an orbit that passes through the object. After repeated collisions the wimp will be thermalized in the core. In the case of supersymmetric dark matter, the trapping rate in the Sun is sufficiently high that the concentration of wimps reaches a steady state where trapping is balanced by either annihilation (for high-mass wimps) or by evaporation (for low-mass wimps). For the Sun, the dividing line between low mass and high mass is $\sim 3 \, \text{GeV}$.

The only annihilation products that can be seen emerging from the Sun or Earth are, of course, neutrinos. The flux of such neutrinos can be calculated for a given wimp candidate and the flux compared with that observed in underground detectors. The observed flux is entirely

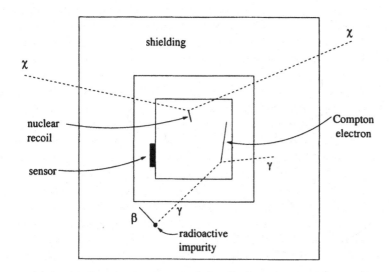

Figure 5. A generic "calorimetric" wimp detector surrounded by its shielding. The galactic wimp χ enters the detector , scatters off a nucleus , and leaves the detector. The recoiling nucleus creates secondary excitations (e.g. scintillation light, ionization, phonons) that can be detected by the sensor. Also shown is a background event due to the ambient radioactivity yielding a Compton electron in the detector.

understood as being due to the decay of cosmic-ray pions and kaons in the Earth's atmosphere (Fig. 6). Certain supersymmetric wimps would give a higher flux and are thus excluded [41, 42].

Because the observed flux of neutrinos is due to an unavoidable background, the only improvements in the limits from these techniques would come from the observation of a small excess of neutrinos coming from the direction of the Sun or center of the Earth. The most reasonable possibility is to search for upward-going muons coming from ν_μ interactions in the rock below a detector. Calculations [43] indicate that a 1 km^2 detector with a muon energy threshold of ~ 10 GeV would be needed to observe a statistically significant solar signal for typical supersymmetric dark matter. Efforts in this direction are underway by instrumenting the Mediterranean [44] or the Antarctic Continental Glacier [45] to observe Cherenkov light produced by muons.

2.2 Axions

Axions [46] are hypothetical light scalar particles invented to prevent CP violation in the strong interactions.[2] They would have been

---

[2] Such a violation would produce a permanent nucleon electric dipole moment in violation of experimental limits [1].

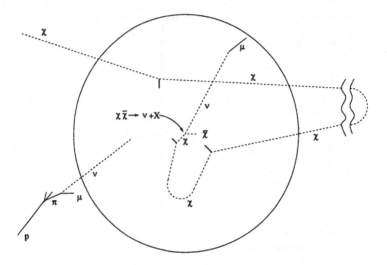

Figure 6.   The capture of a wimp χ in the Earth. If the wimp loses sufficient energy in a collision with a nucleus, the wimp's velocity will drop below the escape velocity placing it in an orbit intersecting the Earth. The subsequent collisions will eventually thermalize the wimp in the center of the Earth. An annihilation with a thermalized anti-wimp may lead to the production of neutrinos that can be detected at the surface. As shown, the background for such events comes from neutrinos produced by decays of pions and kaons produced by cosmic rays in the Earth's atmosphere.

produced in the early Universe via both thermal and non-thermal mechanisms and might produce near-critical relic densities if they have masses in the range $m_a \sim 10^{-5}\,\text{eV}$ to $\sim 10^{-3}\,\text{eV}$. Axions act as cold dark matter and should be present in the galactic halo.

The most popular detection scheme for galactic axions is based on the expected axion-2 photon coupling [47] which allows the axion to "convert" to a photon of frequency $\nu = m_a c^2/h$ in the presence of a magnetic field. If a microwave cavity is tuned to this frequency, the axions will cause an excess power to be absorbed (compared to neighboring frequencies). If the halo is dominated by axions, the predicted power is small, about $10^{-21}\,\text{W}$ for a cavity of volume $3\,\text{m}^3$ and a magnetic field of 10 T. Since the axion mass is not known, it is necessary to scan over the range of interesting frequencies. Pilot experiments [48, 49] have produced limits on the local axion density about a factor 30 above the expected density. Experiments are now in progress to search for axions at the required level of sensitivity [50].

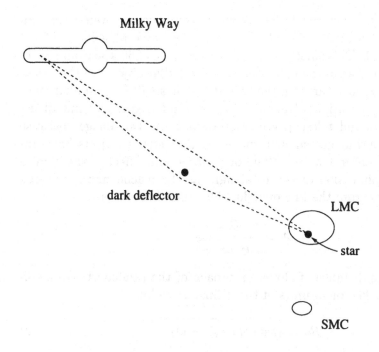

Figure 7. A schematic of the lensing of a star in the Large Magellanic Cloud (LMC) by an unseen object in the galactic halo. While the two images cannot be easily resolved, the combined light from the two images gives a transient amplification of the light from the star as the unseen object passes near the line-of-sight. The light curve for a point source is shown in Fig. 8.

2.3 Baryonic Dark Matter

The theory of primordial nucleosynthesis predicts that the baryon density is about 0.04 times the critical density which suggests that baryons cannot account for all of the dark matter. Nevertheless, baryons could account for *galactic* dark matter if they are in a form that neither absorbs nor emits light in significant quantities. The various possibilities have been reviewed in [51]. The simplest way to hide baryons is to place them in compact objects that either do not burn (e.g. brown dwarfs) or have ceased to burn (e.g. white dwarfs, neutron stars, black holes). Brown dwarfs have masses $< 0.07 M_\odot$ making them too cool to burn hydrogen. They were originally the favored candidates because they completely avoid constraints based on the production of background light or pollution of the interstellar medium with heavy elements through mass-loss or supernova explosions [52].

Dark objects that reside in galactic halos, they are called "machos" for " massive compact halo objects."

Paczyński [53] suggested that machos could be discovered through gravitational lensing of individual stars in the Large Magellanic Cloud (LMC) (Fig. 7). This small galaxy is at a distance of 50 kpc from Earth.

It turns out that in the case of lensing by stellar objects in the galactic halo, the angle separating the two images is small (< 1 milliarcsec). This type of gravitational lensing is therefore referred to as " microlensing." Earth-bound telescopes cannot resolve the two images because atmospheric turbulence smears images so that stellar objects have angular sizes of order 1 arcsec. The only observable effect is therefore a transient amplification of the total light as the macho moves towards and then away from the line-of-sight. The amplification is

$$A = \frac{u^2 + 2}{u\sqrt{u^2 + 4}} , \tag{41}$$

where u is the distance of closest approach of the (undeflected) line-of-sight to the deflector in units of the "Einstein radius"

$$R_{\mathrm{E}} = \sqrt{4GMLx(1-x)/c^2} \tag{42}$$

where L is the observer–source distance, Lx is the observer–deflector distance, and M is the macho mass.

The amplification is greater than 1.34 when the distance to the line-of-sight is less than R_{E}. This amplification corresponds to a reasonable observational threshold since photometry can "easily" be done to better than 10 per cent accuracy. At a given moment, the probability, P, of a given star being amplified by more than a factor 1.34 is just the probability that its undeflected light passes within one Einstein radius of a macho:

$$P \sim n_{\mathrm{macho}} L \pi R_{\mathrm{E}}^2 , \tag{43}$$

where n_{macho} is the mean number density of machos between us and the LMC and L is the distance to the LMC. The density of machos is roughly $n_{\mathrm{macho}} \sim M_{\mathrm{halo}}/ML^3$ where M_{halo} is the total halo mass out to the position of the LMC. Using the expression for the Einstein radius, we find that P is independent of M and determined only by the velocity of the LMC:

$$P \sim \frac{GM_{\mathrm{halo}}}{Lc^2} \sim \frac{v_{\mathrm{LMC}}^2}{c^2} . \tag{44}$$

The LMC is believed to orbit the galaxy with $v_{\mathrm{LMC}} \sim 200\,\mathrm{km\,s^{-1}}$ (corresponding to a flat rotation curve out to the position of the LMC). In this case, P is of order 10^{-6}. More detailed calculations give $P = 0.5 \times 10^{-6}$ [54].

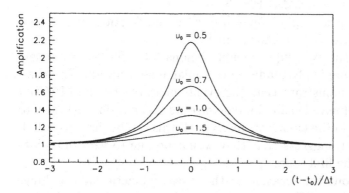

STANDARD MICROLENSING LIGHT CURVE

Figure 8. The microlensing light curve for a point source. The curve is shown for four values of the minimum distance of closest approach (0.5, 0.7, 1.0 and 1.5 Einstein radii). The time is normalized to the mass dependent scale Δt given by (45).

Since the observer, star and deflector are in relative motion a sizeable amplification lasts only as long as the undeflected light beam remains within the Einstein radius. The resulting achromatic, symmetric light curve is shown for various impact parameters in Fig. 8. The time scale of the amplification is the time Δt for the deflecting object to cross one Einstein radius with respect to the observer and source. For the lensing of stars in the LMC by objects in our halo, the relative speeds are of order $200\,\mathrm{km\,s^{-1}}$ and the position of the deflector is roughly midway between the observer and the source ($x \sim 0.5$). The mean Δt is then

$$\Delta t \sim \frac{R_{\mathrm{E}}}{200\,\mathrm{km\,s^{-1}}} \sim 75\,\mathrm{days}\sqrt{\frac{M}{M_\odot}}. \tag{45}$$

The observed time scale distribution can therefore be used to estimate the mass of the machos if one assumes that they are in the galactic halo.

Two groups, the MACHO collaboration and the EROS collaboration, have published results of searches for events in the directions of the LMC and the SMC (the neighboring Small Magellanic Cloud). The lack of events with time scales less than 15 days allowed the two groups to exclude as the dominant halo component objects with masses in the range $10^{-7}M_\odot < M < 10^{-1}M_\odot$ [55]. These limits exclude as a major halo component brown dwarfs of masses $\sim 0.07M_\odot$.

The MACHO collaboration has however observed 13 events of mean duration ~ 50 days [56]. If interpreted as being due to dark lenses in the galactic halo, the rate corresponds to a fraction $f = 0.2$ of the total halo mass being comprised of machos. The time scales correspond to halo objects of mass $\sim 0.4 M_\odot$ (Fig. 9).

EROS has published only upper limits on the fraction of the halo comprised in machos [57], as shown in Fig. 9.

The results of the two experiments demonstrate that it is unlikely that the halo is completely made up of stellar mass objects. The challenge is now to demonstrate that the events observed by the MACHO collaboration are due to lensing by halo objects and not, for instance, to lensing by objects in the clouds themselves. If this can be demonstrated, the estimated mass suggests that they would be due to very old white dwarfs, perhaps the first stars.

Information about the location of the lenses (galactic halo or Magellanic clouds) is very difficult to obtain. It is most easily obtained in very high amplification events, especially events due to binary lenses. In such events, the light curve is modified in a way that depends on the relative distances of the lens and source star [58]. It is also possible to get information on the distance to the lens for very long duration events where the movement of the Earth about the Sun modifies the light curve [59]. In the future, it will also be possible to barely resolve the two microlensing images with space-based interferometric telescopes. Such observations would give enough information to determine the distance to the lenses.

Microlensing searches for dark objects are also being performed for the nearby spiral galaxy M31 [60].

A second way to hide baryons in galactic halos is to place them in small clouds of cold gas comprised of primordial helium and *molecular* hydrogen [13]. The hydrogen must be molecular in order to escape detection via 21 cm emission by atomic hydrogen. The gas must be in clouds because a spatially uniform gas would lead to unobserved absorption of extragalactic sources at molecular transitions [61]. Clouds of a sufficiently high density would be sufficiently rare that most lines-of-sight would have no such absorption.

While this proposal is very efficient in hiding the gas, the plausibility of producing such quantities of molecules is controversial. In galactic disks, molecules are believed to be produced primarily on the surfaces of dust grains and this would not be possible in a primordial mixture of gas.

Limits of the quantity of cold molecular gas clouds near the Milky Way disk have been obtained from limits on the flux of high-energy photons that would be produced by cosmic-ray interactions in the clouds [62]. Limits on the amount in the halo are more difficult to obtain. Under certain conditions, molecular clouds should be observable in microlensing

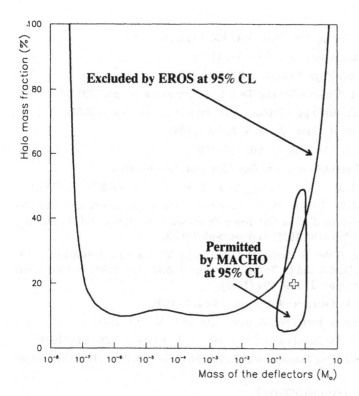

Figure 9. Microlensing constraints on the fraction of the total galactic halo mass that is comprised of machos as a function of the assumed macho mass (in units of $M_\odot$). The figure shows the EROS [57] upper limit using the searches for microlensing in the directions of the Small and Large Magellanic Clouds and the MACHO [56] allowed region centered on a mass fraction of ~ 0.2 and a mass of $\sim 0.5 M_\odot$.

surveys, either in our Galaxy by using the Magellanic Clouds [63] or in galaxy clusters by using background quasars [64].

Acknowledgments

I wish to thank Harrison Prosper organizing a school that was the perfect combination of relaxation and intellectual stimulation.

References

[1] Particle Data Group, Eur. Phys. J. C **15**, 1 (2000).

[2] J. Mather et al., Astrophy. J. **512**, 511 (1999).

[3] Y. Fukuda et al., Phys. Rev. Lett. **81**, 1562 (1998).

[4] D. Tytler et al., Physica Scripta **T85**, 12 (2000) and astro-ph/0001318.

[5] A. Benoit et al., astro-ph/0210306; A.H. Jaffe et al., astro-ph/000733.

[6] S. Perlmutter et al., Astrophy. J. **517**, 565 (1999).

[7] B. Schmidt et al., Astrophy. J. **507**, 46 (1998).

[8] J. Rich, *Fundamentals of Cosmology* (Springer, Berlin,2001).

[9] E. Vangioni-Flam, A. Coc and M. Casse: Astron. Astrophys. **360**, 15 (2000)

[10] S. Burles and D. Tytler: *Stellar Evolution, Stellar Explosions and Galactic Chemical Evolution, Second Oak Ridge Symposium on Atomic and Nuclear Astrophysics* (IOP,Bristol, 1998) and astro-ph/9803071

[11] D. Kirkman D. Tytler, S. Burles, D. Lubin and J. M. O'Meara: Astrophy. J. **529**, 655 (2000); S. Burles and D. Tytler: Astrophy. J. **507**, 732 (1998); S. Burles and D. Tytler: Astrophy. J. **499**, 699 (1998)

[12] R. Cen and J. P. Ostriker: Astrophy. J. **514**, 1 (1999)

[13] Combes, F. and D. Pfenniger: Astron. Astrophys. **V**, 453 (1997)

[14] A.G. Cohen, A. De Rujula and S.L. Glashow: Astrophy. J. **495**, 539 (1998)

[15] N. Bahcall: American Astronomical Society Meeting **194**, 3002 (1999) and astro-ph/9804082.

[16] J. Mohr et al.: astro-ph/0004244 .

[17] K. Griest and J. Silk: Nature **343**, 26 (1990).

[18] B. Carr and C Goymer: astro-ph/0003027.

[19] M. Milgrom: Astrophy. J. **270**, 365 (1983).

[20] J.N. Bahcall et al.: Phys. Rev D **58** 096016 (1998)

[21] P. Smith and J. Lewin: Phys. Rep. **187**, 203 (1990)

[22] L. R. Wang, R. R. Caldwell, J. P. Ostriker and P. J. Steinhardt: Astrophy. J. **530**, 17 (2000); R.R. Caldwell, R. Dave and P.J. Steinhardt, Phys. Rev. Lett. **80**, 1582 (1998); P. Binétruy: Int.J.Theor.Phys. **39**, 1859 (2000)

[23] S.D.M. White, J. F. Navarro, A. E. Evrard and C. S. Frenk: Nature **366**, 429 (1993).

[24] J. Oukbir and A. Blanchard: Astron. Astrophys. **262**, L21 (1992).

[25] J. P. Henry: Astrophy. J. **534**, 565 (2000).

[26] J. Peacock: M.N.R.A.S. **284**, 885 (1997).

[27] Aguirre, A.: Astrophy. J. **525**, 583 (1999).

[28] P. de Bernardis et al.: Nature **404**, 995 (2000).

[29] S. Hanany et al.: Astrophy. J. Lett. **545**, L5 (2000).

[30] Microwave Anisotropy Probe: http://map.gsfc.nasa.gov/; Planck Surveyor: http://sci.esa.int/home/planck/index.cfm

[31] A. G. Riess: Pub. Ast. Soc. Pac. **112**, 1284 (2000).

[32] C. Bennett et al.: Astrophy. J. Lett. **464**, L1 (1996)

[33] U. Seljak and M. Zaldarriaga: CMBFAST, a Microwave Anisotropy Code, http://www.sns.ias.edu/ matiasz/CMBFAST/cmbfast.html.

[34] J. Ellis et al.: Phys.Rev. D **62**, 075010 (2000).

[35] M.W. Goodman and E. Witten: Phys. Rev. D **31**, 3059 (1985).

[36] K. Freese, J. Freedman and A. Gould: Phys. Rev. D **37**, 3388 (1988).

[37] R. Abusaidi et al.: Phys. Rev. Lett. **84**, 5699 (2000).

[38] A. Benoit et al: astro-ph/0206271.

[39] R. Bernabei et al: Phys. Lett. B **480**, 23 (2000).

[40] L. Baudis et al: Phys. Rev. D **59**, 022001 (1999).

[41] M. Mori et al.: Phys. Lett. B **270**, 89 (1991).

[42] M. Mori, et al.: Phys. Lett. B **289**, 463 (1992).

[43] F. Halzen, T. Stelzer and M. Kamionkowski: Phys. Rev. **D45** 4439 (1992).

[44] Antares Collaboration, http://antares.in2p3.fr.

[45] Amanda Collaboration, http://amanda.berkeley.edu.

[46] H. Cheng: Phys. Rep. **158**, 1 (1988).

[47] P. Sikivie: Phys. Rev. Lett. **51**, 1415 (1983).

[48] W. Wuensch et al.: Phys. Rev. D **40**, 3153 (1988).

[49] C. Hagmann et al.: 1990 Phys. Rev. D **42**, 1297 (1990).

[50] C. Hagmann, et al.: Phys. Rev.Lett. **80**, 2043 (1998).

[51] B. Carr: Comments Astrophys. **14**, 257 (1990).

[52] B.D. Fields, K. Freese and D. S. Graff: Astrophy. J. **534**, 265 (2000).

[53] B. Paczyński: Astrophy. J. **304**, 1 (1986).

[54] K. Griest: Astrophy. J. **366**, 412 (1991).

[55] C. Alcock et al.: Astrophy. J. Lett. **499**, L9 (1998).

[56] C. Alcock et al.: Astrophy. J. **542**, 281 (2000).

[57] T. Lasserre et al.: Astron. Astrophys. **355**, L39 (2000); C. Afonso et al., astro-ph/0212176.

[58] C. Afonso et al.: Astrophy. J. **532**, 340 (2000).

[59] C. Alcock et al.: Astrophy. J. Lett. **454**, L125 (1995).

[60] A.P.S. Crotts and A. B. Tomaney: Astrophy. J. Lett. **473**, L87 (1996); S. Paulin-Henriksson et al., astro-ph/0207025.

[61] A. Vidal-Madjar et al.: Astrophy. J. Lett. **538**, L77 (2000).

[62] P. Salati et al.: *Dark matter in astro- and particle physics : (DARK '96)* editors, H.V. Klapdor-Kleingrothaus et al. (World Scientific, 1997).

[63] R. R. Rafikov and B. T. Draine:astro-ph/0006320.

[64] H. Tadros, S.J. Warren and P.C.Hewett, astro-ph/0003422.

COSMOLOGY, INFLATION AND THE PHYSICS OF NOTHING

William H. Kinney
Institute for Strings, Cosmology and Astroparticle Physics
Columbia University
550 W. 120th Street
New York, NY 10027
kinney@physics.columbia.edu

Abstract   These four lectures cover four topics in modern cosmology: the cosmological constant, the cosmic microwave background, inflation, and cosmology at the Planck scale. The underlying theme is that cosmology gives us a unique window on the "physics of nothing," or vacuum energy. The theory of inflation postulates that vacuum energy, or something very much like it, was the dominant force shaping the evolution of the very early universe. Recent astrophysical observations indicate that vacuum energy, or something very much like it, is also the dominant component of the universe today. Therefore cosmology gives us a way to study an important piece of particle physics inaccessible to accelerators. The lectures are oriented toward graduate students with only a passing familiarity with general relativity and knowledge of basic quantum field theory.

1.   Introduction

Cosmology is undergoing an explosive burst of activity, fueled both by new, accurate astrophysical data and by innovative theoretical developments. Cosmological parameters such as the total density of the universe and the rate of cosmological expansion are being precisely measured for the first time, and a consistent standard picture of the universe is beginning to emerge. This is exciting, but why talk about astrophysics at a school for particle physicists? The answer is that over the past twenty years or so, it has become evident that the the story of the universe is really a story of fundamental physics. I will argue that not only should particle physicists care about cosmology, but you should care *a lot*. Recent developments in cosmology indicate that it will be possible to use

H.B. Prosper and M. Danilov (eds.), Techniques and Concepts of High-Energy Physics XII, 189–243.

astrophysics to perform tests of fundamental theory inaccessible to particle accelerators, namely the physics of the vacuum itself. This has proven to be a surprise to cosmologists: the old picture of a universe filled only with matter and light have given way to a picture of a universe whose history is largely written in terms of the quantum-mechanical properties of empty space. It is currently believed that the universe today is dominated by the energy of vacuum, about 70% by weight. In addition, the idea of inflation postulates that the universe at the earliest times in its history was also dominated by vacuum energy, which introduces the intriguing possibility that all structure in the universe, from superclusters to planets, had an ultimately quantum-mechanical origin in the earliest moments of the universe. Furthermore, these ideas are not idle theorizing, but are predictive and subject to meaningful experimental test. Cosmological observations are providing several surprising challenges to fundamental theory.

These lectures are organized as follows. Section 2 provides and introduction to basic cosmology and a description of the surprising recent discovery of the accelerating universe. Section 3 discusses the physics of the cosmic microwave background (CMB), one of the most useful observational tools in modern cosmology. Section 4 discusses some unresolved problems in standard Big-Bang cosmology, and introduces the idea of inflation as a solution to those problems. Section 5 discusses the intriguing (and somewhat speculative) idea of using inflation as a "microscope" to illuminate physics at the very highest energy scales, where effects from quantum gravity are likely to be important. These lectures are geared toward graduate students who are familiar with special relativity and quantum mechanics, and who have at least been introduced to general relativity and quantum field theory. There are many things I will not talk about, such as dark matter and structure formation, which are interesting but do not touch directly on the main theme of the "physics of nothing." I omit many details, but I provide references to texts and review articles where possible.

2. Resurrecting Einstein's greatest blunder.

2.1 Cosmology for beginners

All of modern cosmology stems essentially from an application of the Copernican principle: we are not at the center of the universe. In fact, today we take Copernicus' idea one step further and assert the "cosmological principle": *nobody* is at the center of the universe. The cosmos, viewed from any point, looks the same as when viewed from any other point. This, like other symmetry principles more directly familiar to particle physicists, turns out to be an immensely powerful idea. In particular, it leads to the apparently inescapable conclusion that the universe has a finite age, that there was a beginning of time.

We wish to express the cosmological principle mathematically, as a symmetry. To do this, and to understand the rest of these lectures, we need to talk about metric tensors and General Relativity, at least briefly. A *metric* on a space is simply a generalization of Pythagoras' theorem for the distance ds between two points separated by distances $d\mathbf{x} = (dx, dy, dz)$,

$$ds^2 = |d\mathbf{x}|^2 = dx^2 + dy^2 + dz^2. \tag{1}$$

We can write this as a matrix equation,

$$ds^2 = \sum_{i,j=1,3} \eta_{ij} dx^i dx^j, \tag{2}$$

where η_{ij} is just the unit matrix,

$$\eta_{ij} = \begin{pmatrix} 1 & 0 & 0 \\ 0 & 1 & 0 \\ 0 & 0 & 1 \end{pmatrix}. \tag{3}$$

The matrix η_{ij} is referred to as the *metric* of the space, in this case a three-dimensional Euclidean space. One can define other, non-Euclidean spaces by specifying a different metric. A familiar one is the four-dimensional "Minkowski" space of special relativity, where the proper distance between two points in spacetime is given by

$$ds^2 = dt^2 - d\mathbf{x}^2, \tag{4}$$

corresponding to a metric tensor with indices $\mu, \nu = 0, \ldots, 3$:

$$\eta_{\mu\nu} = \begin{pmatrix} 1 & 0 & 0 & 0 \\ 0 & -1 & 0 & 0 \\ 0 & 0 & -1 & 0 \\ 0 & 0 & 0 & -1 \end{pmatrix}. \tag{5}$$

In a Minkowski space, photons travel on null paths, or *geodesics*, $ds^2 = 0$, and massive particles travel on *timelike* geodesics, $ds^2 > 0$. Note that in both of the examples given above, the metric is time-independent, describing a static space. In General Relativity, the metric becomes a dynamic object, and can in general depend on time and space. The fundamental equation of general relativity is the Einstein field equation,

$$G_{\mu\nu} = 8\pi G T_{\mu\nu}, \tag{6}$$

where $T_{\mu\nu}$ is a *stress energy* tensor describing the distribution of mass in space, G is Newton's gravitational constant and the Einstein Tensor $G_{\mu\nu}$ is a complicated function of the metric and its first and second derivatives. This should be familiar to anyone who has taken a course in electromagnetism, since we can write Maxwell's equations in matrix form as

$$\partial_\nu F^{\mu\nu} = \frac{4\pi}{c} J^\mu, \tag{7}$$

where $F^{\mu\nu}$ is the field tensor and J^μ is the current. Here we use the standard convention that we sum over the repeated indices of four-dimensional spacetime $\nu = 0, 3$. Note the similarity between Eq. (6) and Eq. (7). The similarity is more than formal: both have a charge on the right hand side acting as a source for a field on the left hand side. In the case of Maxwell's equations, the source is electric charge and the field is the electromagnetic field. In the case of Einstein's equations, the source is mass/energy, and the field is the shape of the spacetime, or the metric. An additional feature of the Einstein field equation is that it is *much* more complicated than Maxwell's equations: Eq. (6) represents six independent nonlinear partial differential equations of ten functions, the components of the (symmetric) metric tensor $g_{\mu\nu}(t, \mathbf{x})$. (The other four degrees of freedom are accounted for by invariance under transformations among the four coordinates.)

Clearly, finding a general solution to a set of equations as complex as the Einstein field equations is a hopeless task. Therefore, we do what any good physicist does when faced with an impossible problem: we introduce a symmetry to make the problem simpler. The three simplest symmetries we can apply to the Einstein field equations are: (1) vacuum, (2) spherical symmetry, and (3) homogeneity and isotropy. Each of these symmetries is useful (and should be familiar). The assumption of vacuum is just the case where there's no matter at all:

$$T_{\mu\nu} = 0. \tag{8}$$

In this case, the Einstein field equation reduces to a wave equation, and the solution is gravitational radiation. If we assume that the matter distribution $T_{\mu\nu}$ has spherical symmetry, the solution to the Einstein field equations is the Schwarzschild solution describing a black hole. The third case, homogeneity and isotropy, is the one we will concern ourselves with in more detail here[1]. By homogeneity, we mean that the universe is invariant under spatial translations, and by isotropy we mean that the universe is invariant under rotations. (A universe that is isotropic everywhere is necessarily homogeneous, but a homogeneous universe need not be isotropic: imagine a homogeneous space filled with a uniform electric field!) We will model the contents of the universe as a perfect fluid with density ρ and pressure p, for which the stress-energy tensor is

$$T_{\mu\nu} = \begin{pmatrix} \rho & 0 & 0 & 0 \\ 0 & -p & 0 & 0 \\ 0 & 0 & -p & 0 \\ 0 & 0 & 0 & -p \end{pmatrix}. \tag{9}$$

While this is certainly a poor description of the contents of the universe on small scales, such as the size of people or planets or even galaxies, it is an excellent approximation if we average over extremely large scales in the universe, for which the matter is known observationally to be very smoothly distributed. If the matter in the universe is homogeneous and isotropic, then the metric tensor must also obey the symmetry. The most general line element consistent with homogeneity and isotropy is

$$ds^2 = dt^2 - a^2(t)d\mathbf{x}^2, \tag{10}$$

where the *scale factor* $a^2(t)$ contains all the dynamics of the universe, and the vector product $d\mathbf{x}^2$ contains the geometry of the space, which can be either Euclidian ($d\mathbf{x}^2 = dx^2 + dy^2 + dz^2$) or positively or negatively curved. The metric tensor for the Euclidean case is particularly simple,

$$g_{\mu\nu} = \begin{pmatrix} 1 & 0 & 0 & 0 \\ 0 & -a(t) & 0 & 0 \\ 0 & 0 & -a(t) & 0 \\ 0 & 0 & 0 & -a(t) \end{pmatrix}, \tag{11}$$

which can be compared to the Minkowski metric (5). In this *Friedmann-Robertson-Walker* (FRW) space, spatial distances are multiplied by a dynamical factor $a(t)$ that describes the expansion (or contraction) of the spacetime. With the general metric (10), the Einstein field equations take on a particularly simple form,

$$\left(\frac{\dot{a}}{a}\right)^2 = \frac{8\pi G}{3}\rho - \frac{k}{a^2}, \tag{12}$$

where k is a constant that describes the curvature of the space: $k = 0$ (flat), or $k = \pm 1$ (positive or negative curvature). This is known as the *Friedmann equation*. In addition, we have a second-order equation

$$\frac{\ddot{a}}{a} = -\frac{4\pi G}{3}(\rho + 3p). \tag{13}$$

Note that the second derivative of the scale factor depends on the equation of state of the fluid. The equation of state is frequently given by a parameter w, or $p = w\rho$. Note that for any fluid with positive pressure, $w > 0$, the expansion of the universe is gradually decelerating, $\ddot{a} < 0$: the mutual gravitational attraction of the matter in the universe slows the expansion. This characteristic will be central to the discussion that follows.

2.2   Einstein's "greatest blunder"

General relativity combined with homogeneity and isotropy leads to a startling conclusion: spacetime is dynamic. The universe is not static, but is bound to be either expanding or contracting. In the early 1900's, Einstein applied general relativity to the homogeneous and isotropic case, and upon seeing the consequences, decided that the answer had to be wrong. Since the universe was obviously static, the equations had to be fixed. Einstein's method for fixing the equations involved the evolution of the density ρ with expansion. Returning to our analogy between General Relativity and electromagnetism, we remember that Maxwell's equations (7) do not completely specify the behavior of a system of charges and fields. In order to close the system of equations, we need to add the conservation of charge,

$$\partial_\mu J^\mu = 0, \tag{14}$$

or, in vector notation,

$$\frac{\partial \rho}{\partial t} + \nabla \cdot \mathbf{J} = 0. \tag{15}$$

The general relativistic equivalent to charge conservation is stress-energy conservation,

$$D_\mu T^{\mu\nu} = 0. \tag{16}$$

For a homogeneous fluid with the stress-energy given by Eq. (9), stress-energy conservation takes the form of the *continuity equation*,

$$\frac{d\rho}{dt} + 3H\left(\rho + p\right) = 0, \tag{17}$$

where $H = \dot{a}/a$ is the Hubble parameter from Eq. (12). This equation relates the evolution of the energy density to its equation of state $p = w\rho$. Suppose we have a box whose dimensions are expanding along with the universe, so that the volume of the box is proportional to the cube of the scale factor, $V \propto a^3$, and we fill it with some kind of matter or radiation. For example, ordinary matter in a box of volume V has an energy density inversely proportional to the volume of the box, $\rho \propto V^{-1} \propto a^{-3}$. It is straightforward to show using the continuity equation that this corresponds to zero pressure, $p = 0$. Relativistic particles such as photons have energy density that goes as $\rho \propto V^{-4/3} \propto a^{-4}$, which corresponds to equation of state $p = \rho/3$.

Einstein noticed that if we take the stress-energy $T_{\mu\nu}$ and add a constant Λ, the conservation equation (16) is unchanged:

$$D_\mu T^{\mu\nu} = D_\mu \left(T^{\mu\nu} + \Lambda g^{\mu\nu} \right) = 0. \tag{18}$$

In our analogy with electromagnetism, this is equivalent to adding a constant to the electromagnetic potential, $V'(x) = V(x) + \Lambda$. The constant Λ does not affect local dynamics in any way, but it does affect the cosmology. Since adding this constant adds a constant energy density to the universe, the continuity equation tells us that this is equivalent to a fluid with *negative* pressure, $p_\Lambda = -\rho_\Lambda$. Einstein chose Λ to give a closed, static universe as follows. Take the energy density to consist of matter

$$\begin{aligned} \rho_M &= \frac{k}{4\pi G a^2} \\ p_M &= 0, \end{aligned} \tag{19}$$

and cosmological constant

$$\begin{aligned} \rho_\Lambda &= \frac{k}{8\pi G a^2} \\ p_\Lambda &= -\rho_\Lambda. \end{aligned} \tag{20}$$

It is a simple matter to use the Friedmann equation to show that this combination of matter and cosmological constant leads to a static universe $\dot{a} = \ddot{a} = 0$. In order for the energy densities to be positive, the universe must be closed, $k = +1$. Einstein was able to add a kludge to get the answer he wanted.

Things sometimes happen in science with uncanny timing. In the 1920's, an astronomer named Edwin Hubble undertook a project to measure the distances to the spiral "nebulae" as they had been known, using the 100-inch Mount Wilson telescope. Hubble's method involved using Cepheid variables, named after the star Delta Cephei, the best known member of the class.[1] Cepheid variables have the useful property that the period of their variation, usually 10-100 days, is correlated to their absolute brightness. Therefore, by measuring the apparent brightness and the period of a distant Cepheid, one can determine its absolute brightness and therefore its distance. Hubble applied this method to a

---

[1]Delta Cephei is not, however the nearest Cepheid. That honor goes to Polaris, the north star.[2]

number of nearby galaxies, and determined that almost all of them were receding from the earth. Moreover, the more distant the galaxy was, the faster it was receding, according to a roughly linear relation:

$$v = H_0 d. \tag{21}$$

This is the famous Hubble Law, and the constant H_0 is known as Hubble's constant. Hubble's original value for the constant was something like 500km/sec/Mpc, where one megaparsec Mpc is a bit over 3 million light years.[2] This implied an age for the universe of about a billion years, and contradicted known geological estimates for the age of the earth. Cosmology had its first "age problem": the universe can't be younger than the things in it! Later it was realized that Hubble had failed to account for two distinct types of Cepheids, and once this discrepancy was taken into account, the Hubble constant fell to well under 100 km/s/Mpc. The current best estimate, determined using the Hubble space telescope to resolve Cepheids in galaxies at unprecedented distances, is $H_0 = 71 \pm 6$ km/s/Mpc [4]. In any case, the Hubble law is exactly what one would expect from the Friedmann equation. The expansion of the universe predicted (and rejected) by Einstein had been observationally detected, only a few years after the development of General Relativity. Einstein later referred to the introduction of the cosmological constant as his "greatest blunder".

The expansion of the universe leads to a number of interesting things. One is the cosmological redshift of photons. The usual way to see this is that from the Hubble law, distant objects appear to be receding at a velocity $v = H_0 d$, which means that photons emitted from the body are redshifted due to the recession velocity of the source. There is another way to look at the same effect: because of the expansion of space, the wavelength of a photon increases with the scale factor:

$$\lambda \propto a(t), \tag{22}$$

so that as the universe expands, a photon propagating in the space gets shifted to longer and longer wavelengths. The redshift z of a photon is then given by the ratio of the scale factor today to the scale factor when the photon was emitted:

$$1 + z = \frac{a(t_0)}{a(t_{\text{em}})}. \tag{23}$$

---

[2]The *parsec* is an archaic astronomical unit corresponding to one second of arc of parallax measured from opposite sides of the earth's orbit: 1 pc = 3.26 ly.

Here we have introduced commonly used the convention that a subscript 0 (e.g., t_0 or H_0) indicates the value of a quantity *today*. This redshifting due to expansion applies to particles other than photons as well. For some massive body moving relative to the expansion with some momentum p, the momentum also "redshifts":

$$p \propto \frac{1}{a(t)}. \tag{24}$$

We then have the remarkable result that freely moving bodies in an expanding universe eventually come to rest relative to the expanding coordinate system, the so-called *comoving* frame. The expansion of the universe creates a kind of dynamical friction for everything moving in it. For this reason, it will often be convenient to define comoving variables, which have the effect of expansion factored out. For example, the physical distance between two points in the expanding space is proportional to $a(t)$. We define the comoving distance between two points to be a constant in time:

$$x_{\mathrm{com}} = x_{\mathrm{phys}}/a(t) = \mathrm{const.} \tag{25}$$

Similarly, we define the comoving wavelength of a photon as

$$\lambda_{\mathrm{com}} = \lambda_{\mathrm{phys}}/a(t), \tag{26}$$

and comoving momenta are defined as:

$$p_{\mathrm{com}} \equiv a(t)p_{\mathrm{phys}}. \tag{27}$$

This energy loss with expansion has a predictable effect on systems in thermal equilibrium. If we take some bunch of particles (say, photons with a black-body distribution) in thermal equilibrium with temperature T, the momenta of all these particles will decrease linearly with expansion, and the system will cool.[3] For a gas in thermal equilibrium, the temperature is in fact inversely proportional to the scale factor:

$$T \propto \frac{1}{a(t)}. \tag{28}$$

The current temperature of the universe is about 3 K. Since it has been cooling with expansion, we reach the conclusion that the early universe must have been at a much higher temperature. This is the "Hot Big Bang" picture: a hot, thermal equilibrium universe expanding and

---

[3]It is not hard to convince yourself that a system that starts out as a blackbody stays a blackbody with expansion.

cooling with time. One thing to note is that, although the universe goes to infinite density and infinite temperature at the Big Bang singularity, it does *not* necessarily go to zero size. A flat universe, for example is infinite in spatial extent an infinitesimal amount of time after the Big Bang, which happens *everywhere* in the infinite space simultaneously! The observable universe, as measured by the horizon size, goes to zero size at $t = 0$, but the observable universe represents only a tiny patch of the total space.

2.3   Critical density and the return of the age problem

One of the things that cosmologists most want to measure accurately is the total density ρ of the universe. This is most often expressed in units of the density needed to make the universe flat, or $k = 0$. Taking the Friedmann equation for a $k = 0$ universe,

$$H^2 = \left(\frac{\dot{a}}{a}\right)^2 = \frac{8\pi G}{3}\rho, \qquad (29)$$

we can define a critical density ρ_c,

$$\rho_c \equiv \frac{3H_0^2}{8\pi G}, \qquad (30)$$

which tells us, for a given value of the Hubble constant H_0, the energy density of a Euclidean FRW space. If the energy density is greater than critical, $\rho > \rho_c$, the universe is closed and has a positive curvature ($k = +1$). In this case, the universe also has a finite lifetime, eventually collapsing back on itself in a "big crunch". If $\rho < \rho_c$, the universe is open, with negative curvature, and has an infinite lifetime. This is usually expressed in terms of the density parameter Ω,

$$\Omega \equiv \frac{\rho}{\rho_c} \quad \begin{array}{l} < 1: \text{ Open} \\ = 1: \text{ Flat} \\ > 1: \text{ Closed.} \end{array} \qquad (31)$$

There has long been a debate between theorists and observationalists as to what the value of Ω is in the real universe. Theorists have steadfastly maintained that the only sensible value for Ω is unity, $\Omega = 1$. This prejudice was further strengthened by the development of the theory of inflation, which solves several cosmological puzzles (see Secs. 4.1 and 4.2) and in fact *predicts* that Ω will be exponentially close to unity. Observers, however, have made attempts to measure Ω using a variety of methods, including measuring galactic rotation curves, the velocities of galaxies orbiting in clusters, X-ray measurements of galaxy clusters, the velocities and spatial distribution of galaxies on large scales, and gravitational lensing. These measurements have repeatedly pointed to a value of Ω inconsistent with a flat cosmology, with $\Omega = 0.2 - 0.3$ being a much better fit, indicating an open, negatively curved universe. Until a few years ago, theorists have resorted to cheerfully ignoring the data, taking it almost on faith that $\Omega = 0.7$ in extra stuff would turn up sooner or later. The theorists were right: new observations of the cosmic microwave background definitively favor a flat universe, $\Omega = 1$. Unsurprisingly, the observationalists were also right: only about 1/3 of this density appears to be in the form of ordinary matter.

The first hints that something strange was up with the standard cosmology came from measurements of the colors of stars in globular clusters. Globular clusters are small, dense groups of 10^5 - 10^6 stars which orbit in the halos of most galaxies and are among the oldest objects in the universe. Their ages are determined by observation of stellar populations and models of stellar evolution, and some globular clusters are known to be at least 12 billion years old [3], implying that the universe itself must be at least 12 billion years old. But consider a universe for which $\Omega = 1$ filled with pressureless matter, $\rho \propto a^{-3}$ and $p = 0$. It is straightforward to solve the Friedmann equation (12) with $k = 0$ to show that

$$a(t) \propto t^{2/3}. \tag{32}$$

The Hubble parameter is then given by

$$H = \frac{\dot{a}}{a} = \frac{2}{3}t^{-1}. \tag{33}$$

We therefore have a simple expression for the age of the universe t_0 in terms of the measured Hubble constant H_0,

$$t_0 = \frac{2}{3}H_0^{-1}. \tag{34}$$

The fact that the universe has a finite age introduces the concept of a *horizon*: this is just how far out in space we are capable of seeing at any given time. This distance is finite because photons have only traveled a finite distance since the beginning of the universe. Just as in special relativity, photons travel on paths with proper length $ds^2 = dt^2 - a^2 dx^2 = 0$, so that we can write the physical distance a photon has traveled since the Big Bang, or the *horizon size*, as

$$d_{\mathrm{H}} = a(t_0) \int_0^{t_0} \frac{dt}{a(t)}. \tag{35}$$

(This is in units where the speed of light is set to $c = 1$.) For example, in a flat, matter-dominated universe, $a(t) \propto t^{2/3}$, so that the horizon size is

$$d_H = t_0^{2/3} \int_0^{t_0} t^{-2/3} dt = 3t_0 = 2H_0^{-1}. \tag{36}$$

This form for the horizon distance is true in general: the distance a photon travels in time t is always about $d \sim t$: effects from expansion simply add a numerical factor out front. We will frequently ignore this, and approximate

$$d_H \sim t_0 \sim H_0^{-1}. \tag{37}$$

Measured values of H_0 are quoted in a very strange unit of time, a km/s/Mpc, but it is a simple matter to calculate the dimensionless factor using 1 Mpc $\simeq 3 \times 10^{19}$ km, so that the age of a flat, matter-dominated universe with $H_0 = 71 \pm 6$ km/s/Mpc is

$$t_0 = 8.9^{+0.9}_{-0.7} \times 10^9 \text{ years.} \tag{38}$$

A flat, matter-dominated universe would be younger than the things in it! Something is evidently wrong – either the estimates of globular cluster ages are too big, the measurement of the Hubble constant from from the HST is incorrect, the universe is not flat, or the universe is not matter dominated.

We will take for granted that the measurement of the Hubble constant is correct, and that the models of stellar structure are good enough to produce a reliable estimate of globular cluster ages (as they appear to be), and focus on the last two possibilities. An open universe, $\Omega_0 < 1$ might solve the age problem. Figure 1 shows the age of the universe consistent with the HST Key Project value for H_0 as a function of the density parameter Ω_0. We see that the age determined from H_0 is consistent with globular clusters as old as 12 billion years only for values of Ω_0 less than 0.3 or so. However, as we will see in Sec. 3, recent measurements of the cosmic microwave background (CMB) strongly indicate that we indeed live in a flat ($\Omega = 1$) universe. So while a low-density universe might provide a marginal solution to the age problem, it would conflict with the CMB. We therefore, perhaps reluctantly, are forced to consider that the universe might not be matter dominated. In the next section we will take a detour into quantum field theory seemingly unrelated to these cosmological issues. By the time we are finished, however, we will have in hand a different, and provocative, solution to the age problem consistent with a flat universe.

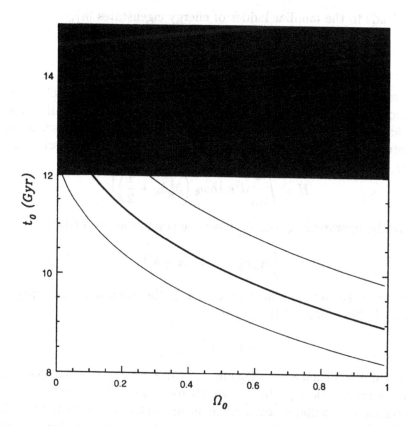

Figure 1. Age of the universe as a function of Ω_0 for a matter-dominated universe. The blue lines show the age t_0 consistent with the HST key project value $H_0 = 72 \pm 6$ km/s/Mpc. The red area is the region consistent with globular cluster ages $t_0 > 12$ Gyr.

2.4 The vacuum in quantum field theory

In this section, we will discuss something that at first glance appears to be entirely unrelated to cosmology: the vacuum in quantum field theory. We will see later, however, that it will in fact be crucially important to cosmology. Let us start with basic quantum mechanics, in the form of the simple harmonic oscillator, with Hamiltonian

$$H = \hbar\omega \left(\hat{a}^\dagger \hat{a} + \frac{1}{2} \right), \tag{39}$$

where $\hat{a}$ and $\hat{a}^\dagger$ are the lowering and raising operators, respectively, with commutation relation

$$\left[\hat{a}, \hat{a}^\dagger \right] = 1. \tag{40}$$

This leads to the familiar ladder of energy eigenstates $|n\rangle$,

$$H |n\rangle = \hbar\omega \left(n + \frac{1}{2} \right) |n\rangle = E_n |n\rangle. \tag{41}$$

The simple harmonic oscillator is pretty much the only problem that physicists know how to solve. Applying the old rule that if all you have is a hammer, everything looks like a nail, we construct a description of quantum fields by placing an infinite number of harmonic oscillators at every point,

$$H = \int_{-\infty}^{\infty} d^3k \left[\hbar\omega_k \left(\hat{a}_k^\dagger \hat{a}_k + \frac{1}{2} \right) \right], \tag{42}$$

where the operators $\hat{a}_k$ and $\hat{a}_k^\dagger$ have the commutation relation

$$\left[\hat{a}_k, \hat{a}_{k'}^\dagger \right] = \delta^3 \left(\mathbf{k} - \mathbf{k'} \right). \tag{43}$$

Here we make the identification that $\mathbf{k}$ is the momentum of a particle, and ω_k is the energy of the particle,

$$\omega_k^2 - |\mathbf{k}|^2 = m^2 \tag{44}$$

for a particle of mass m. Taking $m = 0$ gives the familiar dispersion relation for massless particles like photons. Like the state kets $|n\rangle$ for the harmonic oscillator, each momentum vector $\mathbf{k}$ has an independent ladder of states, with the associated quantum numbers, $|n_k, \ldots, n_{k'}\rangle$. The raising and lowering operators are now interpreted as *creation* and *annihilation* operators, turning a ket with n particles into a ket with $n + 1$ particles, and vice-versa:

$$|n_k = 1\rangle = \hat{a}_k^\dagger |0\rangle, \tag{45}$$

and we call the ground state $|0\rangle$ the *vacuum*, or zero-particle state. But there is one small problem: just like the ground state of a single harmonic oscillator has a nonzero energy $E_0 = (1/2)\hbar\omega$, the vacuum state of the quantum field also has an energy,

$$\begin{aligned}
H |0\rangle &= \int_{-\infty}^{\infty} d^3k \left[\hbar\omega_k \left(\hat{a}_k^\dagger \hat{a}_k + \frac{1}{2} \right) \right] |0\rangle \\
&= \left[\int_{-\infty}^{\infty} d^3k \, \hbar\omega_k / 2 \right] |0\rangle \\
&= \infty.
\end{aligned} \tag{46}$$

The ground state energy diverges! The solution to this apparent paradox is that we expect quantum field theory to break down at very high energy.

We therefore introduce a cutoff on the momentum $\mathbf{k}$ at high energy, so that the integral in Eq. (46) becomes finite. A reasonable physical scale for the cutoff is the scale at which we expect quantum gravitational effects to become relevant, the *Planck scale* m_{Pl}:

$$H\,|0\rangle \sim m_{Pl} \sim 10^{19} \text{ GeV}. \tag{47}$$

Therefore we expect the vacuum everywhere to have a constant energy density, given in units where $\hbar = c = 1$ as

$$\rho \sim m_{Pl}^4 \sim 10^{93} \text{ g/cm}^3. \tag{48}$$

But we have already met up with an energy density that is constant everywhere in space: Einstein's cosmological constant, $\rho_\Lambda = \text{const.}, p_\Lambda = -\rho_\Lambda$. However, the cosmological constant we expect from quantum field theory is more than a hundred twenty orders of magnitude too big. In order for ρ_Λ to be less than the critical density, $\Omega_{\text{Lambda}} < 1$, we must have $\rho_\Lambda < 10^{-120} m_{Pl}^4$! How can we explain this discrepancy? Nobody knows.

2.5 Vacuum energy in cosmology

So what does this have to do with cosmology? The interesting fact about vacuum energy is that is results in accelerated expansion of the universe. From Eq. (13), we can write the acceleration $\ddot{a}$ in terms of the equation of state $p = w\rho$ of the matter in the universe,

$$\frac{\ddot{a}}{a} \propto -(1+3w)\,\rho. \tag{49}$$

For ordinary matter such as pressureless dust $w = 0$ or radiation $w = 1/3$, we see that the gravitational attraction of all the stuff in the universe makes the expansion slow down with time, $\ddot{a} < 0$. But we have seen that a cosmological constant has the odd property of negative pressure, $w = -1$, so that a universe dominated by vacuum energy actually expands faster and faster with time, $\ddot{a} > 0$. It is easy to see that accelerating expansion helps with the age problem: for a standard matter-dominated universe, a larger Hubble constant means a *younger universe*, $t_0 \propto H_0^{-1}$. But if the expansion of the universe is accelerating, this means that H grows with time. For a given age t_0, acceleration means that the Hubble constant we measure will be larger in an accelerating cosmology than in a decelerating one, so we can have our cake and eat it too: an old universe and a high Hubble constant! This also resolves the old dispute between the observers and the theorists. Astronomers measuring the density of the universe use local dynamical measurements such as the orbital velocities of galaxies in a cluster. These measurements are insensitive to a cosmological constant and only measure the *matter* density ρ_M of the universe. However, geometrical tests like the cosmic microwave background which we will discuss in the Sec. 3 are sensitive to the *total* energy density $\rho_M + \rho_\Lambda$. If we take the observational value for the matter density $\Omega_M = 0.2 - 0.3$ and make up the difference with a cosmological constant, $\Omega_\Lambda = 0.7 - 0.8$, we arrive at an age for the universe in excess of 13 Gyr, perfectly consistent with the globular cluster data.

In the mid-1990s, there were some researchers making the argument based on the age problem alone that we needed a cosmological constant [5]. But the case was hardly compelling, given that the CMB results indicating a flat universe had not yet been measured, and a low-density universe presented a simpler alternative, based on a cosmology containing matter alone. However, there was another observation that pointed clearly toward the need for Ω_Λ: Type Ia supernovae (SNIa) measurements. A detailed discussion of these measurements is beyond the scope of these lectures, but the principle is simple: SNeIa represent a *standard candle*, i.e. objects whose intrinsic brightness we know, based on observations of nearby supernovae. They are also extremely bright, so they can be observed at cosmological distances. Two collaborations, the Supernova Cosmology Project [6] and the High-z Supernova Search [7] obtained samples of supernovae at redshifts around $z = 0.5$. This is far enough out that it is possible to measure deviations from the linear Hubble law $v = H_0 d$ due to the time-evolution of the Hubble parameter: the groups were able to *measure* the acceleration or deceleration of the universe directly. If the universe is decelerating, objects at a given redshift will be closer to us, and therefore brighter than we would expect based on a linear Hubble law. Conversely, if the expansion is accelerating, objects at a given redshift will be further away, and therefore dimmer. The result from both groups was that the supernovae were consistently dimmer than expected. Fig. 2 shows the data from the Supernova Cosmology Project [8], who quoted a best fit of $\Omega_M \simeq 0.3$, $\Omega_\Lambda \simeq 0.7$, just what was needed to reconcile the dynamical mass measurements with a flat universe!

So we have arrived at a very curious picture indeed of the universe: matter, including both baryons and the mysterious dark matter (which I will not discuss in any detail in these lectures) makes up only about 30% of the energy density in the universe. The remaining 70% is made of of something that looks very much like Einstein's "greatest blunder", a cosmological constant. This *dark energy* can possibly be identified with the vacuum energy predicted by quantum field theory, except that the energy density is 120 orders of magnitude smaller than one would

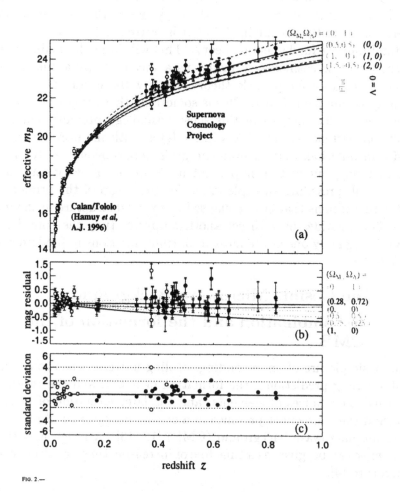

Figure 2. Data from the Supernova Cosmology project. Dimmer objects are higher vertically on the plot. The horizontal axis is redshift. The curves represent different choices of Ω_M and Ω_Λ. A cosmology with $\Omega_M = 1$ and $\Omega_\Lambda = 0$ is ruled out to 99% confidence, while a universe with $\Omega_M = 0.3$ and $\Omega_\Lambda = 0.7$ is a good fit to the data. [8]

expect from a naive analysis. Few, if any, satisfying explanations have been proposed to resolve this discrepancy. For example, some authors have proposed arguments based on the Anthropic Principle [9] to explain the low value of ρ_Λ, but this explanation is controversial to say the least. There is a large body of literature devoted to the idea that the dark energy is something other than the quantum zero-point energy we have considered here, the most popular of which are self-tuning scalar field models dubbed *quintessence* [10]. A review can be found in Ref. [11]. However, it is safe to say that the dark energy that dominates the universe is currently unexplained, but it is of tremendous interest from the standpoint of fundamental theory. This will form the main theme of these lectures: cosmology provides us a way to study a question of central importance for particle theory, namely the nature of the vacuum in quantum field theory. This is something that cannot be studied in particle accelerators, so in this sense cosmology provides a unique window on particle physics. We will see later, with the introduction of the idea of inflation, that vacuum energy is important not only in the universe today, but had an important influence on the very early universe as well, providing an explanation for the origin of the primordial density fluctuations that later collapsed to form all structure in the universe. This provides us with yet another way to study the "physics of nothing", arguably one of the most important questions in fundamental theory today.

3. The Cosmic Microwave Background

3.1 Recombination and the formation of the CMB

The basic picture of an expanding, cooling universe leads to a number of startling predictions: the formation of nuclei, and the resulting primordial abundances of elements, and the later formation of neutral atoms and the consequent presence of a cosmic background of photons, the cosmic microwave background (CMB) [12, 13]. A rough history of the universe can be given as a time line of increasing time and decreasing temperature[14]:

- $T \sim 10^{15}\ K$, $t \sim 10^{-12}$ sec: Primordial soup of fundamental particles.

- $T \sim 10^{13}\ K$, $t \sim 10^{-6}$ sec: Protons and neutrons form.

- $T \sim 10^{10}\ K$, $t \sim 3$ min: Nucleosynthesis: nuclei form.

- $T \sim 3000\ K$, $t \sim 300,000$ years: Atoms form.

- $T \sim 10\ K$, $t \sim 10^{9}$ years: Galaxies form.

- $T \sim 3\ K$, $t \sim 10^{10}$ years: Today.

The epoch at which atoms form, when the universe was at an age of 300,000 years and a temperature of around 3000 K is somewhat oxy-moronically referred to as "recombination", despite the fact that electrons and nuclei had never before "combined" into atoms. The physics is simple: at a temperature of greater than about 3000 K, the universe consisted of an ionized plasma of mostly protons, electrons, and photons, which a few helium nuclei and a tiny trace of Lithium. The important characteristic of this plasma is that it was *opaque*, or more precisely the mean free path of a photon was a great deal smaller than the horizon size of the universe. As the universe cooled and expanded, the plasma "recombined" into neutral atoms, first the helium, then a little later the hydrogen.

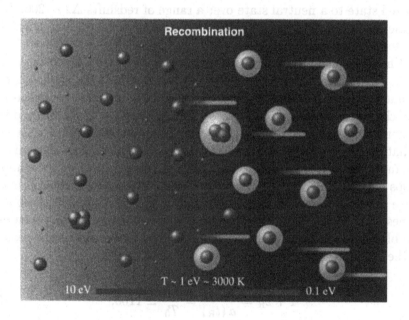

Figure 3. Schematic diagram of recombination.

If we consider hydrogen alone, the process of recombination can be described by the Saha equation for the equilibrium ionization fraction X_e of the hydrogen [15]:

$$\frac{1 - X_e}{X_e^2} = \frac{4\sqrt{2}\zeta(3)}{\sqrt{\pi}}\eta\left(\frac{T}{m_e}\right)^{3/2}\exp\left(\frac{13.6 \text{ eV}}{T}\right). \tag{50}$$

Here m_e is the electron mass and 13.6 eV is the ionization energy of hydrogen. The physically important parameter affecting recombination is the density of protons and electrons compared to photons. This is determined by the *baryon asymmetry*, or the excess of baryons over antibaryons in the universe.[4] This is described as the ratio of baryons to photons:

$$\eta \equiv \frac{n_b - n_{\bar{b}}}{n_\gamma} = 2.68 \times 10^{-8} \left(\Omega_b h^2 \right). \tag{51}$$

Here Ω_b is the baryon density and h is the Hubble constant in units of 100 km/s/Mpc,

$$h \equiv H_0/(100 \text{ km/s/Mpc}). \tag{52}$$

Recombination happens quickly (i.e., in much less than a Hubble time $t \sim H^{-1}$), but is not instantaneous. The universe goes from a completely ionized state to a neutral state over a range of redshifts $\Delta z \sim 200$. If we define recombination as an ionization fraction $X_e = 0.1$, we have that the temperature at recombination $T_R = 0.3$ eV.

What happens to the photons after recombination? Once the gas in the universe is in a neutral state, the mean free path for a photon rises to much larger than the Hubble distance. The universe is then full of a background of freely propagating photons with a black-body distribution of frequencies. At the time of recombination, the background radiation has a temperature of $T = T_R = 0.3eV$, and as the universe expands the photons redshift, so that the temperature of the photons drops with the increase of the scale factor, $T \propto a(t)^{-1}$. We can detect these photons today. Looking at the sky, this background of photons comes to us evenly from all directions, with an observed temperature of the blackbody is $T_0 \simeq 2.73K$. This allows us to determine the redshift of the last scattering surface,

$$1 + z_R = \frac{a(t_0)}{a(t_R)} = \frac{T_R}{T_0} \simeq 1100. \tag{53}$$

This is the cosmic microwave background. Since by looking at higher and higher redshift objects, we are looking further and further back in time, we can view the observation of CMB photons as imaging a uniform "surface of last scattering" at a redshift of 1100. To the extent that recombination happens at the same time and in the same way everywhere, the CMB will be of precisely uniform temperature. In

---

[4] If there were no excess of baryons over antibaryons, there would be no protons and electrons to recombine, and the universe would be just a gas of photons and neutrinos!

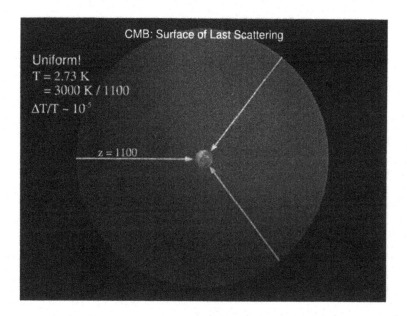

Figure 4. Cartoon of the last scattering surface. From earth, we see microwaves radiated uniformly from all directions, forming a "sphere" at redshift $z = 1100$.

fact the CMB is observed to be of uniform temperature to about 1 part in 10,000! We shall consider the puzzles presented by this curious isotropy of the CMB later.

While the observed CMB is highly isotropic, it is not perfectly so. The largest contribution to the anisotropy of the CMB as seen from earth is simply Doppler shift due to the earth's motion through space. (Put more technically, the motion is the earth's motion relative to a "comoving" cosmological reference frame.) CMB photons are slightly blueshifted in the direction of our motion and slightly redshifted opposite the direction of our motion. This blueshift/redshift shifts the temperature of the CMB so the effect has the characteristic form of a "dipole" temperature anisotropy, shown in Fig. 5. This dipole anisotropy was first observed in the 1970's by David T. Wilkinson and Brian E. Corey at Princeton, and another group consisting of George F. Smoot, Marc V. Gorenstein and Richard A. Muller. They found a dipole variation in the CMB temperature of about 0.003 K, or $(\Delta T/T) \simeq 10^{-3}$, corresponding to a peculiar velocity of the earth of about 600 km/s, roughly in the direction of the constellation Leo.

The dipole anisotropy, however, is a *local* phenomenon. Any intrinsic, or primordial, anisotropy of the CMB is potentially of much greater

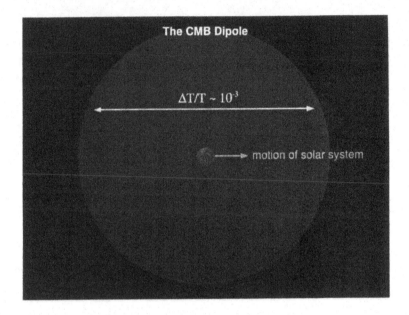

Figure 5.   The CMB dipole due to the earth's peculiar motion.

cosmological interest. To describe the anisotropy of the CMB, we remember that the surface of last scattering appears to us as a spherical surface at a redshift of 1100. Therefore the natural parameters to use to describe the anisotropy of the CMB sky is as an expansion in spherical harmonics $Y_{\ell m}$:

$$\frac{\Delta T}{T} = \sum_{\ell=1}^{\infty} \sum_{m=-\ell}^{\ell} a_{\ell m} Y_{\ell m}(\theta, \phi). \tag{54}$$

Since there is no preferred direction in the universe, the physics is independent of the index m, and we can define

$$C_\ell \equiv \sum_m |a_{\ell m}|^2. \tag{55}$$

The $\ell = 1$ contribution is just the dipole anisotropy,

$$\left(\frac{\Delta T}{T}\right)_{\ell=1} \sim 10^{-3}. \tag{56}$$

It was not until more than a decade after the discovery of the dipole anisotropy that the first observation was made of anisotropy for $\ell \geq 2$, by the differential microwave radiometer aboard the Cosmic Background Explorer (COBE) satellite [16], launched in in 1990. COBE observed that the anisotropy at the quadrupole and higher ℓ was two orders of magnitude smaller than the dipole:

$$\left(\frac{\Delta T}{T}\right)_{\ell>1} \simeq 10^{-5}. \tag{57}$$

Fig. 6 shows the dipole and higher-order CMB anisotropy as measured by COBE. It is believed that this anisotropy represents intrinsic fluctu-

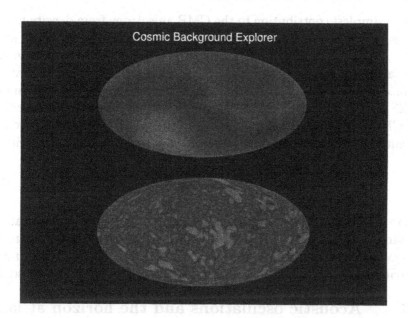

Figure 6. The COBE measurement of the CMB anisotropy [16]. The top oval is a map of the sky showing the dipole anisotropy $\Delta T/T \sim 10^{-3}$. The bottom oval is a similar map with the dipole contribution subtracted, showing the anisotropy for $\ell > 1$, $\Delta T/T \sim 10^{-5}$. (Figure courtesy of the COBE Science Working Group.)

ations in the CMB itself, due to the presence of tiny primordial density fluctuations in the cosmological matter present at the time of recombination. These density fluctuations are of great physical interest, first because these are the fluctuations which later collapsed to form all of the structure in the universe, from superclusters to planets to graduate students. Second, we shall see that within the paradigm of inflation, the form of the primordial density fluctuations forms a powerful probe of the physics of the very early universe. The remainder of this section will be concerned with how primordial density fluctuations create fluctuations in the temperature of the CMB. Later on, I will discuss using the CMB as a tool to probe other physics, especially the physics of inflation.

While the physics of recombination in the homogeneous case is quite simple, the presence of inhomogeneities in the universe makes the situation much more complicated. I will describe some of the major effects qualitatively here, and refer the reader to the literature for a more detailed technical explanation of the relevant physics [12]. The complete linear theory of CMB fluctuations was first worked out by Bertschinger and Ma in 1995 [18].

3.2   Sachs-Wolfe Effect

The simplest contribution to the CMB anisotropy from density fluctuations is just a gravitational redshift, known as the *Sachs-Wolfe effect* [17]. A photon coming from a region which is slightly overdense will have a slightly larger redshift due to the deeper gravitational well at the surface of last scattering. Conversely, a photon coming from an underdense region will have a slightly smaller redshift. Thus we can calculate the CMB temperature anisotropy due to the slightly varying Newtonian potential Φ from density fluctuations at the surface of last scattering:

$$\left(\frac{\Delta T}{T}\right) = \frac{1}{3}\Phi, \tag{58}$$

where the factor $1/3$ is a general relativistic correction. Fluctuations on large angular scales (low multipoles) are actually larger than the horizon at the time of last scattering, so that this essentially kinematic contribution to the CMB anisotropy is dominant on large angular scales.

3.3   Acoustic oscillations and the horizon at last scattering

For fluctuation modes smaller than the horizon size, more complicated physics comes into play. Even a summary of the many effects that determine the precise shape of the CMB multipole spectrum is beyond the scope of these lectures, and the student is referred to Refs. [12] for a more detailed discussion. However, the dominant process that occurs on short wavelengths is important to us. These are *acoustic oscillations* in the baryon/photon plasma. The idea is simple: matter tends to collapse due to gravity onto regions where the density is higher than average, so the baryons "fall" into overdense regions. However, since the baryons and the photons are still strongly coupled, the photons tend to resist this collapse and push the baryons outward. The result is "ringing", or oscillatory modes of compression and rarefaction in the gas to due density fluctuations. The gas heats as it compresses and cools as it expands, and this creates fluctuations in the temperature of the CMB. This manifests itself in the C_ℓ spectrum as a series of bumps (Fig. 8). The specific shape and location of the bumps is created by complicated, although well-understood physics, involving a large number of cosmological parameters. The shape of the CMB multipole spectrum depends, for example, on the baryon density Ω_b, the Hubble constant H_0, the densities of matter and cosmological constant Ω_0, Ω_Λ, and the amplitude of primordial gravitational waves (see Sec. 4.5). This makes interpretation of the spectrum something of a complex undertaking, but it also makes it a sensitive probe of cosmological models. In these lectures, I will primarily focus on the CMB as a probe of inflation, but there is much more to the story.

These oscillations are sound waves in the direct sense: compression waves in the gas. The position of the bumps in ℓ is determined by the oscillation frequency of the mode. The first bump is created by modes that have had time to go through half an oscillation in the age of the universe (compression), the second bump modes that have gone through one full oscillation (rarefaction), and so on. So what is the wavelength of a mode that goes through half an oscillation in a Hubble time? About the horizon size at the time of recombination, 300,000 light years or so! This is an immensely powerful tool: it in essence provides us with a ruler of known length (the wavelength of the oscillation mode, or the horizon size at recombination), situated at a known distance (the distance to the surface of last scattering at $z = 1100$). The angular size of this ruler when viewed at a fixed distance depends on the curvature of the space that lies between us and the surface of last scattering (Fig. 7). If the space is negatively curved, the ruler will subtend a smaller angle

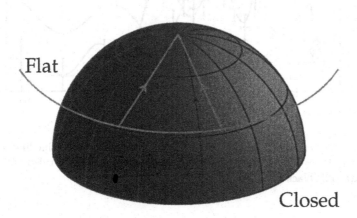

Figure 7. The effect of geometry on angular size. Objects of a given angular size are smaller in a closed space than in a flat space. Conversely, objects of a given angular size are larger in an open space. (Figure courtesy of Wayne Hu [20].)

214

than if the space is flat;[5] if the space is positively curved, the ruler will subtend a larger angle. We can measure the "angular size" of our "ruler" by looking at where the first acoustic peak shows up on the plot of the C_ℓ spectrum of CMB fluctuations. The positions of the peaks are determined by the curvature of the universe.[6]. This is how we measure Ω with the CMB. Fig. 8 shows an $\Omega = 1$ model and an $\Omega = 0.3$ model along with the current data. The data allow us to clearly distinguish between flat and open universes. Figure 9 shows limits from Type Ia supernovae and the CMB in the space of Ω_M and Ω_Λ.

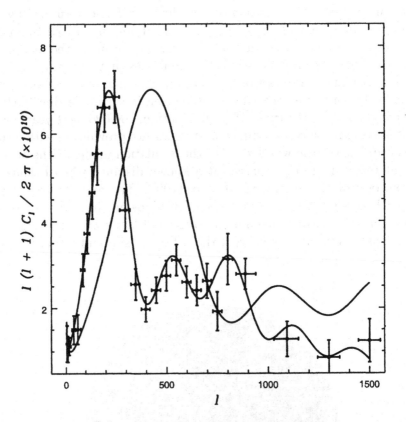

Figure 8. C_ℓ spectra for a universe with $\Omega_M = 0.3$ and $\Omega_\Lambda = 0.7$ (blue line) and for $\Omega_M = 0.3$ and $\Omega_\Lambda = 0$ (red line). The open universe is conclusively ruled out by the current data [19] (black crosses).

---

[5]To paraphrase Gertrude Stein, "there's more *there* there."
[6]Not surprisingly, the real situation is a good deal more complicated than what I have described here. [12]

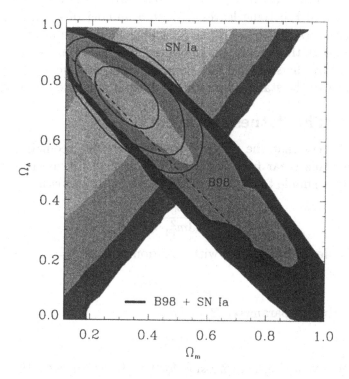

Figure 9. Limits on Ω_M and Ω_Λ from the CMB and from Type Ia supernovae. The two data sets together strongly favor a flat universe (CMB) with a cosmological constant (SNIa). [21]

4. Inflation

The basic picture of big-bang cosmology, a hot, uniform early universe expanding and cooling at late times, is very successful and has (so far) passed a battery of increasingly precise tests. It successfully explains the observed primordial abundances of elements, the observed redshifts of distant galaxies, and the presence of the cosmic microwave background. Observation of the CMB is a field that is currently progressing rapidly, allowing for extremely precise tests of cosmological models. The most striking feature of the CMB is its high degree of uniformity, with inhomogeneities of about one part in 10^5. Recent precision measurements of the tiny anisotropies in the CMB have allowed for constraints on a variety of cosmological parameters. Perhaps most importantly, observations of the first acoustic peak, first accomplished by the Boomerang

[22] and MAXIMA [23] experiments, indicate that the geometry of the universe is flat, with $\Omega_{\text{total}} = 1.02 \pm 0.05$ [21]. However, this success of the standard Big Bang leaves us with a number of vexing puzzles. In particular, how did the universe get so big, so flat, and so uniform? We shall see that these observed characteristics of the universe are poorly explained by the standard Big Bang scenario, and we will need to add something to the standard picture to make it all make sense: inflation.

4.1   The flatness problem

We observe that the universe has a nearly flat geometry, $\Omega_{\text{tot}} \simeq 1$. However, this is far from a natural expectation for an arbitrary FRW space. It is simple to see why. Take the defining expression for Ω,

$$\Omega \equiv \frac{8\pi}{3m_{\text{Pl}}^2} \left(\frac{\rho}{H^2} \right). \tag{59}$$

Here the density of matter with equation of state $p = w\rho$ evolves with expansion as

$$\rho \propto a^{-3(1+w)}, \tag{60}$$

and the Hubble parameter H evolves as

$$H \propto a. \tag{61}$$

We then have a simple expression for how Ω evolves with time:

$$\frac{d\Omega}{d\log a} = (1 + 3w)\Omega\,(\Omega - 1). \tag{62}$$

This is most curious! Note the sign. For an equation of state with $1 + 3w > 0$, which is the case for any kind of "ordinary" matter, a flat universe is an unstable equilibrium:

$$\frac{d\,|\Omega - 1|}{d\log a} > 0, \ 1 + 3w > 0. \tag{63}$$

So if the universe at early times deviates even slightly from a flat geometry, that deviation will grow large at late times. If the universe today is flat to within $\Omega \simeq 1 \pm 0.05$, then Ω the time of recombination was $\Omega = 1 \pm 0.00004$, and at nucleosynthesis $\Omega = 1 \pm 10^{-12}$. This leaves unexplained how the universe got so perfectly flat in the first place. This curious fine-tuning in cosmology is referred to as the *flatness problem*.

4.2   The horizon problem

There is another odd fact about the observed universe: the apparent high degree of uniformity and thermal equilibrium of the CMB. While it might at first seem quite natural for the hot gas of the early universe to be in good thermal equilibrium, this is in fact quite *unnatural* in the standard Big Bang picture. This is because of the presence of a cosmological horizon. We recall that the horizon size of the universe is just how far a photon can have traveled since the Big Bang, $d_H \sim t$ in units where $c = 1$. This defines how large a "patch" of the universe can be in causal contact. Two points in the universe separated by more than a horizon size have no way to reach thermal equilibrium, since cannot have *ever* been in causal contact. Consider two points comoving with respect to the cosmological expansion. The physical distance d between the two points then just increases linearly with the scale factor:

$$d \propto a(t). \tag{64}$$

The horizon size is just proportional to the time, or equivalently to the inverse of the Hubble parameter,

$$d_H \propto H^{-1}. \tag{65}$$

It is straightforward to show that for any FRW space with constant equation of state, there is a conserved quantity given by

$$\left(\frac{d}{d_H}\right)^2 |\Omega - 1| = \text{const.} \tag{66}$$

Proof of this is also left as an exercise for the student. If we again consider the case of "ordinary" matter, with equation of state $1+3w > 0$, we see from Eqs. (62) and (66) that the horizon expands faster than a comoving length:

$$\frac{d}{d \log a}\left(\frac{d}{d_H}\right) < 0, \ 1 + 3w > 0. \tag{67}$$

This means that any two points at rest relative to the expansion of the universe will be causally *disconnected* at early times and causally *connected* at late times. So the natural expectation for the very early universe is that there should be a large number of small, causally disconnected regions that will be in poor thermal equilibrium. This can be seen in a diagram of past light cones in a FRW cosmology: the Big Bang itself forms a surface which cuts off all past light cones. The central question is: just how large was the horizon when the CMB was emitted? Since the universe was about 300,000 years old at recombination, the horizon size then was about 300,000 light years. Each atom at the surface of last scattering could only be in causal contact (and therefore in thermal equilibrium) with with other atoms within a radius

218

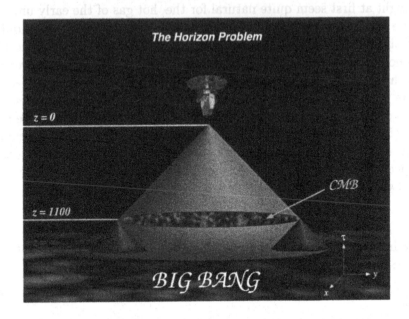

Figure 10. Light cones in a FRW space, plotted in conformal coordinates, $ds^2 = a^2\left(d\tau^2 - dx^2\right)$. The "Big Bang" is a surface at $\tau = 0$. Points in the space can only be in causal contact if their past light cones intersect at the Big Bang.

of about 300,000 light years. As seen from earth, the horizon size at the surface of last scattering subtends an angle on the sky of about a degree. Therefore, two points on the surface of last scattering separated by an angle of more than a degree were out of causal contact at the time the CMB was emitted. However, the CMB is uniform (and therefore in thermal equilibrium) over the *entire sky* to one part in 10^5. How did all of these disconnected regions reach such a high degree of thermal equilibrium?

4.3 Inflation

The flatness and horizon problems have no solutions within the context of a standard matter- or radiation-dominated universe, since for any ordinary matter, the force of gravity causes the expansion of the universe to decelerate. The only available fix would appear to be to invoke initial conditions: the universe simply started out flat, hot, and in thermal equilibrium. While this is certainly a possibility, it hardly a satisfying explanation. It would be preferable to have an explanation for *why* the universe was flat, hot, and in thermal equilibrium. Such an explanation was proposed by Alan Guth in 1980 [24] under the name

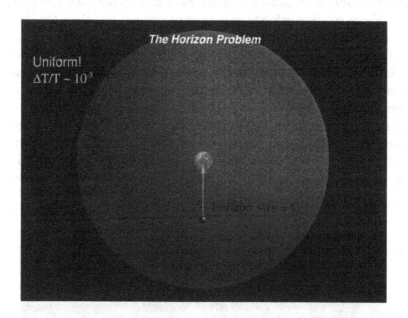

Figure 11. Schematic diagram of the horizon size at the surface of last scattering. The horizon size at the time of recombination was about 300,000 light years. Viewed from earth, this subtends an angle of about a degree on the sky.

of *inflation*. Inflation is the idea that at some very early epoch, the expansion of the universe was accelerating instead of decelerating.

Accelerating expansion turns the horizon and flatness problems on their heads. This is evident from the equation for the acceleration,

$$\frac{\ddot{a}}{a} = -(1 + 3w)\left(\frac{4\pi}{3m_{\rm Pl}^2}\rho\right). \tag{68}$$

We see immediately that the condition for acceleration $\ddot{a} > 0$ is that the equation of state be characterized by negative pressure, $1 + 3w < 0$. This immediately means that the universe evolves *toward* flatness rather than away:

$$\frac{d|\Omega - 1|}{d\log a} < 0, \ 1 + 3w < 0. \tag{69}$$

Similarly, from Eq. (66), we see that comoving scales grow in size more quickly than the horizon,

$$\frac{d}{d\log a}\left(\frac{d}{d_{\rm H}}\right) > 0, \ 1 + 3w < 0. \tag{70}$$

This is a very remarkable behavior. It means that two points that are initially in causal contact ($d < d_{\rm H}$) will expand so rapidly that they will eventually be causally *disconnected.* Put another way, two points in space whose relative velocity due to expansion is less than the speed of light will eventually be flying apart from each other at greater than the speed of light! Note that there is absolutely no violation of the principles of relativity. Relative velocities $v > c$ are allowed in general relativity as long as the observers are sufficiently separated in space.[7] This mechanism provides a neat way to explain the apparent homogeneity of the universe on scales much larger than the horizon size: a tiny region of the universe, initially in some sort of equilibrium, is "blown up" by accelerated expansion to an enormous and causally disconnected scale (Fig. 12.) In typical models of inflation, this process occurs when the

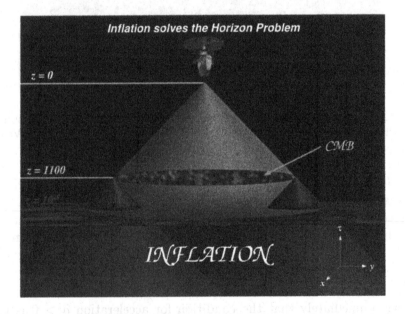

Figure 12. Conformal diagram of light cones in an inflationary space. The end of inflation creates an "apparent" Big Bang at $\tau = 0$ which is at high (but not infinite) redshift. There is, however, no singularity at $\tau = 0$ and the light cones intersect at an earlier time.

universe was around 10^{-42} seconds old, and the scale factor of the universe grows by something more 10^{26} times! In the next section we will

---

[7] An interesting consequence of the currently observed accelerating expansion is that all galaxies except those in our local group will eventually be moving away from us faster than the speed of light and will be invisible to us. The far future universe will be a lonely place, and cosmology will be all but impossible!

talk about how one constructs a model of inflation in particle theory. A more detailed introductory review can be found in Ref. [25].

4.4   Inflation from scalar fields

We have already seen that a cosmological constant due to nonzero vacuum energy results in accelerating cosmological expansion. While this is a good candidate for explaining the observations of Type Ia supernovae, it does not work for explaining inflation at early times for the simple reason that any period of accelerated expansion in the very early universe must end. Therefore the vacuum energy driving inflation must be dynamic. To implement a time-dependent "cosmological constant," we require a field with the same quantum numbers as vacuum, i.e. a scalar. We will consider a scalar field minimally coupled to gravity, with potential $V(\phi)$ and Lagrangian

$$\mathcal{L} = \frac{1}{2} g^{\mu\nu} \partial_\mu \phi \partial_\nu \phi - V(\phi), \tag{71}$$

where we have modified a familiar Minkowski-space field theory by replacing the Minkowski metric $\eta^{\mu nu}$ with the FRW metric $g^{\mu\nu}$. The equation of motion for the field is

$$\ddot{\phi} + 3H\dot{\phi} + p^2 \phi + V'(\phi) = 0. \tag{72}$$

This is the familiar equation for a free scalar field with an extra piece, $3H\dot{\phi}$, that comes from the use of the FRW metric in the Lagrangian. We will be interested in the ground state of the field $p = 0$. This is of interest because the zero mode of the field forms a perfect fluid, with energy density

$$\rho = \frac{1}{2} \dot{\phi}^2 + V(\phi) \tag{73}$$

and pressure

$$p = \frac{1}{2} \dot{\phi}^2 - V(\phi). \tag{74}$$

Note in particular that in the limit $\dot{\phi} \to 0$ we recover a cosmological constant, $p = -\rho$, as long as the potential $V(\phi)$ is nonzero. The Friedmann equation for the dynamics of the cosmology is

$$H^2 = \left(\frac{\dot{a}}{a}\right)^2 = \frac{8\pi}{3m_{\mathrm{Pl}}^2} \left[\frac{1}{2} \dot{\phi}^2 + V(\phi)\right] = 0. \tag{75}$$

In the $\dot{\phi} \to 0$ limit, we have

$$H^2 = \frac{8\pi}{3m_{\mathrm{Pl}}^2} V(\phi) = \text{const.}, \tag{76}$$

so that the universe expands exponentially,

$$a(t) \propto e^{Ht}. \tag{77}$$

This can be generalized to a time-dependent field and a quasi-exponential expansion in a straightforward way. If we have a slowly varying field $(1/2)\dot{\phi}^2 \ll V(\phi)$, we can write the equation of motion of the field as

$$3H\dot{\phi} + V'(\phi) \simeq 0, \tag{78}$$

and the Friedmann equation as

$$H^2(t) \simeq \frac{8\pi}{3m_{\mathrm{Pl}}^2} V[\phi(t)], \tag{79}$$

so that the scale factor evolves as

$$a(t) \propto \exp \int H dt. \tag{80}$$

This is known as the *slow roll* approximation, and corresponds physically to the field evolution being dominated by the "friction" term $3H\dot{\phi}$ in the equation of motion. This will be the case if the potential is sufficiently flat, $V'(\phi) \ll V(\phi)$. It is possible to write the equation of state of the field in the slow roll approximation as

$$p = \left[\frac{2}{3}\epsilon(\phi) - 1 \right] \rho, \tag{81}$$

where the *slow roll parameter* ϵ is given by

$$\epsilon = \frac{16\pi}{m_{\mathrm{Pl}}^2} \left[\frac{V'(\phi)}{V(\phi)} \right]^2. \tag{82}$$

This parameterization is convenient because the condition for accelerating expansion is $\epsilon < 1$

$$\frac{\ddot{a}}{a} = H^2(1 - \epsilon). \tag{83}$$

Specifying a model for inflation is then as simple as selecting a potential $V(\phi)$ and evaluating its behavior as a source of cosmological energy density. Many models have been proposed: Refs. [26, 27] contain extensive reviews of inflationary model building. We will discuss the observational predictions of various models in Section 4.7 below. In the next section, we will discuss one of the central observational predictions of inflation: the generation of primordial density fluctuations.

4.5   Density fluctuations from inflation

So far we have seen that the standard FRW cosmology has two (related) unexplained issues: why is the universe so flat and why is it so smooth? These questions are difficult to answer in the context of a matter- or radiation-dominated cosmology without resorting to simply setting the initial conditions of the universe to match what we see, certainly an unsatisfying solution. The idea of *inflation*, a period of accelerated expansion in the very early universe, provides an elegant solution to these problems. At some early time, the energy density of the universe was dominated by something with an equation of state approximating that of vacuum energy, $p < -(1/3)\rho$. (In the discussion above, we considered a single scalar field as a model for such a fluid.) The useful thing about inflation is that it dynamically drives the universe toward a flat ($\Omega = 1$) geometry, and simultaneously drives causally connected points to non-causal regions, explaining the homogeneity of the universe on scales larger than the horizon. This is an appealing scenario, but it is short on observational consequences: how do we tell whether or not inflation actually happened?

Very soon after the introduction of inflation by Guth, it was realized that inflation had another remarkable property: it could explain the generation of the primordial density fluctuations in the universe. This was first worked out independently by Hawking [28], Starobinsky [29], and by Guth [30]. We will cover the generation of fluctuations in great detail later, but the basic physics is familiar, and is closely related to the generation of Hawking radiation by black holes. It can be explained qualitatively as follows (Fig.   13). In a normal Minkowski space, vacuum fluctuations are interpreted as pairs of virtual particles appearing and then immediately annihilating as a consequence of the Heisenberg uncertainty principle. One qualitative explanation of Hawking radiation at the event horizon of a black hole is that one of the two virtual particles is trapped by the horizon, leaving the other to escape as apparently thermal radiation. A similar process holds in an inflationary spacetime: in inflation, the expansion is so rapid that pairs of virtual particles get "swept up" in the spacetime and are inflated to causally disconnected regions. In essence, they can no longer find each other to annihilate, and the quantum fluctuations become classical modes of the field. Formally, this effect is calculated by considering the equation of motion for a fluctuation in the inflaton field $\delta\phi$,

$$(\ddot{\delta\phi}) + 3H(\dot{\delta\phi}) + \left(\frac{k}{a}\right)^2 (\delta\phi) = 0, \tag{84}$$

Inflation + QM = Fluctuations

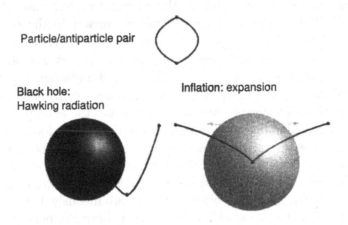

Particle/antiparticle pair

Black hole:
Hawking radiation

Inflation: expansion

Figure 13. Diagrams of virtual particles in a Minkowski space, near the horizon of a black hole, and in an inflationary space. Like the radiation created by one of the virtual pair being lost behind the horizon, virtual particles in inflation are swept out of each others' horizon before they can recombine.

where k is a comoving wavenumber which stays constant with expansion. The physical momentum p of the particle is

$$\mathbf{p} = \frac{\mathbf{k}}{a}. \tag{85}$$

During inflation, the wavelength of a quantum mode is "stretched" by the rapid expansion,

$$\lambda \propto a \propto e^{Ht}. \tag{86}$$

The horizon size, however, remains roughly constant,

$$d_{\mathrm{H}} \sim H^{-1} \simeq \text{const.}. \tag{87}$$

Short-wavelength vacuum fluctuations are then quickly redshifted by the expansion until their wavelengths are larger than the horizon size of the spacetime, and the modes are "frozen" as classical fluctuations. The amplitudes of quantum modes in inflation are conventionally expressed at *horizon crossing*, that is when the wavelength of the mode is equal to the horizon size, or $k/a = H$. The two-point correlation function of the field at horizon crossing is just given by the Hubble parameter:

$$\left\langle \delta\phi^2 \right\rangle_{(k/a)=H} = \left(\frac{H}{2\pi}\right)^2.$$
(88)

Like the horizon of a black hole, the horizon of an inflationary space has a "temperature" $T \sim H^{-1}$. The simplest example of the creation of fluctuations during inflation is the generation of gravitational waves, or *tensor modes*. An arbitrary perturbation to the metric tensor can be expressed as a sum scalar, vector, and tensor fluctuations, depending on how each behaves under coordinate transformations,

$$\delta g_{\mu\nu} = h_{\mu\nu}^{\text{scalar}} + h_{\mu\nu}^{\text{vector}} + h_{\mu\nu}^{\text{tensor}}.$$
(89)

The tensor component can be expressed in general as the superposition of two gravitational wave modes,

$$h_{+,\times}^{\text{tensor}} = \varphi_{+,\times}/m_{\text{Pl}},$$
(90)

where $+, \times$ refer to longitudinal and transverse polarization modes, respectively. Each mode has the equation of motion of a free scalar during inflation. Therefore fluctuations in the field are generated by the rapid expansion,

$$\varphi_{+,\times} = \frac{H}{2\pi}.$$
(91)

The case of scalar fluctuations is more complex, since scalar modes are generated by fluctuations in the inflaton field itself and couple to the curvature of the spacetime. We will simply note the correct expression,

$$\frac{\delta\rho}{\rho} = \frac{\delta N}{\delta\phi}\delta\phi = \frac{H}{2\pi\dot{\phi}},$$
(92)

where N is the number of e-folds of inflation,

$$a \propto e^N \propto \exp \int H dt.$$
(93)

What about vector modes, or primordial vorticity? Since there is no way to source such modes with a single scalar field, primordial vector perturbations vanish, at least in simple models of inflation.

4.6 The primordial power spectrum

Summarizing the results of the last section, inflation predicts not only a flat, smooth universe, but also provides a natural mechanism for the production of primordial density and gravitational wave fluctuations.

The scalar, or density fluctuation amplitude when a mode crosses the horizon is given by

$$\left(\frac{\delta\rho}{\rho}\right)_{k=aH} = \frac{H}{2\pi\dot\phi}, \tag{94}$$

and the gravitational wave amplitude is given by

$$(\varphi_{+,\times})_{k=aH} = \frac{H}{2\pi}. \tag{95}$$

For each of the two polarization modes for the gravitational wave. These are the amplitudes for a single mode when its wavelength (which is changing with time due to expansion) is equal to the horizon size. Note that in the limit that $\dot\phi \to 0$, $H \to$ const., so that fluctuations of all wavelengths have the same amplitude. In the case of slow roll, with $\dot\phi$ small and H slowly varying, modes of different wavelengths will have slightly different amplitudes. If we define the power spectrum as the variance per logarithmic interval,

$$\left(\frac{\delta\rho}{\rho}\right)^2 = \int P_S(k)\,d\log k, \tag{96}$$

inflation generically predicts a power-law form for $P_S(k)$,

$$P_S(k) \propto k^{n-1}, \tag{97}$$

so that the *scale invariant spectrum*, one with equal amplitudes at horizon crossing, is given by $n = 1$. The current observational best fit for the spectral index n is [31]

$$n = 0.91^{+0.15}_{-0.07}. \tag{98}$$

The observations are in agreement with inflation's prediction of a nearly (but not exactly) scale-invariant power spectrum, corresponding to a slowly rolling inflaton field and a slowly varying Hubble parameter during inflation. One can also consider power spectra which deviate from a power law,

$$\frac{d\log P_S(k)}{d\log k} = n + \frac{dn}{d\log k} + \cdots, \tag{99}$$

but inflation predicts the variation in the spectral index $dn/d\log k$ to be small, and we will not consider it further here. Similarly, the tensor fluctuation spectrum in inflationary models is a power-law,

$$P_T(k) \propto k^{n_T}. \tag{100}$$

(Note the unfortunate convention that the scalar spectrum is defined as a power law with index $n - 1$ while the tensor spectrum is defined as a power law with index n_T, so that the scale-invariant limit for tensors is $n_T = 0$!)

It would appear, then, that we have four independent observable quantities to work with: the amplitudes of the scalar and tensor power spectra at some fiducial scale k_*:

$$\begin{aligned} A_S &\equiv P_S(k_*) \\ A_T &\equiv P_T(k_*), \end{aligned} \qquad (101)$$

and the spectral indices n, n_T of the power spectra. In fact, at least within the context of inflation driven by a single scalar field, not all of these parameters are independent. This is because the tensor spectral index is just given by the equation of state parameter ϵ,

$$n_T = -2\epsilon = -\frac{32\pi}{m_{Pl}^2} \left[\frac{V'(\phi)}{V(\phi)} \right]^2 . \qquad (102)$$

However, from Eqs. (94) and (95), and from the equation of motion in the slow-roll approximation (78), we have a simple expression for the ratio r between the amplitudes of tensor and scalar fluctuations:

$$r \equiv \frac{P_T}{P_S} = \dot{\phi} = \epsilon = -2n_T, \qquad (103)$$

so that the tensor/scalar ratio and the tensor spectral index are not independent parameters, but are both determined by the equation of state during inflation, a relation known as the *consistency condition* for slow-roll inflation.[8]

So any simple inflation model gives us three independent parameters to describe the primordial power spectrum: the amplitude of scalar fluctuations A_S, the tensor/scalar ratio r, and the scalar spectral index n. The important point is that these are *observable* parameters, and will allow us to make contact between the physics of very high energies and the world of observational cosmology, in particular the cosmic microwave background.

4.7 Inflationary zoology and the CMB

We have discussed in some detail the generation of fluctuations in inflation, in particular the primordial tensor and scalar power spectra and the relevant parameters A_S, r, and n. However, we have not said much about what kind of potential $V(\phi)$ is suitable for inflation. In

---

[8]In the case of multi-field inflation, this condition relaxes to an inequality, $P_T/P_S \geq -2n_T$.

fact, pretty much *any* potential, with suitable fine-tuning of parameters, will work to drive inflation in the early universe. We wish to come up with a classification scheme for different kinds of scalar field potentials and study how we might find observational constraints on one "type" of inflation versus another. Such a "zoology" of potentials is simple to construct. Figure 14 shows three basic types of potentials: *large field*, where the field is displaced by $\Delta\phi \sim m_{Pl}$ from a stable minimum of a potential, *small field*, where the field is evolving away from an unstable maximum of a potential, and *hybrid*, where the field is evolving toward a potential minimum with nonzero vacuum energy. Large field models

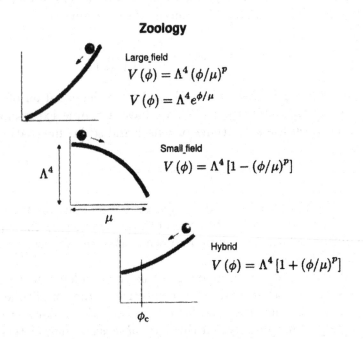

Zoology

Large field
$$V(\phi) = \Lambda^4 (\phi/\mu)^p$$
$$V(\phi) = \Lambda^4 e^{\phi/\mu}$$

Λ^4

μ

Small field
$$V(\phi) = \Lambda^4 [1 - (\phi/\mu)^p]$$

Hybrid
$$V(\phi) = \Lambda^4 [1 + (\phi/\mu)^p]$$

ϕ_c

Figure 14. A "zoology" of inflation models, grouped into *large field*, *small field*, and *hybrid* potentials.

are perhaps the simplest types of potentials. These include potentials such as a simple massive scalar field, $V(\phi) = m^2\phi^2$, or fields with a quartic self-coupling, $V(\phi) = \lambda\phi^4$. A general set of large field polynomial potentials can be written in terms of a "height" Λ and a "width" μ as:

$$V(\phi) = \Lambda^4 \left(\frac{\phi}{\mu}\right)^p, \tag{104}$$

where the particular model is specified by choosing the exponent p. In addition, there is another class of large field potentials which is useful, an exponential potential,

$$V(\phi) = \Lambda^4 \exp(\phi/\mu). \tag{105}$$

Small field models are typical of spontaneous symmetry breaking, in which the field is evolving away from an unstable maximum of the potential. In this case, we need not be concerned with the form of the potential far from the maximum, since all the inflation takes place when the field is very close to the top of the hill. As such, a generic potential of this type can be expressed by the first term in a Taylor expansion about the maximum,

$$V(\phi) \simeq \Lambda^4 \left[1 - (\phi/\mu)^p\right], \tag{106}$$

where the exponent p differs from model to model. In the simplest case of spontaneous symmetry breaking with no special symmetries, the leading term will be a mass term, $p = 2$, and $V(\phi) = \Lambda^4 - m^2\phi^2$. Higher order terms are also possible. The third class of models we will call "hybrid", named after models first proposed by Andrei Linde [32]. In these models, the field evolves toward the minimum of the potential, but the minimum has a nonzero vacuum energy, $V(\phi_{\min}) = \Lambda^4$. In such cases, inflation continues forever unless an auxiliary field is added to bring an end to inflation at some point $\phi = \phi_c$. Here we will treat the effect of this auxiliary field as an additional free parameter. We will consider hybrid models in a similar fashion to large field and small field models,

$$V(\phi) = \Lambda^4 \left[1 + (\phi/\mu)^p\right]. \tag{107}$$

Note that potentials of all three types are parameterized in terms of a height Λ^4, a width μ, and an exponent p, with an additional parameter ϕ_c specifying the end of inflation in hybrid models. This classification of models may seem somewhat arbitrary[9] It is convenient, however, because the different classes of models cover different regions of the (r, n) plane with no overlap (Fig. 15).

How well can we constrain these parameters with observation? Figure 8 shows the current observational constraints on the CMB multipole spectrum. As we saw in Sec. 3, the shape of this spectrum depends on a large number of parameters, among them the shape of the primordial power spectrum. However, uncertainties in other parameters such as the baryon density or the redshift of reionization can confound our measurement of the things we are interested in, namely r and n.

---

[9]The classification might also appear ill-defined, but it can be made more rigorous as a set of inequalities between slow roll parameters. Refs. [33, 34] contain a more detailed discussion.

230

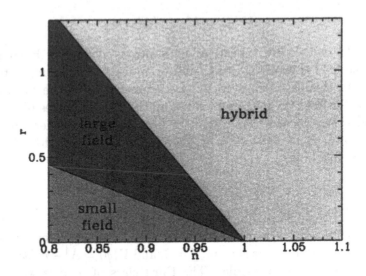

Figure 15.   Classes of inflationary potentials plotted on the (r, n) plane.

Figure 16 shows the current likelihood contours for r and n based on current CMB data [37]. Perhaps the most distinguishing feature of this plot is that the error bars are smaller than the plot itself! The favored model is a model with negligible tensor fluctuations and a slightly "red" spectrum, $n < 1$. Future measurements, in particular the MAP [35] and Planck satellites [36], will provide much more accurate measurements of the C_ℓ spectrum, and will allow correspondingly more precise determination of cosmological parameters, including r and n. (In fact, by the time this article sees print, the first release of MAP data will have happened.) Figure 17 shows the expected errors on the C_ℓ spectrum for MAP and Planck, and Fig. 18 shows the corresponding error bars in the (r, n) plane. Note especially that Planck will make it possible to clearly distinguish between different models for inflation. So all of this apparently esoteric theorizing about the extremely early universe is *not* idle speculation, but real science. Inflation makes a number of specific and observationally testable predictions, most notably the generation of density and gravity-wave fluctuations with a nearly (but not exactly) scale-invariant spectrum, and that these fluctuations are Gaussian and adiabatic. Furthermore, feasible cosmological observations are capable of telling apart different specific models for the inflationary epoch, thus

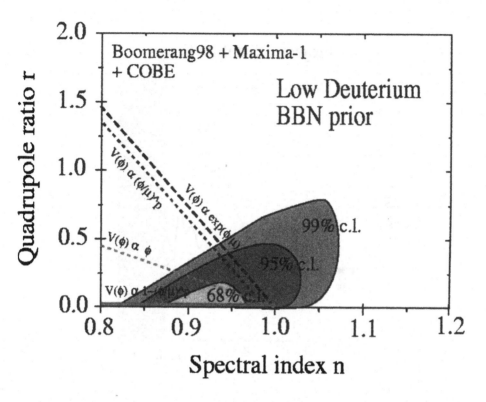

Figure 16.   Error bars in the r, n plane for the Boomerang and MAXIMA data sets. The lines on the plot show the predictions for various potentials. Similar contours for the most current data can be found in Ref. [19]

providing us with real information on physics near the expected scale of Grand Unification, far beyond the reach of existing accelerators. In the next section, we will stretch this idea even further. Instead of using cosmology to test the physics of inflation, we will discuss the more speculative idea that we might be able to use inflation itself as a microscope with which to illuminate physics at the very highest energies, where quantum gravity becomes relevant.

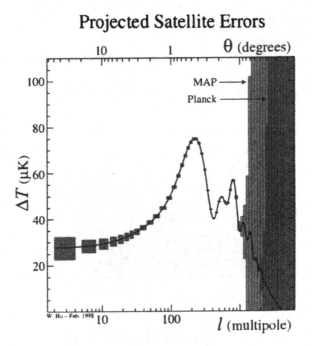

Figure 17. Expected errors in the C_ℓ spectrum for the MAP (light blue) and Planck (dark blue) satellites. [20]

5. Looking for signs of quantum gravity in inflation

We have seen that inflation is an attractive and predictive theory of the physics of the very early universe. Unexplained properties of a standard FRW cosmology, namely the flatness and homogeneity of the universe, are natural outcomes of an inflationary expansion. Furthermore, inflation provides a mechanism for generating the tiny primordial density fluctuations that seed the later formation of structure in the universe. Inflation makes definite predictions, which can be tested by precision observation of fluctuations in the CMB, a program that is already well underway.

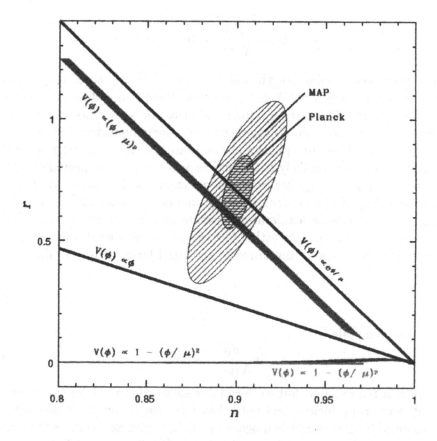

Figure 18. Error bars in the r, n plane for MAP and Planck. These ellipses show the expected $2 - \sigma$ errors. The lines on the plot show the predictions for various potentials.

In this section, we will move beyond looking at inflation as a subject of experimental test and discuss some intriguing new ideas that indicate that inflation might be useful as a tool to illuminate physics at extremely high energies, possibly up to the point where effects from quantum gravity become relevant. This idea is based on a simple observation about scales in the universe. As we discussed in Sec. 2, quantum field theory extended to infinitely high energy scales gives nonsensical (i.e., divergent) results. We therefore expect the theory to break down at high energy, or equivalently at very short lengths. We can estimate the length scale at which quantum mechanical effects from gravity become important by simple dimensional analysis. We define the Planck length ℓ_{Pl} by an appropriate combination of fundamental constants as

$$\ell_{\text{Pl}} \sim \sqrt{\frac{\hbar G}{c^3}} \sim 10^{-35} m. \tag{108}$$

For processes probing length scales shorter than ℓ_{Pl}, such as quantum modes with wavelengths $\lambda < \ell_{\text{Pl}}$, we expect some sort of new physics to be important. There are a number of ideas for what that new physics might be, for example string theory or noncommutative geometry or discrete spacetime, but physics at the Planck scale is currently not well understood. It is unlikely that particle accelerators will provide insight into such high energy scales, since quantum modes with wavelengths less than ℓ_{Pl} will be characterized by energies of order 10^{19} GeV or so, and current particle accelerators operate at energies around 10^3 GeV. However, we note an interesting fact, namely that the ratio between the current horizon size of the universe and the Planck length is about

$$\frac{d_{\text{H}}}{l_{\text{Pl}}} \sim 10^{60}, \tag{109}$$

or, on a log scale,

$$\ln\left(\frac{d_{\text{H}}}{l_{\text{Pl}}}\right) \sim 140. \tag{110}$$

This is a big number, but we recall our earlier discussion of the flatness and horizon problems and note that inflation, in order to adequately explain the flatness and homogeneity of the universe, requires the scale factor to increase by *at least* a factor of e^{60}. Typical models of inflation predict much more expansion, e^{1000} or more. We remember that the wavelength of quantum modes created during the inflationary expansion, such as those responsible for density and gravitational-wave fluctuations, have wavelengths which redshift proportional to the scale factor, so that so that the wavelength λ_i of a mode at early times can be given in terms of its wavelength λ_0 today by

$$\lambda_i \ll e^{-60}\lambda_0. \tag{111}$$

This means that in most models of inflation, fluctuations of order the size of the universe today were smaller than the Planck length during inflation! This suggests the possibility that Plank-scale physics might have been important for the generation of quantum modes in inflation. The effects of such physics might be imprinted in the pattern of cosmological fluctuations we see in the CMB and large-scale structure today.

In what follows, we will look at the generation of quantum fluctuations in inflation in detail, and estimate how large the effect of quantum gravity might be on the primordial power spectrum.

In Sec. 2.4 we saw that the state space for a quantum field theory was a set of states $|n(\mathbf{k}_1), \ldots, n(\mathbf{k}_i)\rangle$ representing the number of particles with momenta $\mathbf{k}_1, \ldots, \mathbf{k}_i$. The creation and annihilation operators $\hat{a}_{\mathbf{k}}^\dagger$ and $\hat{a}_{\mathbf{k}}$ act on these states by adding or subtracting a particle from the state:

$$\begin{aligned}
\hat{a}_{\mathbf{k}}^\dagger |n(\mathbf{k})\rangle &= \sqrt{n+1}\,|n(\mathbf{k})+1\rangle \\
\hat{a}_{\mathbf{k}} |n(\mathbf{k})\rangle &= \sqrt{n}\,|n(\mathbf{k})-1\rangle .
\end{aligned} \tag{112}$$

The ground state, or vacuum state of the space is just the zero particle state,

$$\hat{a}_{\mathbf{k}} |0\rangle = 0. \tag{113}$$

Note in particular that the vacuum state $|0\rangle$ is *not* equivalent to zero. The vacuum is not nothing:

$$|0\rangle \neq 0. \tag{114}$$

To construct a quantum field, we look at the familiar classical wave equation for a scalar field,

$$\frac{\partial^2 \phi}{\partial t^2} - \nabla^2 \phi = 0. \tag{115}$$

To solve this equation, we decompose into Fourier modes $u_{\mathbf{k}}$,

$$\phi = \int d^3 k \left[a_{\mathbf{k}} u_{\mathbf{k}}(t) e^{i\mathbf{k}\cdot\mathbf{x}} + a_{\mathbf{k}}^* u_{\mathbf{k}}^*(t) e^{-i\mathbf{k}\cdot\mathbf{x}} \right], \tag{116}$$

where the mode functions $u_{\mathbf{k}}(t)$ satisfy the ordinary differential equations

$$\ddot{u}_{\mathbf{k}} + k^2 u_{\mathbf{k}} = 0. \tag{117}$$

This is a classical wave equation with a classical solution, and the Fourier coefficients $a_{\mathbf{k}}$ are just complex numbers. The solution for the mode function is

$$u_{\mathbf{k}} \propto e^{-i\omega_k t}, \tag{118}$$

where ω_k satisfies the dispersion relation

$$\omega_k^2 - \mathbf{k}^2 = 0. \tag{119}$$

To turn this into a quantum field, we identify the Fourier coefficients with creation and annihilation operators

$$a_{\mathbf{k}} \to \hat{a}_{\mathbf{k}}, \quad a_{\mathbf{k}}^* \to \hat{a}_{\mathbf{k}}^\dagger, \tag{120}$$

and enforce the commutation relations

$$\left[\hat{a}_{\mathbf{k}}, \hat{a}_{\mathbf{k}'}^{\dagger}\right] = \delta^3 \left(\mathbf{k} - \mathbf{k}'\right). \tag{121}$$

This is the standard quantization of a scalar field in Minkowski space, which should be familiar. But what probably isn't familiar is that this solution has an interesting symmetry. Suppose we define a new mode function $u_{\mathbf{k}}$ which is a rotation of the solution (118):

$$u_{\mathbf{k}} = A(k)e^{-i\omega t + i\mathbf{k}\cdot\mathbf{x}} + B(k)e^{i\omega t - i\mathbf{k}\cdot\mathbf{x}}. \tag{122}$$

This is *also* a perfectly valid solution to the original wave equation (115), since it is just a superposition of the Fourier modes. But we can then re-write the quantum field in terms of our original Fourier modes and new *operators* $\hat{b}_{\mathbf{k}}$ and $\hat{b}_{\mathbf{k}}^{\dagger}$ and the original Fourier modes $e^{i\mathbf{k}\cdot\mathbf{x}}$ as:

$$\phi = \int d^3k \left[\hat{b}_{\mathbf{k}}e^{-i\omega t + i\mathbf{k}\cdot\mathbf{x}} + \hat{b}_{\mathbf{k}}^{\dagger}e^{+i\omega t - i\mathbf{k}\cdot\mathbf{x}}\right], \tag{123}$$

where the new operators $\hat{b}_{\mathbf{k}}$ are given in terms of the old operators $\hat{a}_{\mathbf{k}}$ by

$$\hat{b}_{\mathbf{k}} = A(k)\hat{a}_{\mathbf{k}} + B^*(k)\hat{a}_{\mathbf{k}}^{\dagger}. \tag{124}$$

This is completely equivalent to our original solution (116) as long as the new operators satisfy the same commutation relation as the original operators,

$$\left[\hat{b}_{\mathbf{k}}, \hat{b}_{\mathbf{k}'}^{\dagger}\right] = \delta^3 \left(\mathbf{k} - \mathbf{k}'\right). \tag{125}$$

This can be shown to place a condition on the coefficients A and B,

$$|A|^2 - |B|^2 = 1. \tag{126}$$

Otherwise, we are free to choose A and B as we please.

This is just a standard property of linear differential equations: any linear combination of solutions is itself a solution. But what does it mean physically? In one case, we have an annihilation operator $\hat{a}_{\mathbf{k}}$ which gives zero when acting on a particular state which we call the vacuum state:

$$\hat{a}_{\mathbf{k}} |0_a\rangle = 0. \tag{127}$$

Similarly, our rotated operator $\hat{b}_{\mathbf{k}}$ gives zero when acting on some state

$$\hat{b}_{\mathbf{k}} |0_b\rangle = 0. \tag{128}$$

The point is that the two "vacuum" states are not the same

$$|0_a\rangle \neq |0_b\rangle. \tag{129}$$

From this point of view, we can define any state we wish to be the "vacuum" and build a completely consistent quantum field theory based on this assumption. From another equally valid point of view this state will contain particles. How do we tell which is the *physical* vacuum state? To define the real vacuum, we have to consider the spacetime the field is living in. For example, in regular special relativistic quantum field theory, the "true" vacuum is the zero-particle state as seen by an inertial observer. Another more formal way to state this is that we require the vacuum to be Lorentz symmetric. This fixes our choice of vacuum $|0\rangle$ and defines unambiguously our set of creation and annihilation operators $\hat{a}$ and $\hat{a}^\dagger$. A consequence of this is that an *accelerated* observer in the Minkowski vacuum will think that the space is full of particles, a phenomenon known as the Unruh effect [38]. The zero-particle state for an accelerated observer is different than for an inertial observer.

So far we have been working within the context of field theory in special relativity. What about in an expanding universe? The generalization to a curved spacetime is straightforward, if a bit mysterious. We will replace the metric for special relativity with a Robertson-Walker metric,

$$ds^2 = dt^2 - d\mathbf{x}^2 \rightarrow ds^2 = a^2\left(\tau\right)\left(d\tau^2 - d\mathbf{x}^2\right). \tag{130}$$

Note that we have written the Robertson-Walker metric in a slightly different way than we did in Eq. (10), with the scale factor a multiplying both time- and space-coordinates. This defines a new time variable, called the *conformal time* τ, where

$$dt = a d\tau. \tag{131}$$

This is convenient for doing field theory, because the new spacetime is just a Minkowski space with a time-dependent conformal factor out front. In fact we define the physical vacuum in a similar way to how we did it for special relativity: the vacuum is the zero-particle state as seen by a *geodesic* observer, that is, one in free-fall in the expanding space. This is referred to as the *Bunch-Davies* vacuum.

Now we write down the wave equation for a free field, the equivalent of Eq. (115) in a Robertson-Walker space. This is the usual equation with a new term that comes from the expansion of the universe:

$$\frac{\partial^2 \phi}{\partial \tau^2} + 2\left(\frac{a'(\tau)}{a(\tau)}\right)\frac{\partial \phi}{\partial \tau} - \nabla^2 \phi = 0. \tag{132}$$

Note that the time derivatives are with respect to the conformal time τ, not the coordinate time t. As in the Minkowski case, we Fourier expand the field, but with an extra factor of $a(\tau)$ in the integral:

$$\phi = \int \frac{d^3 k}{a(\tau)} \left[\hat{a}_{\mathbf{k}} u_{\mathbf{k}}(\tau) e^{i\mathbf{k}\cdot\mathbf{x}} + \hat{a}_{\mathbf{k}}^\dagger u_{\mathbf{k}}^*(\tau) e^{-i\mathbf{k}\cdot\mathbf{x}} \right]. \tag{133}$$

Here k is a *comoving* wavenumber (or, equivalently, momentum), which stays constant as the mode redshifts with the expansion $\lambda \propto a$, so that

$$k_{\text{physical}} = k/a. \tag{134}$$

Writing things this way, in terms of conformal time and comoving wavenumber, makes the equation of motion for the mode $u_{\mathbf{k}}(\tau)$ very similar to the mode equation 117 in Minkowski space:

$$u_{\mathbf{k}}'' + \left(k^2 - \frac{a''}{a} \right) u_{\mathbf{k}} = 0, \tag{135}$$

where a prime denotes a derivative with respect to conformal time. All of the effect of the expansion is in the a''/a term. (Be careful not to confuse the scale factor $a(\tau)$ with the creation/annihilation operators $\hat{a}_{\mathbf{k}}$ and $\hat{a}_{\mathbf{k}}^\dagger$!)

This equation is easy to solve. First consider the short wavelength limit, that is large wavenumber k. For $k^2 \gg a''/a$, the mode equation is just what we had for Minkowski space

$$u_{\mathbf{k}}'' + k^2 u_{\mathbf{k}} \simeq 0, \tag{136}$$

except that we are now working with comoving momentum and conformal time, so the space is only quasi-Minkowski. This will be important. The general solution for the mode is

$$u_{\mathbf{k}} = A(k) e^{-ik\tau} + B(k) e^{+ik\tau}. \tag{137}$$

Here is where the definition of the vacuum comes in. Selecting the Bunch-Davies vacuum is equivalent to setting $A = 1$ and $B = 0$, so that the annihilation operator is multiplied by $e^{-ik\tau}$ and not some linear combination of positive and negative frequencies. This is the exact analog of Eq. (122). So the mode function corresponding to the zero-particle state for an observer in free fall is

$$u_{\mathbf{k}} \propto e^{-ik\tau}. \tag{138}$$

What about the long wavelength limit, $k^2 \ll a''/a$? The mode equation becomes trivial:

$$u_{\mathbf{k}}'' - \frac{a''}{a} u_{\mathbf{k}} = 0, \tag{139}$$

with solution

$$\left|\frac{u_{\mathbf{k}}}{a}\right| = \text{const.} \tag{140}$$

The mode is said to be *frozen* at long wavelengths, since the oscillatory behavior is damped. This is precisely the origin of the density and gravity-wave fluctuations in inflation. Modes at short wavelengths are rapidly redshifted by the inflationary expansion so that the wavelength of the mode is larger than the horizon size, Eq. (140). We can plot the mode as a function of its physical wavelength $\lambda = k/a$ divided by the horizon size $d_{\mathrm{H}} = H^{-1}$ (Fig. 19), and find that at long wavelengths, the mode freezes out to a nonzero value. The power spectrum of fluctuations

The Mode Function

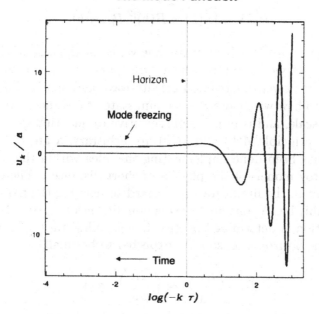

Figure 19. The mode function $u_{\mathbf{k}}/a$ as a function of $d_{\mathrm{H}}/\lambda = k/(aH) = -k\tau$. At short wavelengths, $k \gg aH$, the mode is oscillatory, but "freezes out" to a nonzero value at long wavelengths, $k \ll aH$.

is just given by the two-point correlation function of the field,

$$P(k) \propto \left\langle \phi^2 \right\rangle_{k \gg aH} = \left|\frac{u_{\mathbf{k}}}{a}\right|^2 \neq 0. \tag{141}$$

This means that we have produced *classical* perturbations at long wavelength from quantum fluctuations at short wavelength.

What does any of this have to do with quantum gravity? Remember that we have seen that for an inflationary period that lasts longer than 140 e-folds or so, the fluctuations we see with wavelengths comparable to the horizon size today started out with wavelengths shorter than the Planck length $\ell_{Pl} \sim 10^{-35}$ cm during inflation. For a mode with a wavelength that short, do we really know how to select the "vacuum" state, which we have assumed is given by Eq. (138)? Not necessarily. We do know that once the mode redshifts to a wavelength greater than ℓ_{Pl}, it must be of the form (137), but we know longer know for certain how to select the values of the constants $A(k)$ and $B(k)$. What we have done is mapped the effect of quantum gravity onto a boundary condition for the mode function u_k. In principle, $A(k)$ and $B(k)$ could be anything! If we allow A and B to remain arbitrary, it is simple to calculate the change in the two-point correlation function at long wavelength,

$$P(k) \to |A(k) + B(k)|^2 P_{\text{B}-\text{D}}(k),   (142)$$

where the subscript B – D indicates the value for the case of the "standard" Bunch-Davies vacuum, which corresponds to the choice $A = 1$, $B = 0$. So the power spectrum of gravity-wave and density fluctuations is sensitive to how we choose the vacuum state at distances shorter than the Planck scale, and is in principle sensitive to quantum gravity.

While in principle $A(k)$ and $B(k)$ are arbitrary, a great deal of recent work has been done implementing this idea within the context of reasonable toy models of the physics of short distances. There is some disagreement in the literature with regard to how big the parameter B can reasonably be. As one might expect on dimensional grounds, the size of the rotation is determined by the dimensionless ratio of the Planck length to the horizon size, so it is expected to be small

$$B \sim \left(\frac{l_{Pl}}{d_{\text{H}}}\right)^p \sim \left(\frac{H}{m_{Pl}}\right)^p \ll 1.   (143)$$

Here we have introduced a power p on the ratio, which varies depending on which model of short-distance physics you choose. Several groups have shown an effect linear in the ratio, $p = 1$. Fig. 20 shows the modulation of the power spectrum calculated in the context of one simple model [39]. Others have argued that this is too optimistic, and that a more realistic estimate is $p = 2$ [40] or even smaller [41]. The difference is important: if $p = 1$, the modulation of the power spectrum can be as large as a percent or so, a potentially observable value[42]. Take $p = 2$ and the modulation drops to a hundredth of a percent, far too small to see. Nonetheless, it is almost certainly worth looking for!

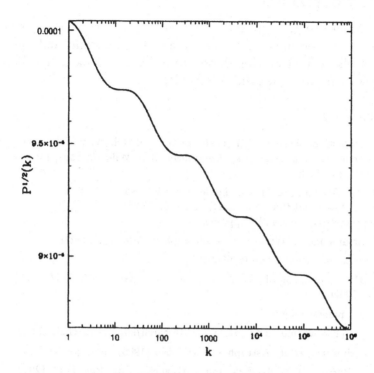

Figure 20. Modulation of the power spectrum of primordial fluctuations for a rotation $B \sim H/m_{Pl}$.

6. Conclusion

We have come a long way in four lectures, from Einstein's misbegotten introduction of the cosmological constant at the beginning of the last century to its triumphant return today. Einstein's blunder is now seen as the key to understanding the very beginning of the universe, as represented by the theory of inflation, as well as the universe today, dominated by the mysterious dark energy that makes up more than two thirds of the entire mass of the cosmos. I have tried to convince you of two things: first, that cosmology *is* particle physics in a very real sense, and second that cosmology is *meaningful* particle physics. Apparently exotic theories of the early universe such as inflation (and perhaps even string theory or some other variant of quantum gravity) are predictive and testable. It is a difficult business, to be sure, compared to the clean physics at, say, an $e^{\pm}$ collider, but what we learn about fundamental theory from cosmology is in many ways complementary to the lessons learned from more traditional particle physics.

Acknowledgments

I would like to give my warmest thanks to Harrison Prosper for organizing a wonderful summer school, and to the bright and motivated students who made lecturing there a rare treat. I thank Richard Easther and Brian Greene for helpful comments.

References

[1] A complete discussion of the basics of General Relativity and cosmology is in S. Weinberg, *Gravitation and Cosmology*, John Wiley & Sons, Inc., New York (1972), pp. 408ff.

[2] M. .W. Feast and R. M. Catchpole, MNRAS **286**, L1 (1997),
M. W. Feast and P.A. Whitelock, astro-ph/9706097,
http://astro.estec.esa.nl/Hipparcos/.

[3] M. Salaris and A. Weiss, Astron. Astrophys., **388**, 492 (2002).

[4] J. R. Mould *et al.*, astro-ph/9909260.

[5] L. M. Krauss and M. S. Turner, Gen. Rel. Grav. **27**, 1137 (1995), astro-ph/9504003.

[6] http://panisse.lbl.gov/.

[7] http://cfa-www.harvard.edu/cfa/oir/Research/supernova/HighZ.html.

[8] S. Perlmutter, *et al.*, Astrophys.J. **517**, 565 (1999), astro-ph/9812133.

[9] J. D. Barrow, H. B. Sandvik, and J. Magueijo, Phys.Rev. D **65** (2002) 123501, astro-ph/0110497,
R. Kallosh and A. Linde, hep-th/0208157,
J. Garriga and A. Vilenkin, astro-ph/0210358.

[10] R. R. Caldwell, R. Dave, P. J. Steinhardt, Phys. Rev. Lett. **80**, 1582 (1998), astro-ph/9708069,
L. Wang, R. R. Caldwell, J. P. Ostriker, and P. J. Steinhardt, Astrophys. J. **530**, 17 (2000), astro-ph/9901388.

[11] P. J. E. Peebles and B. Ratra, astro-ph/0207347.

[12] For a review of the physics of the CMB, see: http://background.uchicago.edu/,
http://www.hep.upenn.edu/ max/cmb/experiments.html,
W. Hu, astro-ph/0210696,
W. Hu and S. Dodelson, Ann. Rev. Astron. Astrophys. **40**, 171 (2002), astro-ph/0110414.

[13] For an extensive list of bibliographic references relevant to the CMB, see: M. White and J. D. Cohn, *TACMB-1: The Theory of Anisotropies in the Cosmic Microwave Background*, AJP/AAPT Bibliographic Resource letter (2002), astro-ph/0203120.

[14] A good discussion of thermodynamics in the early universe can be found in E. W. Kolb and M. S. Turner, *The Early Universe*, Addison-Wesley Publishing Company, Reading, MA (1990), Ch. 3.

[15] E. W. Kolb and M. S. Turner, *Ibid.*, p. 77.

[16] C. L. Bennett, *et al.*, Astrophys. J. **464**, L1 (1996).

[17] R. K. Sachs and A. M. Wolfe, Astrophys. J. **147**, 73 (1967).

[18] E. Bertschinger and C. P. Ma, Astrophys. J. **455**, 7 (1995), astro-ph/9506072.

[19] X. Wang, M. Tegmark, B. Jain, and M. Zaldarriaga, astro-ph/0212417.

[20] Figure courtesy of Wayne Hu, http://background.uchicago.edu/.

[21] P. de Bernardis *et al.*, Astrophys. J. **564**, 559 (2002), astro-ph/0105296.

[22] P. de Bernardis *et al.*, Nature **404**, 955 (2000), astro-ph/0004404, A. E. Lange *et al.*, Phys.Rev. D63 (2001) 042001, astro-ph/0005004.

[23] S. Hanany *et al.*, Astrophys.J. **545**, L5 (2000), astro-ph/0005123, A. Balbi *et al.*, Astrophys. J. **545**, L1 (2000); Erratum-ibid. **558**, L145 (2001), astro-ph/0005124.

[24] A. Guth, Phys. Rev. D **23**, 347 (1981).

[25] A. Liddle, astro-ph/9901124.

[26] A. Linde, *Particle Physics and Inflationary Cosmology*, Harwood Academic Publishers, London (1990).

[27] A. Riotto and D. H. Lyth, Phys. Rept. **314**, 1 (1999), hep-ph/9807278 .

[28] S. Hawking, Phys. Lett. **B115**, 295 (1982).

[29] A. Starobinsky, Phys. Lett. **B117**, 175 (1982).

[30] A. Guth, Phys. Rev. Lett. **49**, 1110 (1982).

[31] X. Wang, M. Tegmark, and M. Zaldarriaga, Phys. Rev. D **65**, 123001 (2002), astro-ph/0105091 .

[32] A. Linde, Phys. Rev. D **49**, 748 (1994), astro-ph/9307002.

[33] S. Dodelson, W. H. Kinney, and E. W. Kolb, Phys. Rev. D **56**, 3207 (1997), astro-ph/9702166.

[34] W. H. Kinney, Phys. Rev. D **58**, 123506 (1998), astro-ph/9806259.

[35] http://map.gsfc.nasa.gov/.

[36] http://astro.estec.esa.nl/SA-general/Projects/Planck/.

[37] W. H. Kinney, A. Melchiorri, A. Riotto, Phys. Rev. D **63**, 023505 (2001), astro-ph/0007375.

[38] W. G. Unruh, Phys. Rev. D **14**, 870 (1976).

[39] U. H. Danielsson, Phys .Rev. D **66**, 023511 (2002), hep-th/0203198, R. Easther, B. R. Greene, W. H. Kinney, and G. Shiu, Phys. Rev. D **66**, 023518 (2002), hep-th/0204129.

[40] N. Kaloper, M. Kleban, A. Lawrence, and S. Shenker, hep-th/0201158, N. Kaloper, M. Kleban, A. Lawrence, S. Shenker, and L. Susskind, hep-th/0209231.

[41] J. C. Niemeyer, R. Parentani, and D. Campo, Phys. Rev. D **66**, 083510 (2002), hep-th/0206149.

[42] L. Bergstrom and U.H. Danielsson, hep-th/0211006.

SILICON DETECTORS

Alan Honma

CERN, EP Division

1211 Genève 23, Switzerland

alan.honma@cern.ch

Abstract

These lectures review silicon strip detectors and their use in high energy physics experiments. The principles of operation, their performance, the effect of radiation damage, methods of fabrication, their construction into silicon detector modules, their read-out electronics and their usage in current experiments are described. In addition a short discussion of other types of silicon detectors is included.

1.   Introduction

The first usage of silicon detectors as precision position measurement devices dates from the late 1970's, where the pioneering work was done for vertex detection in fixed-target charm physics experiments. These early devices used segmented sensors (strips) with fine pitch compared to the other direct read-out position measuring devices in use at the time (wire read-out gas detectors). The first usage in colliding beam experiments started in the late 1980's with the Mark II experiment at the SLC collider at SLAC followed by all four LEP experiments at CERN. These were used for precision vertex finding and identification of secondary vertices. The first pixel-type silicon device was used in high energy physics in the early 1980's, also for a fixed-target charm experiment. The reasons that silicon detectors were not used prior to this were three-fold: the cost of the relatively new silicon processing technology, the small signal size of silicon devices, and the need for miniaturized read-out electronics. Advances in microelectronics starting in the 1970's progressively overcame all these problems.

In the present day experiments and those planned for the near future (LHC), the usage of silicon detectors is increasing. The hadron colliders such as the Tevatron at Fermilab and the LHC at CERN have a severe

H.B. Prosper and M. Danilov (eds.), Techniques and Concepts of High-Energy Physics XII, 245–271.

radiation environment which is often beyond that tolerated by existing gas detectors. Thus more and more of the inner tracking devices are based on silicon which can be made to support the higher radiation levels.

These lectures will concentrate mainly on silicon strip devices which are the "mainstay" of charged particle silicon detectors used in high energy physics. Other types of silicon detectors: pixel devices, pad devices, and drift devices, will be briefly discussed at the end of these lectures. Note that there are also many types of photon detecting silicon-based devices but these will not be discussed.

2.   Silicon Strip Detector Principles of Operation

The primary motivation for using a silicon strip detector is to obtain a position measurement of the charged particle passage. This uses the ionization signal (dE/dx) from the particle passage which in the case of a semiconductor, comes from the creation of electron-hole (e-h) pairs. This is completely analogous to a gas drift chamber where the ionization produces electron-ion pairs and a strong electric field is used to drift the electrons and ions to the oppositely charged electrodes where the charge can be amplified and measured. In silicon, one needs 3.6eV to produce one e-h pair. In the absence of an electric field the pair recombines quickly and thus one needs a means of preventing the recombination. The method used to achieve the electric field which drifts the electrons and holes to opposite faces of the silicon device, exploits the properties of the p-n junction.

2.1   The p-n Junction (Diode)

Recall that a p-type semiconductor has positive free charge carriers (holes) that are usually obtained by adding impurities of acceptor ions like boron. Boron is a type III element in the periodic table having 3 valence electrons, therefore in the crystal lattice it has one electron acceptor site. An n-type semiconductor has negative carriers (electrons) obtained with impurities of donor ions like phosphorus. Phosphorus is a type V element having 5 valence electrons and thus one electron donor site. The addition of impurities to silicon to obtain a p- or n-type semiconductor is called doping or implanting.

When the two types of semiconductor are brought into contact to form a junction, there appears a gradient of electron and hole densities resulting in a diffusive migration of majority carriers across the junction (see Fig. 1a). The migration leaves a region of net charge of opposite sign on each side, called the space-charge region or depletion region (depleted of charge carriers). The electric field set up in the region prevents further migration of carriers. In the depletion region, e-h pairs won't as easily recombine but will drift away from each other due to the field. If we make the p-n junction at the surface of a silicon wafer with the bulk being n-type (you could also do it the opposite way), we then need to extend the depletion region throughout the n bulk to get maximum charge collection. This can be achieved by applying a reverse bias voltage. Note that since this is a diode, a forward bias would result in current flow.

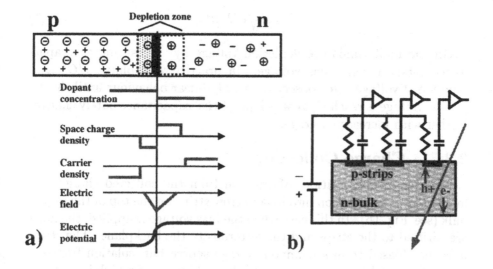

Figure 1. a) schematic view of p-n junction. b) basic strip detector.

2.2 Properties of the Depletion Region

The depth w of the depletion region is a function of the bulk resistivity, the charge carrier mobility μ and the magnitude of the reverse bias voltage V_b:

$$w = \sqrt{2\epsilon\rho\mu V_b}, \quad \text{where } \rho = 1/e\mu N, \text{ and} \qquad (1)$$

N is the doping concentration, e is the charge of the electron and ϵ is the dielectric constant. One easily finds that to fully deplete a device of thickness d, one needs a voltage, known as the depletion voltage V_d of:

$$V_d = d^2/(2\epsilon\mu\rho). \tag{2}$$

Thus one needs a higher voltage to fully deplete a low resistivity material. One also sees that a higher voltage is needed for a p-type bulk since the carrier mobility of holes is lower than for electrons (450 vs. 1350 cm^2/Vs). The capacitance is simply the parallel plate capacity of the depletion zone.

$$C = A\sqrt{\epsilon/2\rho\mu V_b} \tag{3}$$

One normally finds the depletion voltage by measuring the capacitance versus reverse bias voltage. In practice, one plots $1/C^2$ as a function of voltage. One observes a nearly linear behaviour until the depletion voltage is reached, at which point the capacitance barely changes further with increased voltage.

2.3   Charge Collection

In order to get position information from the collected charge, one makes the p- type silicon into long narrow strips on the top of the n-type bulk (see Fig. 1b). In this case when the bias voltage is applied, the holes are drifted to the strips and the electrons to the backplane. The strips must be biased from a common voltage source but isolation from the other strips and from the source is achieved by use of a high impedance bias connection (resistor or equivalent). In addition, one usually wants to AC couple the input amplifier to avoid large DC input currents. Both the biasing structure and AC coupling are often integrated directly on the silicon sensor. Bias resistors are made via deposition of doped polysilicon, and capacitors via metal readout lines over the implants but separated by an insulating dielectric layer (SiO$_2$ or Si$_3$N$_4$).

The magnitude of the collected charge is usually specified in terms of minimum ionizing charge deposition. The mean dE/dX loss in silicon is 3.88 MeV/cm, thus for 300μm thickness one expects 116 keV energy loss. Charge collection in silicon detectors is normally given in terms of the most probable energy loss since this value is easily determined from an energy loss histogram. The most probable loss for the expected Landau distribution is about 0.7 of the mean, thus in our example, about 81 keV. Since 3.6eV is needed to create an electron- hole pair, this would give about 22500 electrons (e) of collected charge.

The drift velocity of the charge carriers is $v = \mu E$, and thus the charge collection time $t_d = d/\mu E$. Using typical values for E of 2.5kV/cm and a thickness, $d = 300\mu m$, gives collection times of about 9ns for electrons and 27ns for holes. Note that one can reduce the collection time by simply increasing the bias voltage. One can also estimate the charge radius which is proportional to the square root of the charge collection time. In this example the radius works out to be about $6\mu m$. The diffusion of the charge could be exploited to obtain better position resolution by means of centroid finding if the charge would be spread over 2 or more strips. This would require keeping long drift times, hence low drift fields which implies silicon with low depletion voltage. Note that although the charge collection times differ for electrons and holes, their diffusion radius is the same.

2.4 Double-sided Detectors

The fact that electrons and holes drift to opposite sides of the silicon device means that a second coordinate measurement can be obtained by using the electron charge. This is not completely trivial because, unlike the p-strips which naturally separate the charge collection on that face, on the opposite face one has only the n-bulk with no structures that separate the electron charge on that face. One can implant n-type strips on the back side, with the strip orientation orthogonal to that of the p-strips. However, this is not sufficient to ensure that the electron charge remains confined to the region of its deposition. The reason for this is that there will be a SiO_2 layer on the back face between the n-strips and the electrons will tend to accumulate there. The mobility of the electrons between the strips at this interface allows the charge to disperse. Two methods are commonly employed to "break" this electron accumulation layer, the first uses implanted p-strips between the n-strips to provide a potential barrier, the second uses "field plates" (metallized strips over oxide over the n-strips) which are biased slightly negative with respect to the n-strips such that the electrons are repelled from the interstrip region.

2.5 Leakage Currents, Breakdown, Defects

Leakage currents are not desired since they can lead to increased readout noise. The main source of unwanted current flow in a silicon strip device is known as the generation current which comes from charge generated in the depletion zone (bulk) by defects or contaminants. The current has an exponential dependence on temperature due to the thermal dependence of e-h pair creation by defects. Surface currents are another type of undesired leakage current driven by defects occuring typically at the top edge surfaces of the device. Unlike in the bulk where the potential varies uniformly from p-strip to backplane, the cut edges and top corner of a sensor are at the backplane potential and hence the voltage drop occurs on the top surface between the p-strip bias ring and the top edges (see Fig. 2a). Typically, n-type implants are put along the top side edges and one or more guard rings (usually floating) are placed between this edge n implant and the p-strip bias ring (Fig. 2b). This structure makes for well determined fields over the top edge surfaces and a more continuous potential drop in the critical high field regions.

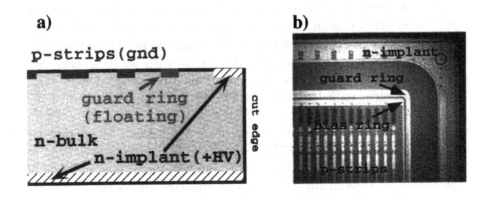

Figure 2. a) Cross-sectional view of silicon strip device showing edge structures. b) Top view of edge and corner of a CMS tracker silicon strip device.

Another potential undesirable effect is breakdown. This is a sharp increase in leakage current on one or more strips with increased bias voltage. Obviously this is most worrisome when it occurs below the normal operating voltage of the device. Breakdown occurs when the field is high enough to initiate avalanche multiplication, the point at which the charge carriers have enough energy to produce more e-h pairs.

This usually occurs around $30V/\mu m$ (compared to a typical operating field of $<1V/\mu m$). Local defects and inhomogeneities can result in fields approaching the breakdown point. Breakdown can occur through the bulk, in the edge regions (usually surface breakdown), and also between any structures with sufficiently different potentials.

There are also a number of different types of strip defect that can degrade the performance of a silicon strip device. Noisy strips are usually due to high DC leakage current but sometimes are seen as large fluctuations in current. This can be due to defects, damage or incorrect processing. Shorts can occur between strips and from strip to other structures (biasing ring). Open strips are interruptions of a strip. Other important properties of strips include the total capacitance which is the sum of the back-plane and inter-strip capacitances (normally kept low to avoid charge loss and increased noise), interstrip resistance (normally kept high to avoid charge loss and increased noise), line resistance of metal and implant (normally kept low to avoid increased noise). The quality of both the device design and the fabrication plays a critical role in reducing all the above-mentioned undesirable effects.

3. Silicon Strip Detector Performance

3.1 Position Resolution

If one treats the detected charge on each strip in a binary way (threshold discrimination), the resolution is simply: $\sigma = p/\sqrt{12}$, where p is the strip pitch. So for a typical strip pitch of $50\mu m$, $\sigma = 14.4\mu m$. As mentioned earlier, if the charge distribution is shared between adjacent strips, can use centroid finding to improve this resolution. However, since typical charge distribution sizes are of order 5-10 μm this implies quite fine strip pitch. In fact it is not practical to make sensors of pitch less than $20\mu m$ and most are greater.

Test devices have been made that have achieved $\sigma < 3\mu m$ (using a read-out pitch of 25 μm), this is near the limit on precision determined by diffusion and statistical fluctuations of the ionizing energy deposition. Read-out electronics pitch limit is around $40\mu m$ and time/cost constraints often argue for even larger read-out pitches. Fortunately there is a way to preserve resolution with larger read-out pitch. If one reads out only every n^{th} strip but preserves the signal magnitude, the charge gets shared such that the centroid resolution is nearly that obtained by reading out every strip. This technique, known as capacitive charge division, has limitations in that some signal is lost (capacitive coupling to the backplane), the noise is a bit higher (more input capacitance) and one loses some two track separation capability. However, this is clearly an economic solution in the case of low occupancy and has been used extensively. It does require a good signal/noise ratio, which is the next topic.

3.2   Signal to Noise Ratio (S/N)

The charge signal has a Landau distribution with a significant low energy tail which becomes even lower with noise broadening as shown in Fig. 3a. One usually has low occupancy in silicon sensors thus most channels have no signal. One doesn't want noise to produce fake hits so one needs to cut high above the noise tail to define good hits. But if one cuts too high one loses efficiency for real signals. The centroid determination is also degraded by poor signal to noise.

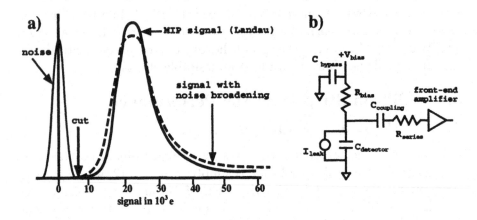

Figure 3.   a) Signal and noise distributions. b) Typical input load to front-end amplifier.

As was determined previously, the basic signal produced is about 22500e. Typical losses of 5-10% can be expected depending on the nature of the chosen electrical network at the input (AC coupling capacitor, stray capacitances and resistances, see Fig. 3b) and the front-end electronics. The noise is usually expressed as equivalent noise charge (ENC), also in units of electron charge e. Here we assume the use of a CR-RC amplifier shaper circuit as is most commonly used. The main sources of noise are:

- capacitive load, C_d (often the major source, the feedback mechanism of most amplifiers makes the amplifier internal noise dependent on input capacitive load), ENC $\propto C_d$.

- sensor leakage current (shot noise), ENC $\propto \sqrt{I}$.

- parallel resistance of bias resistor (thermal noise), ENC $\propto \sqrt{kT/R}$.

The total noise is generally expressed in the form: ENC $= a + b \cdot C_d$, where the last two sources are absorbed into the constant term a.

One must consider the noise implications on the detector design. For example, the strip length, device quality, choice of bias method will affect noise. Temperature is important for both the leakage current noise (current doubles for an increase of 7°C) and for the bias resistor contribution. Noise is also very frequency dependent, thus it will be dependent on the choice of read-out method.

A typical noise value for a LEP silicon strip module (OPAL) is: ENC $= 500 + 15 \ C_d$. The typical strip capacitance is about 1.5pF/cm, strip length is 18cm, so C_d=27pF. Thus, ENC = 900e, and since S=22500e, the S/N ≈ 25. Note that for a typical LHC silicon strip module (CMS): ENC $= 425 + 64 \ C_d$. The much larger capacitive term is largely due to the much faster shaping time needed in the LHC (nearly three orders of magnitude shorter bunch crossing time). In order that the device performance does not degrade the physics results, a S/N of at least 10 is generally needed.

4. Radiation Damage to Silicon Devices

In many of the operating experiments and those in construction, the radiation environment will be such that the properties of the silicon strip detectors will be directly altered by radiation damage. Most radiation issues apply to all types of silicon detectors not just strips.

4.1 Damage from Ionization Energy Loss

The silicon bulk damage from ionization energy loss is negligible (see below), however, surface damage can be significant. The main effect in silicon is increased positive charge accumulation at the oxide interfaces. This increases interstrip capacitance (more noise), can degrade breakdown performance and affects the behaviour of other structures dependent on near-surface effects (back-side isolation strips, biasing structures).

4.2 Non-ionizing Energy Loss (NIEL)

Non-ionizing energy loss in silicon strip devices causes lattice displacement damage of silicon atoms (atomic displacement, nuclear interactions). For ionizing radiation one needs E >250KeV to produce a displacement (low probability) and thus when such a rare event occurs, one typically produces a point defect. For neutrons and charged hadrons, however, damage starts at low energy (E >200eV for a displacement) and can cause a large region of damage. Thus hadron collider (FNAL, RHIC, LHC) and high intensity hadron beam fixed target experiments are most at risk. NIEL damage creates large numbers of new donor and acceptor states. This changes the charge density in the space-charge (depletion) region. The increased numbers of generation-recombination centres leads to increased leakage current. New energy levels are available to trap the carriers leading to lowered mobility.

NIEL damage causes n-type silicon to become less n-type (reduces the negative carrier density) until the substrate undergoes type inversion (becomes p-type), and then becomes increasingly p-type (increases the positive carrier density) with further irradiation. Note that if one starts with p-type silicon, it just becomes more p-type with irradiation. The depletion voltage for the silicon device goes as $1/\rho \propto N_{eff}$, where N_{eff} is the effective doping concentration, the carrier density in the depletion region. The change of N_{eff} as a function of the fluence Φ (integrated dose) is shown in Fig. 4. Therefore the depletion voltage also decreases until inversion and then increases. Note that Φ is usually quoted as the fluence giving a damage equivalent to the fluence of 1 MeV neutrons (in particles per unit area) since particle types have different levels of damage. Thus one must take into account the damage factor corresponding to the type of radiation received.

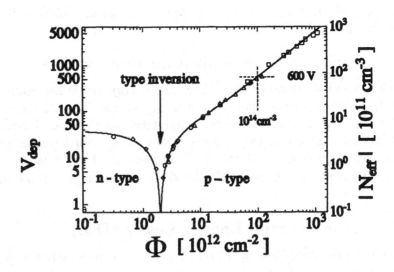

Figure 4. Change of N_{eff} and V_{dep} as a function of fluence (data from Ref. [1]).

An interesting effect observed after irradiation is that the damage is found to diminish, the rate of this "healing" is highly temperature dependent. This effect is called annealing and is in part due to true annealing, the repair of lattice defects. For N_{eff}, two annealing effects are observed: "beneficial" with a short time scale, and "reverse" annealing with a long time scale as shown in Fig. 5. Reverse annealing actually causes the damage to increase. Annealing occurs faster at higher temperatures and can be halted at low temperatures. Annealing effects lead to the following situation for a running silicon detector in LHC: one must keep the detector cold (- 10°C or less) most of the time to avoid reverse annealing. However, one can allow short periods at warmer temperatures periodically to allow for beneficial annealing. Note that these bulk damage effects are only significant at fluences of $> 10^{12}$ n/cm^2.

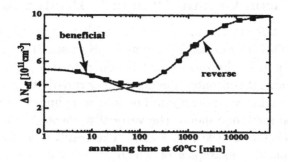

Figure 5. Change of N_{eff} as a function of time after irradiation.

NIEL damage also results in increased leakage current as well as reduced charge collection efficiency. The change of leakage current with fluence is given by $\Delta I_{leak} = \alpha\Phi$, where α is known as the damage constant. For I_{leak} there is only beneficial and no reverse annealing effects. A bad side-effect of the increased leakage current is the danger of thermal runaway. This effect consists of the irradiated device heating up from resistive heating due to the increased leakage current. If this heat cannot be evacuated efficiently, the increased heat will lead to higher leakage current (since it is strongly dependent on the temperature). This can quickly run away uncontrollably. The increased number of trapping states in the depletion region from NIEL damage also leads to a decrease of charge collection. The magnitude of the effect is device dependent but typically can be a 10% loss of signal for $\Phi = 2 \times 10^{14}$ n/cm^2.

In summary, the main radiation damage effects for sensors are an increase of interstrip capacitance (higher noise), an increase of leakage current (higher noise, risk of thermal runaway), a decrease of charge collection (less signal), and a decrease followed by an increase of depletion voltage (possible breakdown). Thus one should choose device parameters appropriately and one must also consider the evolution of the device with radiation.

5.   Silicon Sensor Fabrication

The production steps required to produce a silicon strip sensor[1] are very similar to those used to produce many of the integrated circuits used in everyday electronics (memory chips, processors). For the steps listed below, many are done by separate companies, although in some cases a single company has the in-house capability for all steps.

5.1   Silicon Crystal ("ingot") Production

Silicon crystals are produced starting with very pure quartzite sand, cleaned and further purified by chemical processes. The sand is melted and the phosphorus (boron) dopant is added to make n(p) type silicon (remember, dopant concentration determines resistivity). This is poured in a mold to make a polycrystalline silicon cylinder. Using a single silicon crystal seed, one melts the vertically oriented cylinder onto the seed using RF power (see Fig. 6a,b). The result is a single crystal of silicon or "ingot", examples of which are shown in Fig. 6 c and d. This is known as the float zone (FZ) method. An alternative method (widely used for microelectronics grade silicon wafers) is called the CZ method, invented by Czochralski. Here one "draws" out the single crystal from a crucible of liquid silicon using a seed. However, high resistivity silicon (as is needed for silicon strip sensors) is more reliably obtained with the FZ method. The CZ method can more easily obtain larger diameter ingots. The current largest diameter silicon wafer for the microelectronics industry is now 12 inches (30cm), whereas for high energy physics detectors, the largest wafer size is 6 inches (15cm). Other factors that limit the physics detector wafer size are the device yield (difficult to produce large area, defect-free sensors) and the thickness (large area wafers need to be thicker to survive the processing, whereas thin detectors are preferred for material budget reasons).

---

[1] The term "sensor" will denote a silicon strip device made from a single wafer, as the term "detector" often refers to the ensemble of sensors used in an experiment.

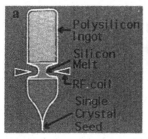

Figure 6. a,b) FZ method for growing a single silicon crystal (ingot). c,d) photographs of ingots. (drawings and photos from Ref. [2])

The ingot is then sliced into wafers of thickness 300-500μm with diamond encrusted wire or disc saws. Lapping (grinding away large imperfections), etching (more removal of impurities and imperfections), and polishing are needed to attain the desired wafer thickness and to ensure a surface with minimal defects.

5.2 Wafer Processing

The following processing steps shown in schematically in Fig. 7 are needed to make the most simple DC-coupled silicon strip detector:

1. Start with n-doped silicon wafer, $\rho \approx$ 1-10 kΩ cm

2. Oxidation at 800 - 1200°C

3. Photolithography (= mask align + photo-resist layer + developing) followed by etching to make windows in oxide

4. Doping by ion implantation (or by diffusion)

5. Annealing (healing of crystal lattice) at 600°C

6. Photolithography followed by Al metallization over implanted strips and over backplane usually by evaporation or sputtering.

For double-sided silicon sensors one must start with wafers polished on both sides. Then the following additional steps are needed:

- Implant n strips instead of a full backplane.

- Add the blocking p-strips.

- Al metallization over implanted back-side read-out strips.

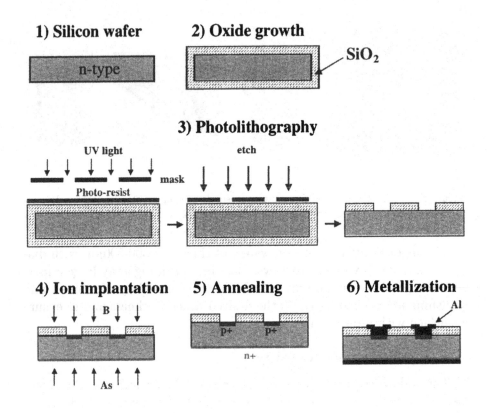

1) Silicon wafer

n-type

2) Oxide growth

SiO$_2$

3) Photolithography

UV light   etch

mask

Photo-resist

4) Ion implantation

B

As

5) Annealing

p+   p+

n+

6) Metallization

Al

Figure 7.   The main processing steps for the simplest single-sided silicon strip device.

The above three implantations each require a set of photolithography and etching steps. If one also wants integrated biasing structures one can choose from one of these three methods (which must be performed on both sides in the case of double-sided devices):

- Polysilicon resistor biasing: add doped polysilicon (non-single crystal) using length to width aspect ratio to get desired resistance.

- "Punch-through" biasing: make a gap between end of p-strip and a p-type implanted bias ring. A voltage difference between two p implants is needed for current flow (typically a few volts). The effective series resistance is very high for low currents.

- FoxFET (Field oxide FET) biasing: same as punch-through but with a gate electrode to control voltage properties of the gap.

If in addition one wants integrated AC coupling capacitors one makes a strip of aluminium over the oxide layer over the p-strip to make the capacitor. It turns out that an oxide thickness of 0.1-0.2μm is required. The same technique can be used on the backside of a double-sided device. It turns out to be difficult to make a perfect oxide insulator over such a large surface. The most common defects are called "pinholes", a short (or low resistive connection) through the oxide. A technique of putting an additional very thin layer of silicon nitride (Si_3N_4) has been used successfully to overcome this problem.

It is sometimes the case that one wants to route the signals differently from the metal lines on or over the strips, for instance to read out the signal of the n-strips of a double-sided sensor which are oriented orthogonal to the p- strips. To route those strips to the same edge of the sensor one could make another metal layer with orthogonally oriented strips above an insulation layer. One then needs "vias", electrical contact holes in the insulation layers just where the appropriate lines cross. This technique is known as a "double metal layer" and is quite achievable in principle. In practice one has to worry about "pinholes", failed vias, the increased capacitance, and cross-talk to all other channels.

5.3   Final Steps

This completes most of the more complex processing steps needed to make a typical high energy physics type silicon sensor. What is left are the final finishing steps:

- **Passivation** is the application of a layer of SiO_2 or other suitable material (polyimide is very common) to the external surfaces. This acts to protect the surfaces that are not needed for electrical contact from physical damage, chemical interactions, and other environmental effects (humidity).

- **Cleaning** is usually performed to remove any residual chemicals left from the processing steps.

- **Testing** is then performed to determine the quality of the devices on the wafer. This is often done prior to cutting out the individual devices because if many devices are made on a single wafer, much time will be spent aligning each device on the probe station. Test structures are often included on the wafer design in order to test specific properties of the processing and design.

- **Dicing** involves cutting the sensor(s) from the wafer using a diamond disc saw. The width of the cut is about 50 microns. A "scribe and break" method has also been used. As dicing should not effect most structures (except for the edges), only limited (e.g. total leakage current) or no further testing is usually performed. A final cleaning is needed after dicing.

6. Construction of Detector Modules

The "module" is the basic building block of a silicon tracking detector. In a modular design one tries to make many identical sub-units so as to be able to use interchangeable components. A typical module consists of a mechanical support structure, the silicon strip sensors, the front-end electronics and the signal routing (see Fig. 8).

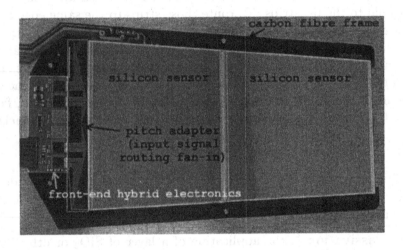

Figure 8. A typical LHC experiment silicon strip detector module (CMS).

Many constraints are often imposed on the module design: low mass (minimize multiple scattering), rigidity, strength, low coefficients of thermal expansion (CTE), good thermal conduction, compactness (restricted space), low cost, radiation hardness, proper functioning over a large temperature range.

6.1 Mechanical Support Structure (Frame)

Exotic materials such as carbon-fibre or graphite composites are often needed to meet the requirements of low mass, high strength, high thermal conductivity, and low CTE. Honeycomb structures and foams are used with the composite sheets to provide rigidity. For applications where the support infrastructure is in the active detection volume (all collider experiments and some fixed target) one needs to minimize the material budget in terms of radiation lengths. Low Z metals (beryllium, aluminium) for beam pipe, support fixtures, thermal contacts and cooling system are used when possible. The various components are usually glued together to avoid metal fasteners.

6.2 Sensor Design Choices

Sensor design must first follow the physics requirements, yet there remain many options: geometrical shape, thickness, read-out and implant pitch, p or n bulk silicon, resistivity, double-sided or single-sided, type of biasing structure, AC or DC coupling, double-metal read-out. In many cases there are conflicting design trade-offs between these choices. One finds that economics (limited budget) often forces the design decision. An example is that a double-sided sensor gives two coordinate position information for half the silicon material budget compared to using single-sided sensors. However, the cost of processing a double-sided sensor is about three times that of a single-sided, thus the total sensor cost can be less for the single-sided choice.

6.3 Module Dront-end Electronics and Connectivity

In a module, often several sensors have their strips connected together in series to make multi-sensor modules. This saves on electronics channels, and has little or no impact on the detector performance if there is still sufficient S/N and the track occupancy is low. To connect sensor to read-out chips, one often uses a "pitch adapter", a fan-out circuit on glass or other substrate which matches the sensor pitch to that of the read-out chip. The connections to and from the pitch adapter as well as those between sensors are done by wire bonding. The front-end read-out electronics usually consist of ASIC (application specific integrated circuits) chips mounted on a hybrid circuit which provides power lines, control signals and read-out connectivity. Modules are typically mounted onto a low-mass structure which must provide good contact to the cooling system as well as a stable and precise positioning. For connection of front-end to back-end electronics (signals, controls, power, monitoring), one usually uses standard copper conductor cables. If this cabling is in the active volume, it can be the largest contribution to the material budget as low noise electronics need good grounding and low impedance power connections implying thick conductors.

7. Front-end Electronics

The final performance result of the silicon detector system is directly linked to the optimal choice of amplifier and read-out method as much as to the sensor design. The read-out electronics is often more difficult to fabricate than the sensors, thus it is of interest to examine in more detail some of the more relevant design choices and methods of implementation of the front-end electronics.

7.1   Choice of Front-end Chip Design

Recall that the signal into the front-end (FE) electronics has a typical duration of about 30ns and a magnitude of about 22500e for one MIP (minimum ionizing particle). Noise contributions come from the total input capacitance, bias resistance and sensor leakage current. A typical amplifier circuit consists of a preamplifier stage followed by an CR-RC shaper which acts as a differentiator followed by an integrator as shown in Fig. 9.

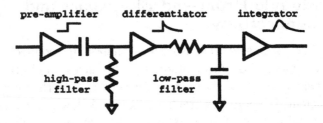

Figure 9.   Schematic of pre-amplifier followed by CR-RC shaper circuit.

The choice of amplifier design will be determined by a number of factors. The bunch crossing rate will determine the differentiation and integration time constants needed. The types and trajectories of particles to be detected (MIP depositions or more?) as well as the type of data transmission and downstream electronics will determine the dynamic range and gains needed. Other choices concerning the design of the front-end chip include the possibility of incorporating multiplexing, pipelining, zero suppression, pedestal subtraction, and waveform processing.

After signal amplification one must choose a read-out architecture: binary (hit / no hit comparator circuit), digital (fast ADC conversion of input charge), or analogue (analogue output proportional to input charge). Each has advantages and disadvantages in terms of ease of fabrication, redundancy, robustness (including radiation hardness), data volume and cost. Advantages of the binary read-out are that it can use a fast bipolar technology and it has minimal data output volume. Disadvantages are that it a non-standard process, is hard to debug and the pulse information is lost. Advantages of the digital read-out are the absence of further S/N loss of analogue signals, pulse information can be kept and the data volume is moderate. Disadvantages are the complexity of the electronics needed, the power consumption and the time needed for conversion to digital. Advantages of analogue are that this involves the least manipulation of the data and pulse information can be available. Disadvantages are the S/N loss of analogue signal transport and the high data volume.

The technological choices in the implementation of front-end chips have been following the rapid changes which are typical of the micro-electronics industry. For example, the recent 0.25 μm CMOS technology (the 0.25 μm refers to the minimum feature size) has been proven to be sufficiently radiation hard for LHC usage, a major impediment in the past for CMOS devices (see Fig. 10). This, coupled with the fact it is a microelectronics industry standard process has made it more than cost competitive with bipolar technology. Still there are trade-offs between CMOS and bipolar in terms of power consumption, speed and manufacturability.

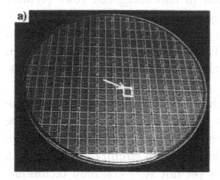

Figure 10. a) Over 400 ASIC front-end chips (for CMS) are fabricated on a single 8 inch wafer using IBMs standard 0.25μm technology. b) The ATLAS front-end hybrid (outlined) mounted directly over one of the two silicon sensors on the detector module.

7.2 Choice of Hybrid Circuit Technology

Standard hybrid circuit technology options include conventional printed circuit boards made on conventional FR4 or G10 glass fibre substrates (cheap, many producers but limited by relatively large feature size), thick film gold on ceramic (expensive but reliable and with small feature size), copper/Kapton flex circuits (smaller feature sizes than with FR4, getting very inexpensive due to consumer electronics), or a mix of Kapton and FR4 (easy to merge these technologies). The hybrid circuit must also adapt to the mechanical and thermal properties of the module as well as the requirement of low material budget which in many cases leads to exotic materials and highly custom designs. Aluminium in place of copper as a conductor or beryllium oxide in place of aluminium oxide as a ceramic substrate can benefit both the thermal performance and reduce the material budget. The ATLAS hybrid circuit (Fig. 10b) is a good example of usage of low mass materials with a Cu/Kapton flex circuit mounted on a carbon composite substrate.

7.3   Connectivity Technology

For fine pitch applications such as connecting sensors to each other and to the FE chips ultrasonic wedge wire bonding is the standard method. It is also usually employed for all connections of the front-end chips and bare die ASICs. A "mature" technology which has been around for about 40 years, this technique uses ultrasonic power to vibrate a needle-like tool on top of the wire. Friction welds the wire to metallized substrate underneath (see Fig. 11). One can easily achieve $80\mu m$ pitch in a single row and $40\mu m$ in two staggered rows (typical FE chip input pitch is $44\mu m$). One generally uses $25\mu m$ diameter aluminium wire and bond to aluminium pads (chips) or gold pads (hybrid substrates). A variant known as gold ball wire bonding is heavily used in industry (PC processors) but usually not with such thin wire or fine pitch.

Other more exotic connection techniques include TAB (tape automated bonding). This is a variant to wire bonding, using the same machines but different tools to make the "bond". In this case a thin metallized Kapton finger (or other thin flexible circuit) is welded to a substrate. This is a fairly new and as yet not very common technique. Another technique known as bump bonding, or "flip-chip", is similar to soldering but with very fine pitch connections. One chip is "welded" to another face to face with points of connections having "bumps" of solder-like metal or conductive adhesive. This technique, shown in Fig. 12, is essential for pixel detector connections where a dense 2-D array of connections is required. It is also used to connect cables directly to chips.

To provide connections between the various types of electronic components, a number of different technologies are available. High density

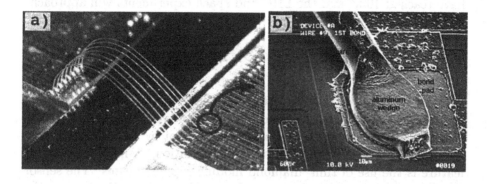

Figure 11. a) Microscope view of a row of 25 μm diameter wire bonds connecting pitch adapter to silicon sensor. b) Electron photomicrograph showing a bond "foot (welded end of wire).

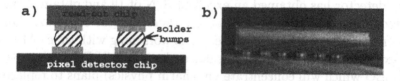

Figure 12. a) Schematic of bump bonding technique. b) Photo of bump bonding between pixel detector chip and read-out chip.

interconnects (HDI) include industry standard and custom cables, usually flexible Kapton/copper with miniature connectors. Soldering is still the standard for surface mount components, packaged chips and some cables. Conductive adhesives are often a viable low temperature alternative, especially for delicate substrates.

8. Silicon Detectors in Experiments

Silicon detectors are in use in nearly all high energy physics experiments today. The experiments at the Tevatron collider, HERA, the B-factories, RHIC, and LHC all contain silicon strip detectors and several have pixel detectors as well. A number of fixed target and space based experiments also are using strip detectors. In addition, drift and pad detectors are in use or are foreseen for many of the RHIC and LHC experiments. The size (in total surface area) of strip detectors has also increased significantly with time. Areas of around 1m^2 were common for the LEP and B-factory experiments but the upgraded FNAL detectors have reached 10m^2 and the LHC and space experiments will approach and exceed 100m^2.

8.1   A Recent Silicon Strip Detector: CLEO at CESR

The CLEO silicon strip detector is a 4-layer device with 61 modules, and 447 silicon sensors. It uses double-sided n-type DC coupled sensors with the p-side giving the z coordinate read-out by means of a double-metal layer. The n-side uses p-strips for isolation. A novel mechanical feature is the use of thin diamond module support beams. The sensor is AC coupled to the read-out and has resistor biasing by means of a separate circuit which also serves as a pitch adapter. A high density Cu/Kapton flex circuit is used to connect the sensors to the AC coupling/biasing circuit. The front-end electronics consists of an amplifier chip followed by a separate ADC chip, hence is an example of digital read-out. A schematic overview of the CLEO module is shown in Fig. 13.

The detector has obtained an excellent S/N of 19 and obtained the expected position resolution. However, there is clear radiation damage occurring in the silicon and the effect is worsening with time. Although the exact nature of the damage is not understood, CLEO-c (the upgrade of CLEO which will concentrate on charm physics) plans to replace the silicon detector with a low mass drift chamber.

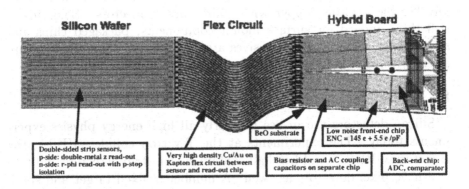

Figure 13.   Schematic of CLEO silicon strip detector module.

8.2 An LHC Silicon Strip Detector: CMS

The initials of the CMS experiment could stand for "Currently the Most Silicon" as it plans to use more than 200m^2 of silicon strip sensors in 10 tracking layers. In addition, inside the strip detector will be 3 layers of the silicon pixel detector. The LHC radiation environment (max $\Phi = 1.6 \cdot 10^{14}$ n/cm^2) governs many of the design choices for the silicon as well as the other components. The need to maximize the silicon lifetime implies a cold (-10°C) environment which imposes further requirements especially on the mechanical design. The CMS silicon strip detector will use simple p-strip on n-bulk single-sided AC coupled sensors with integrated polysilicon bias resistors. The FE chip used is implemented in 0.25μm CMOS technology and is an example of analogue read-out having an analogue pipeline memory and waveform processing. More than 15,000 modules are required and a total of more than 26,000,000 wire bonds will be needed to connect the sensors and electronics. Simplicity of the module design is essential since to construct so many modules will require a maximum of automation, including a robotic module assembly machine for placing and gluing the components of the module (see Fig. 14). Module construction is just beginning and it is expected to take about 2 years to make all the modules.

Figure 14. a) CMS module assembly robot applying glue to carbon fibre frames. b) Measuring silicon sensor placement accuracy on three modules after gluing.

9. Other Types of Silicon Detector

9.1 Silicon Pixel Devices

The benefit of pixel detectors is quite obvious in a high multiplicity environment, the combinatorial problem of coordinate matching in single coordinate devices quickly gets out of hand. The first silicon pixel devices for high energy physics were built from CCD chips. CCDs pixel devices now exist in different varieties but have been joined by two new technologies: the monolithic active pixel sensors (MAPS) and the hybrid active pixel sensors (HAPS). Pixel devices must survive extreme radiation levels in hadron colliders. This is often possible owing to their thin active layer and lower operating voltages. However, in some cases it is expected to replace periodically the pixel detectors damaged by the radiation.

CCD pixel detectors are derived from the CCD technology used for cameras, but they need more active thickness (cameras need just a micron depth to get a good signal). One needs to grow additional layers of silicon on the substrate (called epitaxial layers) with certain properties in order to get deeper charge collection. Still the active depth is usually quite small (typically 15μm) so the ionization signal is small. The charge is kept isolated in the pixel and then shifted along a row to the edge cell. Due to the small signal, the noise must be kept low which can be achieved by cooling to about 200°K. When the charge packets arrive at the edge of the chip, one can buffer and send the signal out to other chips for further manipulation (convert to binary, sparsification, etc). Note that the wells are always alive so you are always subject to collecting charge even if it comes "out of time". Also the clocking speed is limited so one cannot clear out the "event" faster than a certain time. This limitation makes CCDs unsuitable for high rate applications like general purpose vertex detection at LHC. However, for lower rate applications (linear colliders, some fixed target experiments) this is ideal. The SLD silicon pixel vertex detector (Fig. 15) was the first in a collider experiment. It had 20μm x 20μm pixels and achieved about 4μm resolution.

Figure 15. The SLD CCD pixel vertex detector.

Monolithic active pixel sensors (MAPS) use a thin active layer like the CCD (epitaxial layer). Unlike a CCD, this does not shift the charge but collects it directly on a circuit grown on the pixel surface. Only minimal logic is in the pixel, so on-chip digitization and sparsification must be done at the edge of the chip (edge logic). Pixel size can be as small as 20μm x 20μm. This is a fairly new technology under active development.

Hybrid active pixel sensors (HAPS) are basically a strip sensor with the strips further segmented into pixels. In order to read-out the collected charge, the read-out chip must have the same "pixellation" as the sensor. The read-out chip is then bump-bonded onto the pixel sensor, the number of connections (bumps) equal to the number of pixels. The size of the read-out circuit (and likely limitations in power and heat dissipation) limit the pixel size to about 150μm x 150μm or larger. Both CMS and ATLAS are building pixel detectors following the HAPS scheme.

9.2 Silicon Pad Devices

These use the same principle as strip detectors but with wider and shorter strips (see Fig. 16a). Often the biasing and read-out lines are routed to the edge using a double metal layer. A concern is the large input capacitance (high noise) if one uses large pad sizes. However, these devices are often used for calorimeter read-out (e.g. ALEPH Si-W calorimeter) or high accuracy electromagnetic calorimeter pre-shower detectors (CMS) where the signals are very large. Other uses are in situations not needing high precision but having unusual geometries (annular) or environments (high radiation) that may not be possible with other tracking methods (gas detectors). The reduced cost of silicon has made this more attractive.

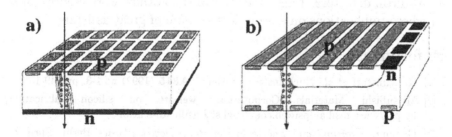

Figure 16. a) Schematic drawing of a generic silicon pad device. b) Schematic drawing of a silicon drift device.

9.3 Silicon Drift Devices

In silicon drift devices (see Fig. 16b) one implants p-strips on both faces and they are put to negative potential. Electrons drift parallel to substrate surface to an n-pad anode. One needs a voltage divider network for the p-strips (like in drift chambers) to provide a uniform drift field. One needs to ensure good material uniformity, low defect rates, good drift field homogeneity, precise voltage dividing on p-strips and good temperature control. Spatial resolutions of 40-50μm from the drift time and 20-30μm from the anode segmentation have been achieved. These devices have not been used much in high energy physics but have recently been implemented for STAR at RHIC and for ALICE at LHC.

10. Conclusions

Silicon strip detectors, based on simple p-n junction diode principle, are now a "mature" technology. Their widespread use and low cost are largely thanks to advances in the microelectronics industry. Their versatility and flexibility allows for many options and design possibilities. Pixel detectors have replaced strips in region closest to interaction point and have become the "hottest" area of development in HEP silicon. Pads devices are still useful in calorimetry and lower resolution tracking. Drift devices are now beginning to be used, we will have to wait and see how well they work. Silicon detectors remain an exciting and interesting field of development and application for high energy physics experiments.

Acknowledgments

I wish to thank Harrison Prosper, Kathy Mork, and all the students for an enjoyable experience at the school. I would also like to thank my colleagues in OPAL and CMS for the opportunity to learn and work with silicon detectors. I am indebted to the authors of references [3] to [8] whose publications on silicon devices were of great assistance.

References

[1] R. Wunstorf *et al.*, Nucl. Instr. and Meth., **A315** (1992) no.1-3, pp.149-155.

[2] Mitsubishi Materials Corporation, website on silicon fabrication, http://www.msil.ab.psiweb.com/english/msilhist0-e.html

[3] G. Lutz, *Semiconductor Radiation Detectors: Device Physics*, Berlin, Springer, 1999.

[4] C. Damerell, *Vertex detectors: The State of the Art and Future Prospects*, 23rd SLAC Summer Institute on Particle Physics: the Top Quark and the Electroweak Interaction, Stanford, CA, USA, 10 - 21 Jul 1995 - SLAC, Stanford, CA, 1997.

[5] A. Peisert, *Silicon Microstrip Detectors*, Instrumentation in High Energy Physics, F.Sauli (ed), World Scientific, (1992).

[6] H. Spieler, *Semiconductor Detectors*, UCB Berkeley Physics 198 course notes, http://www-physics.lbl.gov/~spieler

[7] Vertex 2000 conference proceedings. Nucl. Instr. and Meth., **A473** (2001), No.1 and 2.

[8] S.M. Sze, *Semiconductor Sensors*, New York, Springer, 1994.

GASEOUS DETECTORS: THEN AND NOW

Archana Sharma
CERN
Geneva, Switzerland
Archana.Sharma@cern.ch

Abstract

I give a broad survey of the physics and operation of gas detectors with an emphasis on the practical issues that arise in the design and operation of these most versatile of position-sensitive detectors.

1 Introduction and Historical Overview

The subject of gaseous detectors for detecting ionizing radiation has always been a fascinating one since the last several decades. Radiation is a far-reaching tool to study the fine structure of matter and the forces between elementary particles. Instruments to detect radiation find their applications in several fields of research. To investigate matter, three types of radiation are available: photons, neutrons and charged particles. Each of them has its own way to interact with material, and with the sensitive volume of the detector. The interaction mechanisms have been well studied and are documented in several references [1]. Photons in the meV-eV range permit spectroscopy, i.e. to probe the energy level of atoms, molecules or solids. The penetrating power of X-ray photons is exploited in radiography, tomography and imaging, for example, of tumors, in hospitals and medical or biological studies. In engineering, similar techniques aid in the verification of the quality and composition of materials, for example to detect defects. High energy charged particles are used as projectiles in nuclear and high energy physics (HEP) experiments to investigate the elementary building blocks and forces of nature. The large physics experiments involve very large international collaborations that make use of accelerators to accelerate and bring these particles into collision. Overviews of HEP experiments may be found in numerous references, of which some are quoted here [2].

The first documented gas radiation detector was the gold leaf electroscope, of the eighteenth century: two thin gold leaves inside a volume of air fly away from each other when like charges are placed on

H.B. Prosper and M. Danilov (eds.), Techniques and Concepts of High-Energy Physics XII, 273–339.

them, measuring the total charge deposited on them from the outside. This was followed by the ionization chamber of Rutherford and Geiger which, in essence, was the first proportional counter since it produces a pulse of charge for each incident radiation in the gas inside. Visualization of high density tracks came with the Cloud Chamber in which saturated air with water vapor was contained inside a chamber fitted with a movable piston which, when moved, allows the expansion of the air producing water droplets along the tracks of any ionizing particle passing through the chamber. This technique suffered the drawback that it did not offer any time information about the ionizing radiation.

Charged particle detection was revolutionized by the invention in 1968 of the Multiwire Proportional Chamber (MWPC) by Georges Charpak [3]. Its advanced derivatives are high accuracy drift chambers (DC) and Time Projection Chambers (TPC). Despite their various shortcomings, namely modest rate capabilities and limited two-track resolutions, they have been exploited impressively in a large number of particle physics experiments providing excellent tracking and momentum analysis over the last three decades. Their impact in other fields has been non-negligible, for example in X-ray and medical imaging, UV photon detection, neutron, and crystal diffraction and other material science studies, astronomy, etc. An example of radiography shown in Fig. 1 shows the radiography of Charpak's hand made with a digital x-ray imaging apparatus based on the MWPC [4] called the Siberian Digital Radiography System, installed in several hospitals. A large volume detector with wires oriented towards a narrow fanned beam, it is successfully competing with traditional film X-ray imagers. The device also delivers a considerably reduced dose to patients. More details on medical applications may be found in [5].

In these lectures I will present some basics and state-of-the-art of multiwire proportional chambers, and their derivatives for large area and volume detection of incident radiation. Eventually, their limitations and long term operation will be discussed.[1]

---

[1] The reader may note, since this review is based on lectures, it is a commentary on the figures provided. All further details may be obtained from the quoted literature.

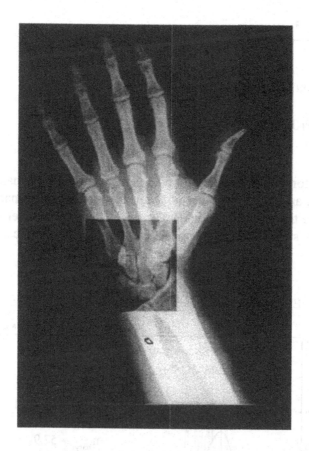

Fig. 1 A radiograph of G. Charpak's hand with the Siberian Digital Radiography System installed in the Hospital St. Vincent, Paris.

2 Single Wire Proportional Chamber (SWPC)

2.2 Primary and Total Ionization

When ionizing radiation passes through a gas filled SWPC, it releases a number, N_p, of electron-ion pairs, called primary ionization, by collisions with the atoms of the gas. Generally, the resulting primary electron will have sufficient kinetic energy to ionize other atoms. The total number of created electron–ion pairs, N_t, can be much higher than the primary ionization.

The actual number of primary electron-ion pairs is Poisson distributed

276

$$P(m) = \frac{\bar{n}^m e^{-\bar{n}}}{m!}$$

The detection efficiency is hence limited to

$$\varepsilon_{det} = (1 - P(0)) = 1 - e^{-\bar{n}}$$

Therefore, for thin layers this may be significantly lower than unity. For example, in an Argon gas layer of 1mm, the number of primary electrons is 2.3, hence the detection efficiency is 0.92. The total number of electron-ion pairs is subject to large, non-Poissonian fluctuations called Landau tails.

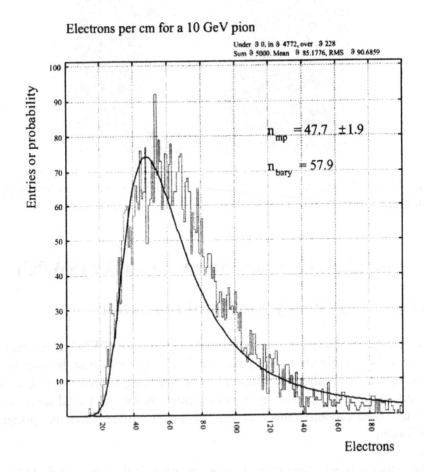

Fig. 2

Fig. 2 shows the number of electrons per cm in propane, most probable being ~ 48 for this gas. The Poissonian nature of the distribution is well reproduced by measurement as well as simulation, the most probable value being 48 for this gas. GARFIELD [6] is a simulation program for studying gas detectors. In this program the sub program HEED [7] takes care of the energy loss and cluster production; this plot has been made using this program.

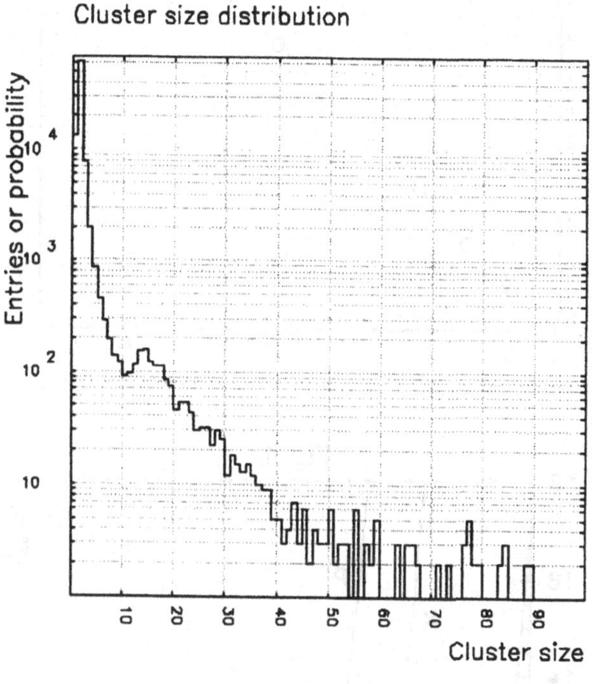

Fig. 3

The cluster size distribution shown in Fig. 3 for propane, after the rapid initial drop, and the bumps that correspond to interactions in the K, L, M etc shells and for large cluster sizes, an approximately $1/n^2$ drop. These large clusters are in fact delta electrons and Heed is able to reproduce their spatial extent. The most probable energy loss depends on the ambient conditions. Fig. 4 shows its dependence on pressure.

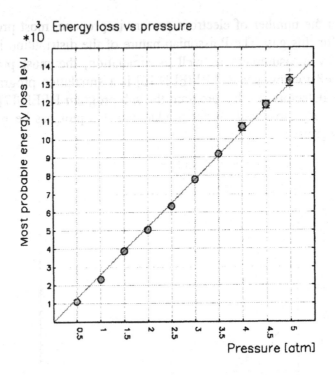

Fig. 4

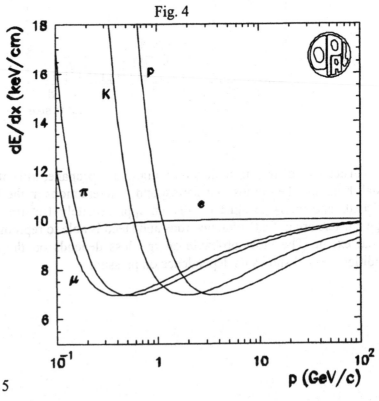

Fig. 5

Table 1

Gas	Ex [eV]	Ei [eV]	wi [eV]	(dE/dX) mip [keV/cm]	N_p (# of primary electrons) [1/cm]	N_t (# of total electrons) [1/cm]	Radiation Length [m]
He	19.8	24.5	41	0.32	4.2	8	5299
Ar	11.6	15.7	26	2.44	23	94	110
Ne	16.67	21.56	36.3	1.56	12	43	345
Xe	8.4	12.1	22	6.76	44	307	15
CF_4	12.5	15.9	54	7	51	100	92.4
DME	6.4	10.0	23.9	3.9	55	160	222
CO_2	5.2	13.7	33	3.01	35.5	91	183
CH_4	9.8	15.2	28	1.48	25	53	646
C_2H_6	8.7	11.7	27	1.15	41	111	340
$i-C_4H_{10}$	6.5	10.6	23	5.93	84	195	169

The detectors capable of identifying primary clusters, typically provide a better energy resolution. Energy loss is also a technique for particle identification, as exemplified in a measurement from Ref. [8], shown in Fig 5. Table 1 gives the basic properties of some typical wire chamber gases obtained from various sources; see bibliography.

Even 100 electron-ion pairs are not easy to detect since, typically, the noise of an amplifier is equivalent to ~ 1000 electrons. Therefore, one needs to increase the number of electron-ion pairs in order to increase the signal to noise ratio. This is achieved by "gas amplification."

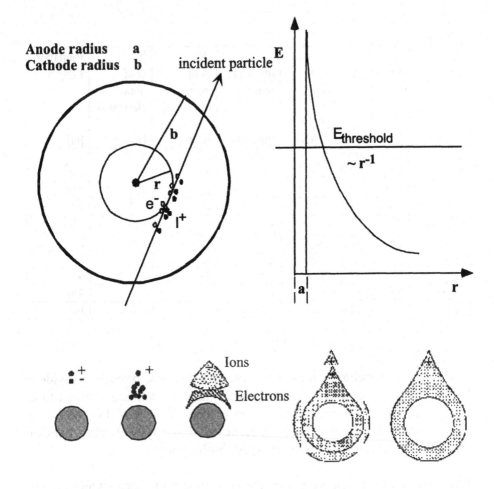

$$E(r) = \frac{CV_0}{2\pi\varepsilon_0}\frac{1}{r} \qquad V(r) = \frac{CV_0}{2\pi\varepsilon_0}ln\frac{r}{a} \qquad V_0 = V(b)$$

$$C = \frac{2\pi\varepsilon_0}{ln\dfrac{b}{a}} \qquad V(a) = 0$$

Fig. 6 Schematics of the operation of a Single Wire Proportional Chamber [9]. For a counter with wire radius a =10 µm, cathode radius b =10 mm: $C \approx 8$ pF/m.

Principle of Operation – Gas Amplification

Figure 6 shows a schematic of the operation of a single wire proportional counter. In the cylindrical field geometry the electrons drift towards the anode wire in an increasing electric field. Close to the anode wire, the field, $E_{threshold}$, is sufficiently high (several kV/cm) to result in further ionization by electrons released in the gas volume. The number N at a given time will be given by:

$$N = N_0 e^{\alpha E(x)}$$

Where α is the First Townsend Coefficient, that is, the number of ionizing collisions (number of electron-ions pairs created) per cm of drift, and is given by

$$\alpha = \frac{1}{\lambda}$$

where λ is the mean free path between collisions. For low energies,

$$\alpha \approx kN\varepsilon$$

where ε is the characteristic energy of the electron. The multiplication factor or gain may be given as

$$M = \frac{N}{N_0} = \exp\left[\int_a^{r_c} \alpha(r)dr\right]$$

More details on calculations and parameterization of gain may be found references [10].

Figure 7 shows the cross section of different kinds of collision that an electron can experience in argon. One sees that the energy dissipation is mainly by ionization; noble gases have high specific ionization. De-excitation of noble gas atoms is only possible via emission of photons. In argon, the photons are 11.6 eV, which is above the ionization threshold of metals, for example in Cu it is 7.7 eV. Photo-extraction from the cathode results in new avalanches (Malter Effect) and eventually a state of permanent discharge is reached.

One solution is to add polyatomic gases as quenchers. Figure 8 shows the cross section of electrons in carbon dioxide. There are many vibrational and rotational energy levels, thus facilitating the absorption of photons over a large energy range. Energy dissipation is also possible by dissociation into smaller molecules. In methane for example the absorption band is 7.9 – 14.5 eV.

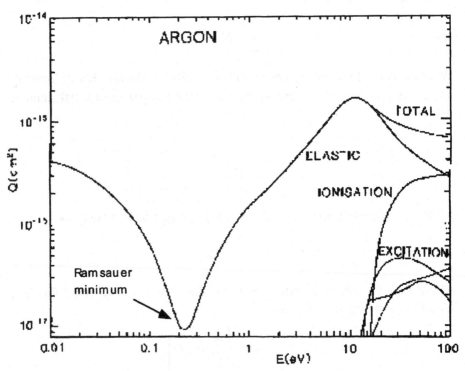

Fig. 7 Cross Section of electron collisions in argon

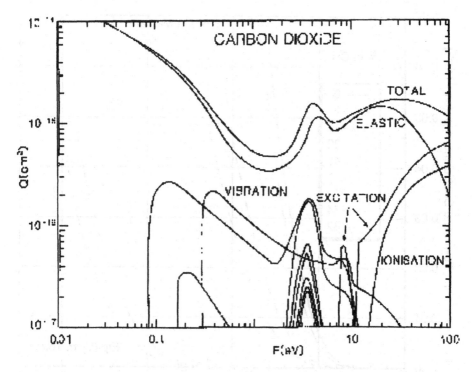

Fig. 8 Cross Section of electron collisions in carbon-dioxide

284

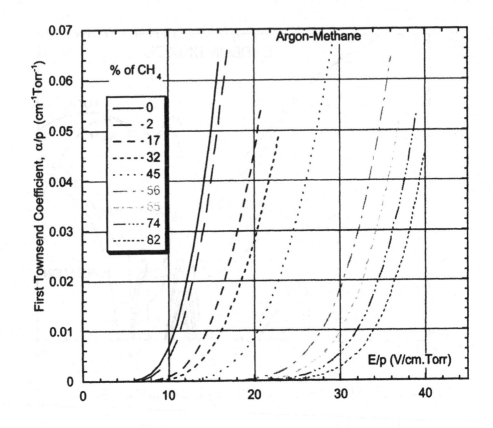

Fig. 9 Townsend Coefficient α, measured for argon-methane mixtures [11].

Figure 9 shows the measured values of the Townsend Coefficient in some argon-methane gas mixtures. The present version[2] of Garfield has incorporated in it a Monte Carlo version of the electron transport program MAGBOLTZ (NMONTE, IMONTE) which gives good estimates of the Townsend coefficient. As sketched in Fig. 6 the electrons advancing towards the anodes multiply and within a few wire radii form an avalanche within < 1ns. A signal is induced on both the anode and cathode due to the moving charges, both electrons and ions, their speeds determined by the local electric fields:

[2] GARFIELD v 8.33 Nov. 1999

$$dv = \frac{Q}{lCV_0}\frac{dV}{dr}dr$$

Electrons are collected by the anode wire and dr is very small ~ few μm; hence electrons contribute very little to the signal, with a time constant such that it is impossible to have only an electron signal. The ions have to drift back to the cathode, since most of them are created in the avalanche close to the anode, hence dr is big, and the major contribution to the signal is from the ions as sketched in Fig. 10. The signal duration being limited by total ion drift time, electronic signal differentiation is needed to limit dead time.

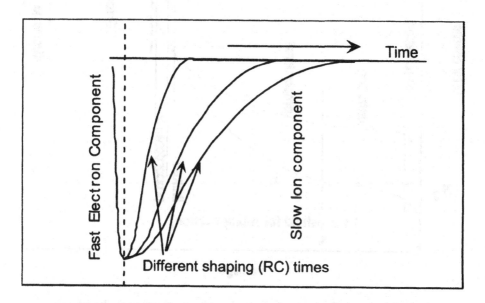

Fig. 10 Schematic of electron and ion component of signal

Modes of Operation

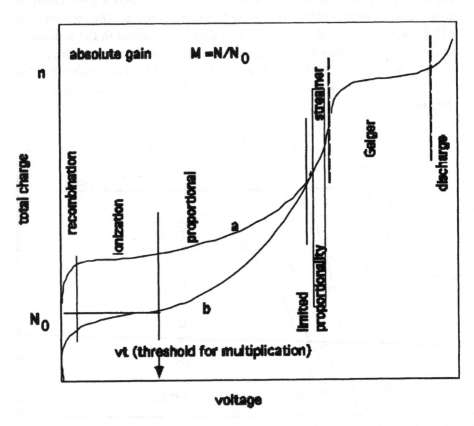

Fig. 11 shows the operation modes of a proportional chamber.

As the voltage on the anode wire is increased from zero, various modes of operation of the wire chamber are encountered. Referring to Fig. 11, zone I is the recombination mode where the field is not high enough for the primary electrons to drift to the anode, and they are lost by recombination to the positive ions. The second region (II) is the ionization mode where the field is high enough for the electrons to reach the anode, but still not enough to multiply. Nevertheless, all of them are collected.

Region III is called the proportional mode region; here, above the critical or threshold voltage multiplication starts and the detected signal is proportional to the original ionization, thus the chamber can be used for an energy measurement (dE/dx). However, the secondary avalanches have to be quenched since gains can reach up to 10^5 in the proportional mode.

Increasing the voltage further one encounters the region of limited proportionality (IV) where multiplication reaches saturation. The gain no longer increases with voltage but the avalanche transits into a streamer as shown in Fig.11. In this region of operation there is a strong probability of photoemission from the avalanches where secondary avalanches merge with the original avalanche creating a streamer (V). This mode of operation has the advantage of very large gains (10^{10}), hence simple electronics. Nevertheless, the gas mixtures require a strong quencher as one of its components. This mode of operation has been applied in several experiments [12] where cheap large area coverage is needed.

Increasing the voltage still further one reaches the Geiger or discharge mode which is dominated by tremendous amount of photoemission where the full length of the anode wire is arrested by the avalanche. To operate in this mode the HV needs to be pulsed and very strong quenchers are needed.

Energy Resolution and Escape Peak

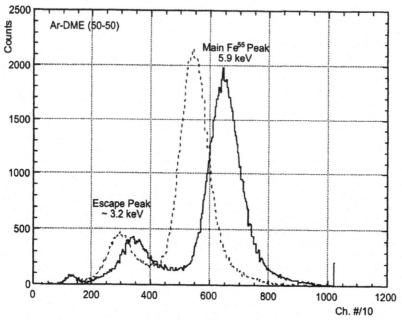

Fig. 12

288

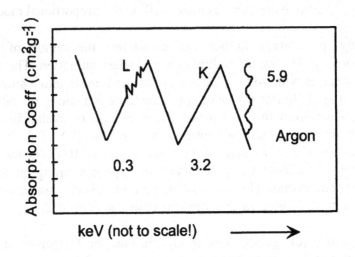

Fig. 13 (a) Absorption in Argon

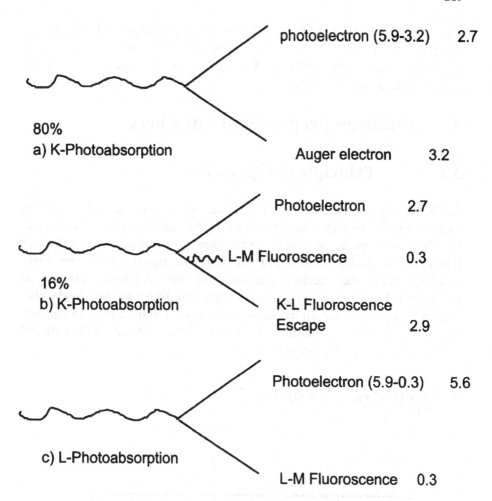

Fig. 13 (b) Main and escape peaks in argon mixtures

A typical measurement of pulse height spectra from an Fe[55] source is shown in Fig. 12, where the main peak corresponds to 5.9 keV, accompanied by a small escape peak. Increasing the voltage, the peak shifts in proportion to the gain achieved as a result of the increased electric field in the counter.

Figures 13 (a) and (b) show a schematic of the formation of the main and escape peaks in argon mixtures. The sum of a and b are responsible for the main 5.9 keV peak while the K-L Fluorescence is responsible for the escape peak, which is ~ 15%.

3 Multiwire Proportional Chambers

3.1 Principle of Operation

An MWPC is sketched in Fig. 14. It consists of a plane of sense wires running parallel to each other enveloped by two cathode planes, which can also be a wire plane or simple metal sheet. The capacitive coupling of non-screened parallel wires and the negative signal on all the wires resulting from the electron motion and the avalanche process is compensated by the positive signal induction from the ion avalanche. Typical parameters are an anode spacing s, of 1 mm, anode to cathode distance of 5 mm, and diameter of anode wires ~ 20 μm. With a digital read-out the spatial resolution is limited to

$$\sigma_x \approx \frac{s}{\sqrt{12}}\,(s = 1mm, \sigma_x = 300\,\mu m)$$

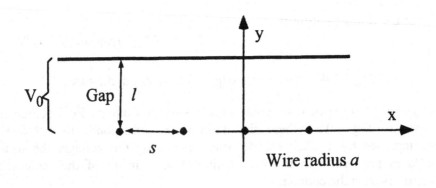

Fig. 14

Electric field and potential in the MWPC are:

$$E(x,y) = \frac{CV_0}{2\varepsilon_0 s}\left\{1 + \tan^2\frac{\pi x}{s}\tanh^2\frac{\pi y}{s}\right\}^{\frac{1}{2}}\left\{\tan^2\frac{\pi x}{s} + \tanh^2\frac{\pi y}{s}\right\}^{-\frac{1}{2}}$$

$$V(x,y) = \frac{CV_0}{4\pi\varepsilon_0}\left\{\frac{2\pi l}{s} - \ln\left[4\left(\sin^2\frac{\pi x}{s} + \sinh^2\frac{\pi y}{s}\right)\right]\right\}$$

Capacity per unit length: $\qquad C = \dfrac{2\pi\varepsilon_0}{\dfrac{\pi l}{s} - \ln\dfrac{2\pi a}{s}} \qquad V(a)=V_0 \qquad V(l)=0$

Along the axis (centered on the anode):

$$E_y = E(0,y) = \frac{CV_0}{2\varepsilon_0 s}\coth\frac{\pi y}{s} \qquad E_x = E(x,0) = \frac{CV_0}{2\varepsilon_0 s}\cot\frac{\pi x}{s}$$

Since a<<s, $\qquad C < \dfrac{2\varepsilon_0 s}{l}$

For y<<s: $\quad E(x,y) \cong \dfrac{CV_0}{2\varepsilon_0}\dfrac{1}{r} \qquad r = \sqrt{x^2 + y^2}$ (cylindrical counter)

For y>s: $\qquad \coth\dfrac{\pi y}{s} \cong 1 \quad E_y \cong \dfrac{CV_0}{2\varepsilon_0 s}$

For a parallel plate chamber of same geometry:

$$E_{//} = \frac{V_0}{l} = \frac{C_{//}V_0}{2\varepsilon_0 s} \qquad E_y = \frac{C}{C_{//}}\frac{V_0}{l}$$

Typical values of the capacitance per unit length C (in pF/m):

l (mm)	$2a$ (μm)	s(mm)		
		1	2	5
4	20	3.6	5.7	8.1
8	20	2.0	3.5	5.9

Angular distribution of avalanche in a MWPC and charge induction

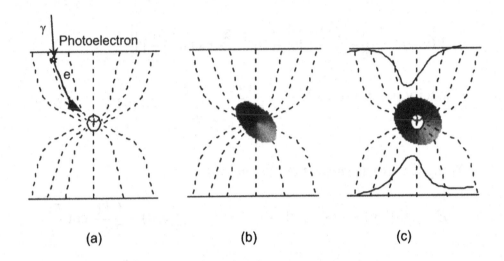

(a)   (b)   (c)

Fig.15

The avalanche process is not uniformly distributed around the circumference of the anode. As illustrated in Fig.15 it is concentrated towards the direction where the primary ionization has taken place. This asymmetry results in the angularly localized positive ion distribution, the angular charge density being more concentrated in the direction of the primary ionization.

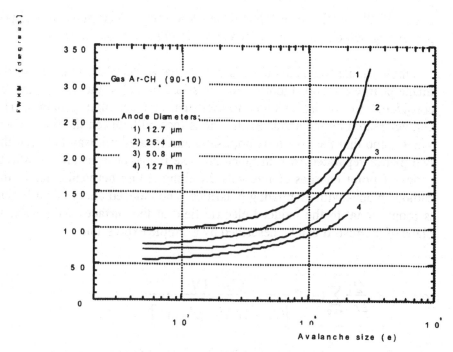

Fig. 16 Avalanche Asymmetry [13]

The motion of the positive ions induces charge on both cathode planes and the asymmetry in the ion distribution causes changes in the centroid and the width of the induced charge distribution. The angular distribution of the avalanche is dependent on the anode wire diameter, its voltage, i.e. avalanche size, gas mixture and the type of radiation. Figure 16 shows some results of the angular distribution of the avalanche around the anode wire in an argon-methane mixture with different anode wire diameters [13]. The measurements were performed with an azimuthally segmented cylindrical proportional chamber.

3.2   Performance of a Position Sensitive MWPC

Position resolution is defined as the standard deviation (sigma) or the full width at half maximum (FWHM) of the reconstructed position distribution for a given incident particle or photon, the position of which is a delta function. The position resolution of a detector system is determined by many factors, such as the signal to noise ratio, the physical spread of the primary ionization, the spread of the avalanche and the width and incident angle of the track. It is also depends on the electronics and signal processing, which however are beyond the scope of these lectures.

A simple model to calculate the induced charge distribution is to use the image charge method. Ignoring the presence of wires, the system can be considered to consist of two infinitely parallel conducting planes. The charge created by the anode avalanche is located between the two planes. We assume that the charge is point-like and located midway between the two planes separated by a distance 2h. The image charges are an infinite series of point charges at intervals 2h along a line perpendicular to the cathode planes with alternating polarities. The induced charge distribution is proportional to the induced electric field at the surface of the cathode plane of interest. It can be described by a simple expression:

$$\rho(x,y) = \frac{-Q_A}{2\pi} \sum_{n=0}^{\infty} (-1)^n \frac{(2n+1)l}{\left[(2n+1)^2 h^2 + x^2 + y^2\right]^{\frac{3}{2}}}$$

where Q_A is the avalanche size. The charge distribution along any one dimension can be derived by integrating the above equation over the other dimension:

$$\rho(x) = \int_{-\infty}^{\infty} \rho(x,y)dy$$

A single parameter formula is given here, see [14-17] for details:

$$\frac{\rho(\lambda)}{Q_A} = K_1 \frac{1 - \tanh^2(K_2\lambda)}{1 + K_3 \tanh^2(K_2\lambda)}$$

where

$$K_1 = \frac{K_2\sqrt{K_3}}{4\tan^{-1}\sqrt{K_3}}$$

$$K2 = \frac{\pi}{2}\left(1 - \frac{1}{2}\sqrt{K_3}\right)$$

K_3 is a parameter, the value of which depends on the geometry of the chamber, namely the anode-cathode gap and inter-anode gap. This is important for designing readout chambers for large detectors, as we will see in a later section.

3.3   Read-out Methods of the MWPC

A single plane of a conventional MWPC gives positional information of the incident particle in one direction, and the position resolution is given by the anode wire pitch (σ=s/sqrt12). An interpolating method is one in which signals from a few read-out channels (adjacent pads) can be processed to obtain position information much finer than the read-out spacing. These methods have been used in several detectors, and they are mainly of three kinds: resistive charge division, geometrical charge division, and capacitive charge division

3.3.1   Charge Division with a Resistive Anode Wire

Resistive electrodes have been used in many forms in radiation detectors. The most commonly used position sensitive detectors use the charge division method. Usually a position resolution σ of 0.4 % of the anode wire length is achieved. The charge division with resistive electrodes is independent of the capacitance and resistance of the electrode. It is well described in the Refs. [15].

3.3.2   Resistively Coupled Cathode Strips and Wires

Another charge division scheme uses the induced charge on the cathode of an MWPC. As illustrated in Fig.17, one or both cathode planes can be subdivided into strips interconnected by resistors. Certain strips at regular intervals are connected to charge sensitive preamplifiers. The centroid of the induced charge on the cathode (in the direction across the strips) can be evaluated by using signals only from a few read-out channels. This method equally applies to the special cases of cathode strips individual or a group of interconnected cathode wires. The position resolution of such a system is primarily determined by the signal/noise ratio (s/n), and the read-out spacing. An optimal performance in position resolution with a minimum number of electronic channels is achieved by the subdivided charge division method. Typically, no more than three read-out channels are used to perform centroid computation to minimize the noise contribution from the preamplifiers.

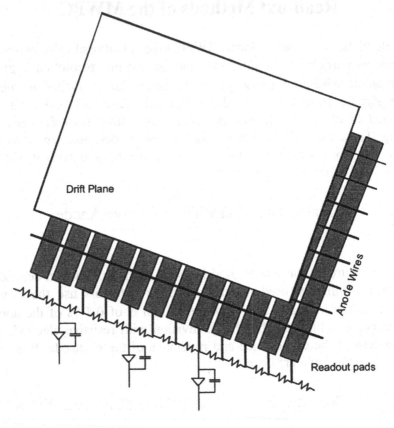

Drift Plane

Anode Wires

Readout pads

Fig. 17

The position linearity of the system is determined by the uniformity of the resistance, width of the individual strip, and matching of time constants of the electrode and the electronics. Because the centroid finding is performed locally, the position resolution is independent of the dimensions of the detector. The induced charge on the cathode plane from an avalanche is spread over a localized area. The amount of charge induced depends on the position of the avalanche, which will therefore be determined with good accuracy.

3.3.3   Strip Cathode

The simplest geometry uses cathode strips adjacent to each other with strips widths slightly less than the anode cathode spacing to sample the induced charge. Each of the strips can be directly connected to a preamplifier; closely spaced wires can be substituted as illustrated in Fig. 18. With an inter-anode spacing of 1 mm, σ_x=300 μm. On the other hand with a center of gravity method the position resolution can be of the order of 100 μm.

The second co-ordinate may be obtained in several ways. Using crossed or small angle stereo wire-planes can give the second co-ordinate, though this is limited to low multiplicity of primary ionizations. Charge division with resistive wires and reading both ends of the anode wires give accuracies up to 0.4% of its length. The timing difference between the two ends of the anode wire can also be measured and a timing resolution $\sigma(dt)$ ~ 100ps gives measurement accuracies $\sigma(y)$ ~ 4cm over a few meters. Coarse measurements over long distances [16] have been made using these techniques. The analog read-out of a wire plane combined with one or two segmented (stripped) cathode planes afford very high accuracy tracking with MWPCs.

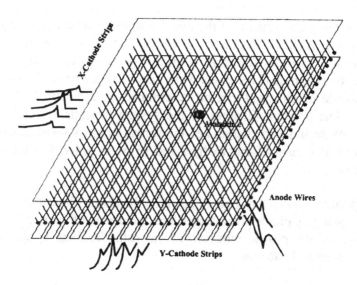

Fig. 18 High accuracy two-dimensional information from a multiwire proportional chamber.

3.3.4   Wedges and Strip Cathode

The wedge and strip cathode is a classical example of the geometrical charge division method [17] and has evolved into several other variations.

As shown in Fig. 19, electrodes A and B are wedge shaped. Their width varies linearly along the y- direction.

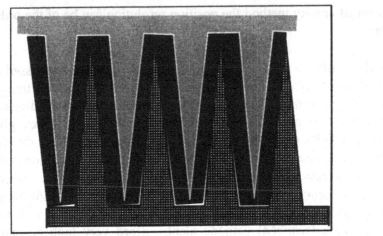

Fig. 19

Therefore the y co-ordinate of the event centroid can be determined from the ratio A-B/A+B or A/A+B , where A and B are the electrical signals collected on electrodes A and B, respectively. Along the x-direction, the width of the electrode C increases while that of D decreases. Therefore the x-co-ordinate of an event can be determined by the ratio C-D/C+D or C/C+D. The position linearity of the system depends critically on the footprint of the induced charge, i.e., the spread of the induced charge must cover several groups of the wedges and strips in order to achieve good position linearity.

It is important to keep the spacing between the anode plane and the wedge and strip pattern cathode plane larger than the period of the pattern, so that the image of the charge induced on the cathode plane is spread over several wedge and patterns.

This is a true two-dimensional interpolating method. However, the reconstructed position in the direction across the anode wires suffers large modulations at a period of the anode wire spacing. This is due to the design of the MWPC in which the avalanches are localized near the anode wires. This limits the position resolution to the anode pitch in the direction across the anode wires.

The large electrode capacitance limits the maximum dimension of these electrodes due to the electronic noise. The minimum size of such electrodes is also limited by etching techniques.

3.3.5   Chevron Pads

Recently a lot of progress has been made in the so-called Chevron pads, sketched in Fig. 20, both in the manufacturing techniques and in the read-out algorithms [18].

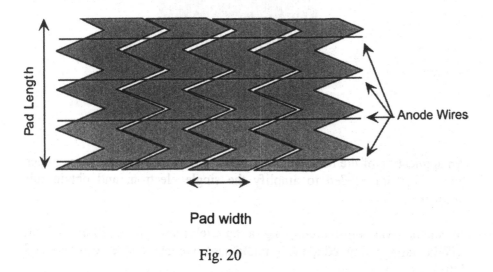

Fig. 20

3.4   Simple Derivatives of the MWPC

The multistep chamber (MSC)

Photon feedback being a nasty problem in the stable operation of gaseous detectors with high gain, an important breakthrough came with the introduction of two step devices [19]. A derivative of the MWPC, the MSC separates the amplification into parts as shown schematically in Fig. 21 with a low field transfer region, such that an overall high gain is obtained in two sub-critical steps. Total gains of $\sim 10^6$ to 10^8 have been achieved by reducing the photon feedback problem. The electron transparency of wire meshes depends on the field ratio E_D/E_M.

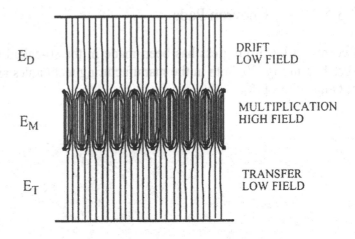

E_D DRIFT
LOW FIELD

E_M MULTIPLICATION
HIGH FIELD

E_T TRANSFER
LOW FIELD

Fig. 21

An application of the MSC is single photo electron detection where rather high gains are needed to amplify the single electron, and obtain full efficiency.

In another variation, combining a parallel plate preamplifier and an MWPC large combined gains in photo-sensitive gases have been obtained [20].

Thin Gap Chambers are MWPCs with specialized cathode planes of given resistivities developed to be very fast with ~ 2 ns rise time, and large signals 10^6 gain and robust. [21]. Some variations, e.g., honeycomb chambers, may be found in literature [22].

4 Drift and Diffusion of Charges in Gases Under the Influence of an Electric Field

Drift and diffusion has been studied in great detail, both through measurement [23] and simulation [24], by several workers in the field.

4.1 Magnetic Field, B=0

In the absence of any external field, the electrons and ions will lose their energy through collisions with gas atoms, eventually reaching thermalization with

$$\varepsilon = \frac{3}{2}kT \approx 40meV$$

An originally localized ensemble of electrons/ions will diffuse, undergoing multiple collisions with a diffusion coefficient D and σ_x:

$$\frac{dN}{N} = \frac{1}{\sqrt{4\pi DT}} e^{(x^2/4Dt)} dx \qquad D = \frac{\sigma_x^2(t)}{2t} \qquad \sigma_x = \sqrt{2Dt}$$

LONGITUDINAL AND TRANSVERSE DIFFUSION

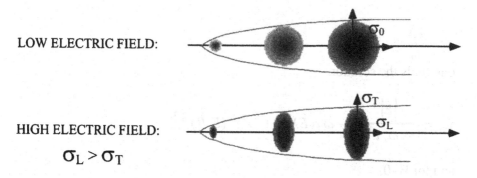

LOW ELECTRIC FIELD:

σ_0

HIGH ELECTRIC FIELD:

$\sigma_L > \sigma_T$

σ_T

σ_L

Fig. 22 Schematic of the transverse and longitudinal diffusion of the electrons [9]

Figure 22 sketches the movement of electrons in low and high electric field and the difference in the transverse and longitudinal components of diffusion. In the presence of electric and magnetic fields the equation of motion is

$$m\frac{d\vec{v}}{dt} = e\vec{E} + e(\vec{v}x\vec{B}) + \vec{Q}(t)$$

Q(t) represents a stochastic force resulting from the multiple scattering. We are interested only in the time averaged solution, namely, velocity v ~ constant. Q(t) can then be replaced by a friction term

$$\left(\frac{d\vec{v}}{dt}\right) = 0 = e\vec{E} + e(\vec{v}x\vec{B}) - \frac{m}{\tau}\vec{v}_D$$

$$\rightarrow$$

$$\frac{m}{\tau}\vec{v}_D$$

With mobility, μ, defined as

$$\mu = \frac{e\tau}{m}$$

and cylcotron frequency ω

$$\omega = \frac{e\vec{B}}{m}$$

one finds the solution:

$$\vec{v}_D = \frac{\mu|\vec{E}|}{1+\omega^2\tau^2}\left[\vec{E} + \omega\tau(\vec{E}x\vec{B}) + \omega^2\tau^2(\vec{E}.\vec{B})\vec{B}\right]$$

and for B=0,

$$\vec{v}_D = \mu\vec{E}$$

For the case when B is not zero, the drift velocity has three components, one parallel to the E-field, another parallel to the B-field and one parallel to the ExB component. In case E and B are not perpendicular there can be two scenarios: If the electrons in the drift gas have $\omega\tau \ll 1$, that is, either they experience a small B field or a have short collision time, they will follow the E-field. If, however, $\omega t \gg 1$ namely, large B field or longer time between collisions, they will follow the B-field. The angle that the electron makes relative to the direction of the E field, in non-parallel E and B fields, is called the Lorentz angle, α, and may be approximated by

$$\tan \alpha = \omega\tau$$

Drift velocity in Ar/DME mixtures

Drift velocity in Ar/DME mixtures

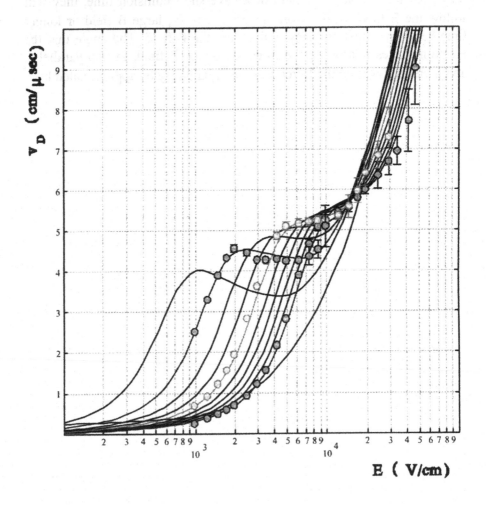

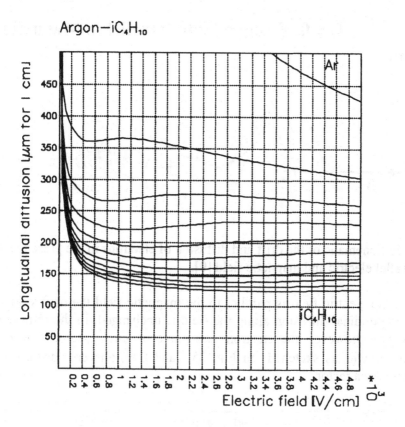

Fig. 24 Longitudinal Diffusion in Argon-Isobutane gas mixtures moving in percentages by 10.

Mobilities of Ions

Tables and compilations exist in several references [25], the important point being that they are rather slow as compared to electrons (~1000 times). Most "foreign" ions, i.e. ions of other molecules than Argon, usually the noble component of the operational gas mixture, move faster than singly ionised Argon ions do. Lighter ions accelerate quickly and tend to scatter with minor loss of energy, Heavier ions need more time to accelerate, but are less deviated in collisions with the drift gas. We will return to this issue when encountering drift of ions in a TPC.

4.2   B ≠ 0, Electric Field Parallel to Magnetic Field

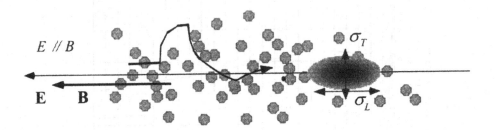

Fig. 25 Schematic of the drift and diffusion of electrons in the presence of a parallel electric and magnetic field [9].

When the electric field is parallel to the magnetic field the drift velocity and longitudinal diffusion along the field is unchanged. In the transverse projection, however, the electrons are move on circle segments with a radius of v/ω, as sketched in Fig. 25, and the transverse diffusion is reduced by a factor

$$\sqrt{1+\omega^2\tau^2}$$

This fact has been used in designing Time Projection Chambers (TPCs) to improve measurement accuracies over large distances.

Figure 26 shows the drift velocity, transverse and longitudinal diffusion per cm of drift with and without magnetic field. Typical drift velocities are $\sim$ 5cm/µs while the ion drift velocities are almost 1000 times smaller. Compilations of transport parameters are available in several references [11].

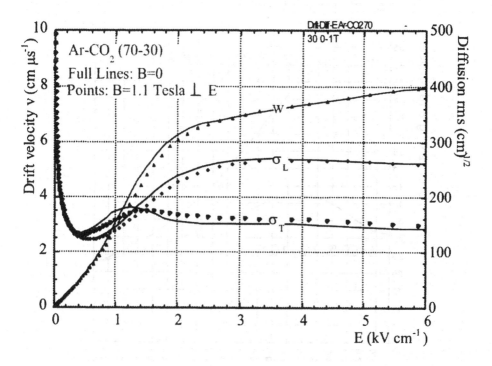

Fig. 26 Transport parameters of argon-carbon dioxide

The diffusion coefficient is in general a function of the electron energy and hence of the applied external fields. In "cool" gases like carbon dioxide and di-methyl ether (DME), electrons stay thermal up to high electric fields (~2 kV/cm), and one expects a small diffusion, while in "hot" gases, for example argon, the electrons become non-thermal at relatively low electric fields (~ few V/cm), which results in large non-isotropic diffusion.

For cool gases the drift velocities are small, and hence the Lorentz angle too is small. (Compare results for the two gases shown in Fig. 27.) The Lorentz angles for four different values of magnetic field perpendicular to the electric field in a detector are shown for a typical Ar-CO_2 gas mixture in Fig. 28.

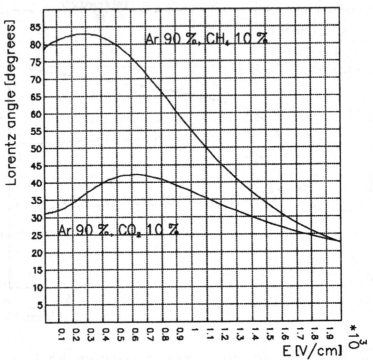

Fig. 27. Lorentz angle at B=1 T for Ar-CO2(90-10) and Ar-CH4 (90-10)

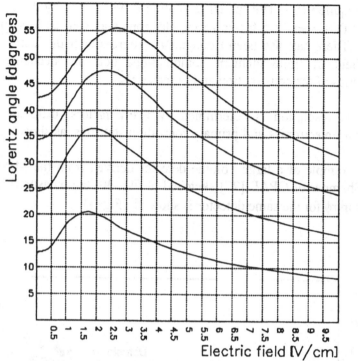

Fig. 28. Lorentz angles for Ar_CO2 at different magnetic fields.

5   Large Volume Tracking

5.1   Drift Chambers (DC)

High Accuracy Drift Chambers [26] have been built and used by several experiments, in a simple extrapolation of MWPCs by measuring the arrival time of electrons at the sense wire relative to a start time t_0. In these chambers the high field regions of the anodes is interspersed with low field regions, by introducing field wires in between the anodes. This allows the electrons to drift to nearby anodes in a relatively low field, chosen to be uniform. Anodes typically consist of doublets such that the left-right ambiguity can be solved.

To design drift chambers capable of operating over large areas in big experiments, the geometry is optimized so that a uniform electric field results, thus yielding little or no dependence of the drift velocity. This results in linear space-time relations. Gas mixtures are chosen conforming to these requirements. The spatial resolution is not limited by the cell size as in MWPCs, there are fewer wires, electronics as well as less support structure. There has also been tremendous progress over the last decades in the field of finite element computations both in mechanics and electrostatics of the construction and in materials.

The spatial resolution is determined by many factors: diffusion, path fluctuations, electronics and primary ionization statistics, as shown in Fig. 29.
Another example of a large volume detector derived from the MWPC is the Jet Chamber of OPAL [27] sketched in Fig. 30; the measured spatial accuracy in rφ for the same detector is shown in Fig. 31.

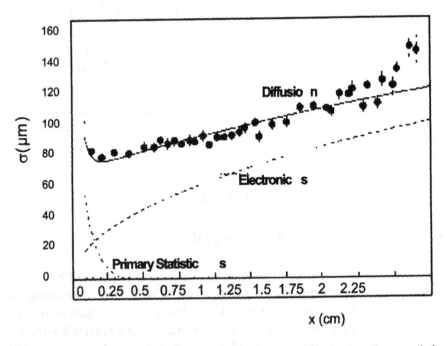

Fig. 29 Measurement of spatial accuracy in a drift chamber and its components [26].

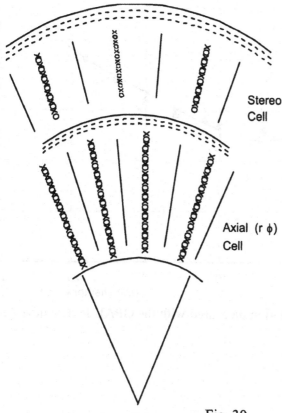

Stereo
Cell

Axial (r φ)
Cell

Fig. 30

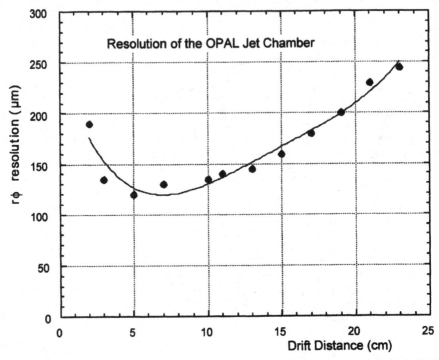

Fig. 31 Spatial (rφ)resolution measured with the OPAL Jet Chamber [27]

ExB Effect

This effect consists of the twisting of electron drift lines from their straight paths towards the anode. This effect is more pronounced at the cathode plane since the field near the cathode wires is lower as compared with the anode, and as one sees from Fig. 27 the Lorentz angles are rather large at lower electric fields [28]. This effect is visible in the plot below, Fig. 32, where maximum distortion in the electron drift lines is apparent at the cathode grid (above–thicker wires) while no effect is seen at the anode (below- thinner wire).

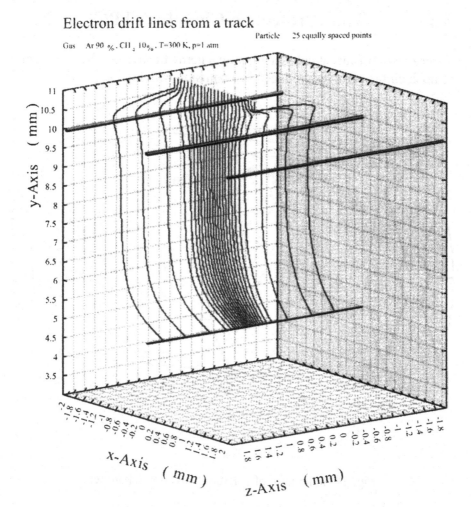

Electron drift lines from a track

Gas   Ar 90 %, CH₄ 10%, T=300 K, p=1 atm

Particle   25 equally spaced points

Fig. 32 ExB Effect

5.2   Time Projection Chamber (TPC)

Typical experiments at colliders require a hermetic 4π sensitivity in order to track all the particles produced in a collision with high accuracy.

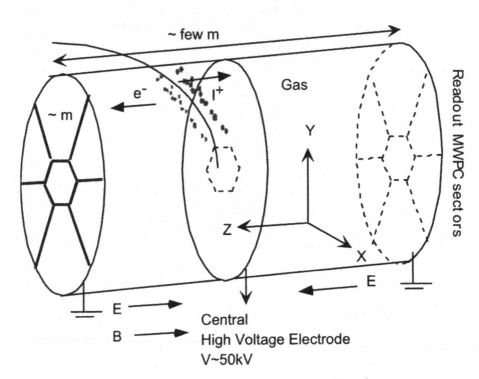

Fig. 33 Schematic of a Time Projection Chamber

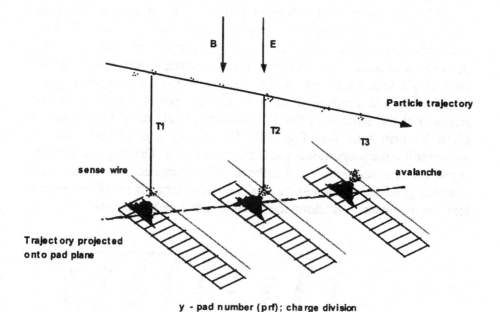

Fig. 34 Principle of operation of a TPC

A TPC, conceived by D. Nygren [29], is the ultimate form of the application of the multiwire chamber. It gives a full 3D track reconstruction as sketched in Fig. 33. A central electrode is held at a high potential with respect to the endplates thus ensuring electron drift towards them in a low and uniform electric field. The X-Y information is given from wires and segmented cathode of the MWPC read-out sectors arranged at the endcaps of the TPC cylinder, while the Z information is given by drift time (see Fig. 34). The energy loss of each track is measured by the proportional gains at the MWPCs. The magnitude of the signal on each sense wire and cathode pad is sampled in ~ 100 ns time interval; good resolution (~3 % rms) on the determination of the energy loss can be obtained by taking the mean of many measurements.

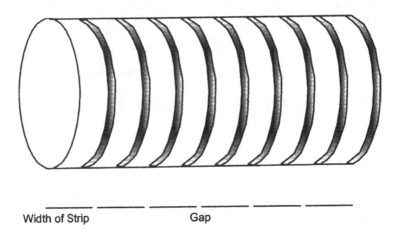

Width of Strip   Gap

Fig. 35 Field Cage in a TPC

Description of field cages and their design considerations may be found in many references. The high voltage electrode (see Fig. 33) needs to be perfectly parallel to the endplanes so that the relative field distortions are ~ 10^{-4} in the drift volume. (Distortions translate into the drift volume ~ the dimension of the object producing it). Finite element analysis programs like MAXWELL [30] are used to compute these distortions in the designing stage of the experiment. A field cage or potential degrader is necessary to permit the uniform drop of the voltage from the very high (~50kV) potential of the HV plane to the grounded endplane. Strips of metal are interspersed with insulator to form cylinders of field cages (see Fig. 35). The NA49 experiment has pioneered a new technique of building a field cage without insulator in between, by using several columns of support with ribbons of aluminized mylar wound around. The insulator can sometimes be a potential source of discharges when it charges up in hot spots of radiation in the experiment. The electric field near the field cages is usually distorted due to the finite sizes of the strips. The distortions can be well studied in the designing stages also with simulations [31]. An example is shown in Fig. 36, of the equipotentials in the region close to the strips of the field cage. Distortions usually exist near the edges of the drift volume.

With the help of lasers, ionization is created in regions of the drift volume and tracked thereby mapping the electric field with the help of well known drift velocity of electrons. More details may be found in references [32]. Distortions due to the magnetic field are minimal since the electric field is parallel to it; though local regions may still exist. Dependence on ambient conditions can also be non-negligible.

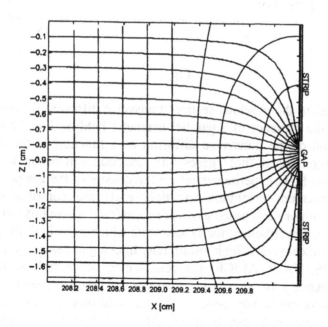

Fig. 36

Using the same principle a radial TPC has been built at CERN [NA45], with a central cylinder as the high voltage electrode and MWPCs arranged in modules around it at a distance of almost 1m. Here the electron drifts away from the central electrode in a 1/r field, contrary to the almost uniform field in the traditional TPC (for example as in ALEPH [33]). In

318

this cylindrical TPC, the end plates are covered by Kapton "field cages" that follow the 1/r field for the potentials of their strips and close the drift volume electrically.

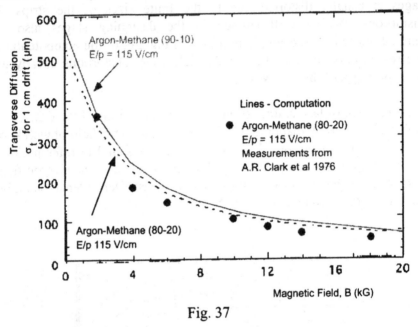

Fig. 37

As discussed in the previous section, transverse diffusion is significantly reduced in the presence of a parallel magnetic field. In Fig. 37 is shown the reduction of the transverse diffusion as a function of the magnetic field. In argon based "hot" gases this reduction can be up to 7 times depending on B. In experiments where the density of the gas plays a role in the momentum resolution (multiple scattering contribution being higher), helium or neon based gases are considered where this reduction is not dramatic. Nevertheless, for experiments which must look for particles with low momenta (as in high multiplicity tracking in heavy ion collisions [34] CERES, NA49, ALICE) the contribution of multiple scattering is large, and hence a low mass gas mixture is the best compromise. Several hundred tracks can be reconstructed over several meters with a resolution of ~ 200 µm with the help of the TPC and very sophisticated reconstruction algorithms [35]. The references should be studied for details.

5.2.1   Effect of Pollutants and Contaminants on Drift

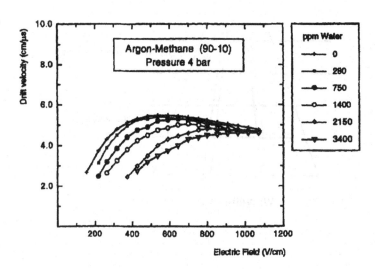

Fig. 38 Effect of the addition of water on the drift velocity in Ar-CH$_4$, as a function of electric field.

Figure 38 shows measurements of drift velocity [36] exemplifying the effect of addition of water to an argon-methane (90-10) mixture, at ~ 100 V/cm. Adding 1000 ppm of water changes the drift velocity by 20%, while at higher fields the change is negligible. The change is more at higher fields for a richer noble gas mixture. An electron capture (attachment) phenomenon has also a non-negligible electron detachment probability, and the transport parameters may also be sensitive to this electron slowing down mechanism. Incidentally, the first paper by Biagi [24] where the computation program was introduced exemplified the very good agreement between these data and the calculations.

The presence of pollutants in the operating gas mixture affects the gas detector operation. There are two effects: one is the modification of transport parameters and the second is electron loss by capture due to electro-negative pollutants.

5.2.2   Pad Response Function

The geometry of the MWPC is optimized in such a way that one maximizes the signal wire coupling to the pad from the avalanche on the anode wire. This depends on several parameters. Field wires, inter-electrode distances and the angle of the ion drift.

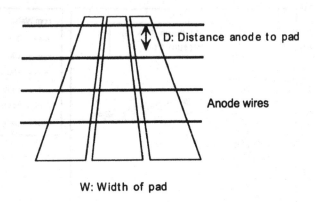

Fig. 39

The Pad Response Function is defined as the integrated charge over the area of the strip. As sketched in fig 39, W/D is chosen carefully such that the signal ratio on neighboring strips, r is high. Meaurements show when: W/D = 1 r ~ 60%, W/D = 0.5   r ~ 20%. The signal is also dependent on the position of the avalanche along the anode, namely whether it was at the edge of a facing pad or at the center of it [37]. An exercise of signal formation is included in the following to demonstrate how the choice of a readout geometry is dictated.

5.2.3 Measurement Accuracy of TPCs

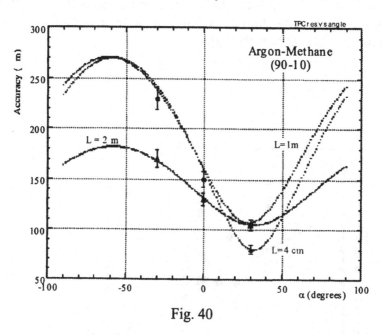

Fig. 40

As mentioned before, the z-coordinate in a TPC is given by the measurement of the drift time. This may be estimated by simulation when designing a TPC. The major factors on which it depends are transverse and longitudinal diffusion of the electrons in the gas, number and average distance between the primary ionization clusters, angle of incident track with respect to the anode wire and read-out pad. In principle the electric and magnetic fields are parallel but there can be localized regions where this is not so, in which case the ExB effects come into play. In addition, close to the anode wire, where the field is radial and the approaching electron is constantly changing its angle with respect to the magnetic field, a gas mixture with a small Lorentz angle is optimal. Depending on the cluster size and the distance of the drift length, an effect called "de-clustering" is observed, namely the obliteration of the primary clusters as they drift towards the end caps. This improves the accuracy the larger the drift distance! Lastly, there is also a component of the electronics noise to the measurement accuracy. All these components were evaluated and the accuracy was computed for an argon-methane mixture operational in ALEPH [38]. Plotted in Fig. 40 is the measured and computed accuracy. There is also a contribution due to the ambient conditions namely temperature and pressure which is why in these systems they are monitored very carefully.

The dE/dx resolution is improved by a long path length due to the increased number of clusters from the primary ionization. Figure 41 shows the dE/dx resolution as a function of total normalized detector depth for several experiments [39].

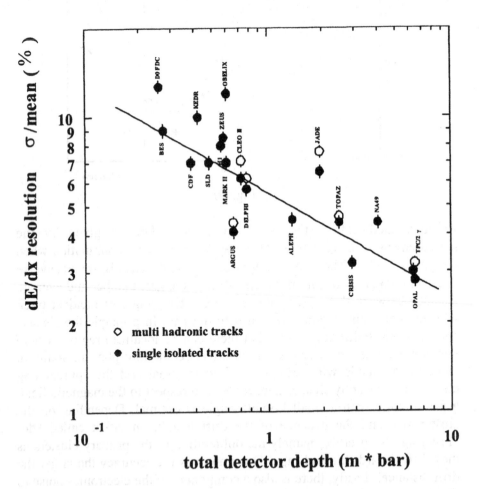

Fig. 41 dE/dx resolution of large detectors as a function of the drift depth

5.2.4   Positive Ion Feedback Problem and Distortions: Gating

As seen before, avalanches at the read-out MWPC end-cap of the TPC produce positive ions, which in the case of a TPC, recede towards the "cathode," namely the high voltage central electrode. Typically, these detectors operate at gains of several 10^3 to 10^4, yielding many ions drifting back per electron. These moving positive charges have a lower velocity; typical collection times are ~ 100 ms/m of drift. Slowly, there is a rate dependent space charge that builds up in the drift volume, which eventually distorts the electric field both in its magnitude and direction, depending on the drift length and the drift field, the distortion being directly and inversely proportional, respectively, to the two quantities [40].

To avoid this drawback, a simple grid of wires is placed in between the cathode of the MWPC read-out, and the drift region with alternating potential between adjacent wires. No positive ions can pass through to the drift region, nor can any electrons be admitted into the amplification region in between interesting events. This prevents the positive space charge build up in the drift region. Details may be found in [41].

The ExB effect is particularly important and has to be studied in the case of the TPC readout. As seen before magnetic deflection is largest when E and B are perpendicular and when in addition E is small. The most pronounced region of this kind is in the proximity of the cathode wires. One can try to smooth the field transition at the cathode wires; vary the number of cathode wires per cell. Smoothing the field transition is accomplished by varying the offset potential of the gate wires such that the field in the region between gate wires and cathode wires becomes larger. When doing this, the cathode wires cause smaller magnetic deflections, but the gate wires are no longer neutral, and they start to cause deflections.

The Ring Cathode Chamber (RCC) [42] is a new kind of wire chamber developed with the state-of-the-art technology and materials. This incorporates a readout scheme with ring shaped cathode elements thus permitting a good coupling of the signal at the anode to the readout pad.

Here we demonstrate the gating properties with an RCC [31]. The gate here consists of strips arranged between two cells as demonstrated in Figs.

42 and 43. The offset voltage applied such that no electrons can pass into (or positive ions can pass out of) the active cell.

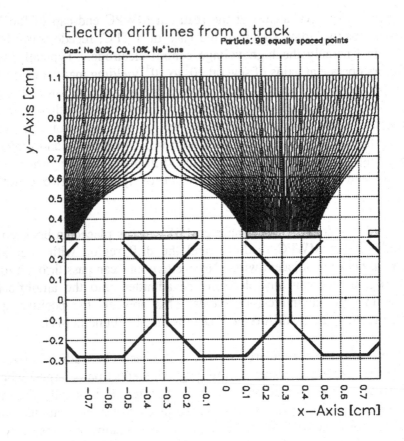

Figs 42, Closed Ring cathode Cell

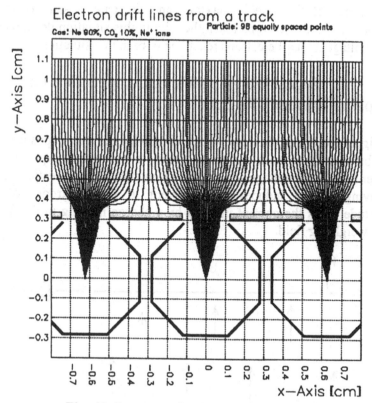

Fig. 43 Gate open for electrons in RCC

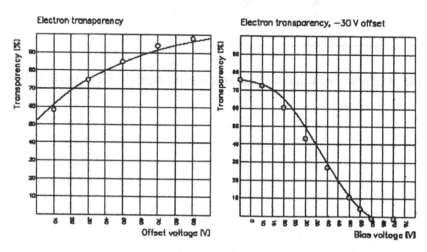

Fig 44. Electron Transparencies versus offset and bias gate voltage.

In order to open and close the entrance gate of the RCC, a positive or a negative bias voltage is applied to alternating the gating electrodes on top of a constant offset voltage. When the bias voltage is set to 0 V, the transparency of the gate is maximal and the gate is said to be open. Whether or not all electrons reach the wire in this situation depends on the offset voltage. Not all electrons reach the wire in this situation.

5.2.5 Signal Weighting Field: Computations for Comparing Different Readout

The signal weighting field is a vector field which is a superposition of the contributions from each electrode. Being a geometrical quantity, it is independent of the voltage settings. Its orientation is the direction in which a moving charge induces most current in the electrode, while its magnitude gives the current induced by a charge moving in the direction of the weighting field.

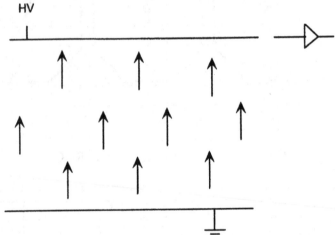

Fig. 45 Demonstration of the vector weighting field for a parallel plate

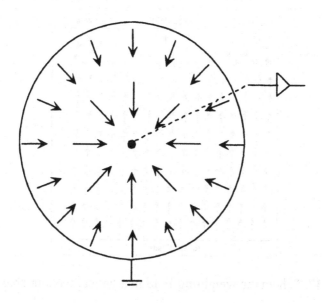

Fig. 46. Demonstration of the vector weighting field for an SWPC

The mechanism to compute a signal consists in placing a charge in the sensitive volume of the detector, thus causing a voltage shift on all the electrodes (prevented by the voltage supply which reshuffles the charges)[3]. A moving charge causes currents to flow between the electrodes given by:

$$I = -Q\mathbf{E_w} \cdot \mathbf{v_d}$$

---

[3] Note that the total charge on the electrodes remains constant

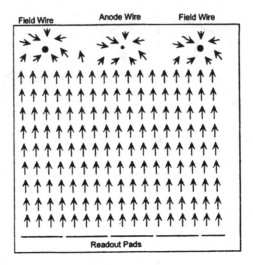

Fig. 47. Schematic weighting field in a typical readout chamber

Where E_w, the weighting field, given by geometry gives the direction of the maximal current, and v_d is the speed.

Both electrons and ions are moving along the electric field (assuming B=0 and σ=0) with a speed which is determined by the local E-field. The drift velocity and ion mobilities have been discussed in other sections, and are known for many chamber gas mixtures.

Consider a traditional readout chamber geometry with anode wires in between field wires and a simple pad readout. The weighting field is sketched in Fig. 47.

The various signals that contribute to the shape of the final signal are sketched in Fig. 48. Unipolar signals arise from ions moving from an electrode that is not read: the readout pads. Bi-polar signals are due to the ions moving between the pads, while secondary peaks correspond to the arrival of ions on field and cathode wires, not on pads.

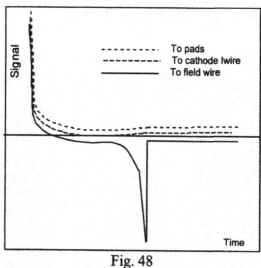

Fig. 48

Charge sharing between the various electrodes can also be studied, especially when designing a readout such that the charge is maximized on the readout electrode. A summary of one such study is shown in Table 2 for several different experiments [43]. The two columns show that charge coupling to the readout element integrated within 100 ns, a typical shaping time for the readout electronics (also shown for 0 ns). One can clearly see the reason for the circular pad of the RCC readout.

Table 2

LAYOUT	t = 0 ns (%)	T=100 ns (%)
ALEPH	22	17
ALEPH (with field wires)	37	33
NA 49	27	18
NA 49 (with Field Wires)	40	33
CERES	44	37
STAR Inner	45	37
STAR Outer	36	32
RCC	95	94

Passive gates have also been studied [44] that depend on a high magnetic field (hence large Lorentz angle at low E-field) for their operation; they have yet to find a real application.

5.3 Gaseous Detector Aging

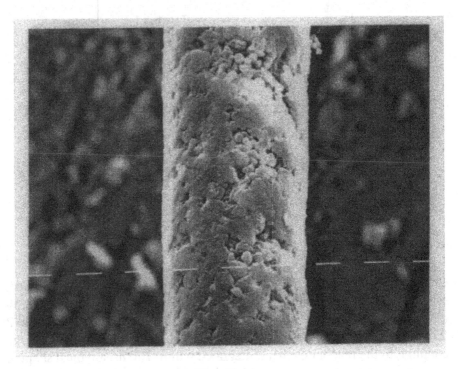

Fig. 49 (a)

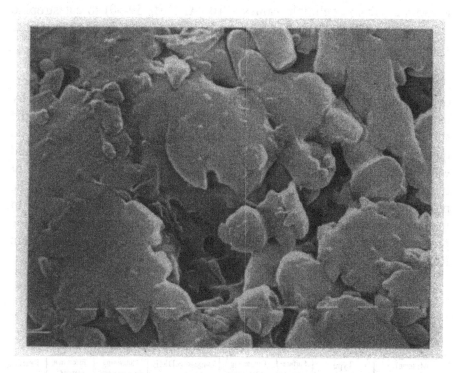

Fig. 49 (b) Examples of aged wires with deposits [45]

Gas detector aging is a very complex process and has been the subject of intensive research over the last few decades [46]. During the avalanche formation, a micro plasma discharge results inside the gas volume in the high fields close to the wire. This may result in one or a combination of the following: decomposition of chamber gas and possible contaminants, highly active radicals with dipole moments, polymerization and resistive carbon or silicon based deposits on the anodes and cathodes. On the anode this results in a non-uniform increase of diameter, with the electrons stranded on its surface. The resulting field around the anode is much reduced and inhomogeneous; as a consequence the gain is dependent on the rate. On the cathode strong dipoles are formed by the ions deposited on top of the surface deposits and their mirror charges resulting in a Malter effect, namely the field emission of electrons and higher dark current. This, in combination with the anode deposits, can lead to sparking which in turn can lead to corrosion and eventually even evaporation of the electrodes. The aging rate is a function of the relative amplitude loss: Specific Aging R, (in %/C/cm) can be estimated from: $\Delta A/A = RQ/l$ where A is the gain of the detector, Q the accumulated charge and l the length of the wire.

R can vary from negligible values ~ 10 (Ar/C_2H_6 50-50) to catastrophic values > 10^6 (CH4 + 0.1% TMAE). An R < 100 is considered as moderate aging.

Aging can be minimized by careful material selection and cleanliness during chamber construction, chamber geometry and operation and by using aging resistant gases. A small admixture of water increases the surface conductivity of the anode; alcohols like methylal have good quenching capabilities. They reduce lateral avalanche propagation and effectively suppress polymerization. Clean gas systems with no PVC or teflon tubes or silicon oil bubblers are also helpful in reducing the problems related to aging. Many references [46-48] give a list of materials/contaminants and their effects on gas detector operation; the properties of some materials tested are reproduced in Table 3, taken from Ref. [49].

Table 3 Materials and epoxies tested for outgassing

Material	Type Curing	Surface (cm^2)	Outgassing ARGON	Outgassing DME	Outgassing Ar–DME	Effect in a SWPC	Result
Duralco 4525	Epoxy Room T	156			Yes Room T	Gain loss	Bad
Duralco 4461	Epoxy Room T	156			Yes Room T	Gain loss	Bad
Hexcel EPO 93L	Epoxy - room T	150	No	No			OK
	- 16 h@23°C + 4h@45°C	75	No	No			OK
Hexcel A40	Epoxy room T	150	Yes T>40°C	Yes T>40°C			Bad
Technicoll 8862 Hard. Tech. 8263	Epoxy room T	115	Yes Room T	- Benzene comps. - Xylene			Bad
Stycast 1926 Hard. 266	Epoxy room T	150	No	No			OK
Loctite 330	Epoxy	20	Yes Room T				Bad
Amicon 125	Epoxy 1.5 h @ 80°C	150	No	No			OK
Araldit AW106 Hard.HV 953 U	Epoxy 2 days @ 70°C	176			Yes Room T	Gain loss	Bad
Araldit 2014	Epoxy	150		Yes Room T			Bad

Epotecny E505.SIT	Epoxy - 30 min@ 80°C - 24 h @ 50°C	131 75		No	Yes room T	No effect	OK OK
Epotecny 503	Epoxy 60 min. @ 65°C	185	Yes Room T	- Silicon			Bad
Epo–Tek E905	Epoxy room T	150	Yes T>45°C	Yes T>45°C			Bad
Epo–Tek H72	Epoxy 1.5 h @ 65°C	150	Fast	Fast			Bad
Polyimide Dupont 2545	Epoxy 2h @ 200°C	110	No	No			OK
Norland NEA 155	Epoxy	150	Yes Room T	Yes Room T			Bad
Norland NEA 123	Epoxy UV light	150	Yes Room T	Yes Room T			Bad
Norland UVS 91	Epoxy UV light+50°C	150	Yes Room T	Yes Room T			Bad
Stesalit 4411W	fibreglass	380			Yes T~ 75°C	No effect	OK
Vectra C150	Liquid crystal polymer	142			Yes Room T	No effect	OK
Ryton R4	polysulphur phenylene	336			Yes Room T	Gain loss	Bad
PEEK (crystalline)	Poly ether ether ketone	260	No	No			OK
PEEK (amorphous)	Poly ether ether ketone	2m pipe 3mm ø		Yes Room T			Bad
PEE		60		No			OK
EPDM	Copolymer ethylene propylene	300	Yes Room T	Yes Room T			Bad
PVDF	Fluorinated polyvinylidene	500	Yes Room T	Yes Room T			Bad
Viton	Fluorinated copolymer	412	Yes Room T	Yes Room T			Bad
Pipe	Polyeurethane	2m 3 mmØ (ext.)		No @40°C No			OK

Stycast 2	R T Curing	150		No 50 °C No 80 °C No			OK
Araldit 2	Durcisseur 951 C. T. 23 °C	150		No 80°C No			OK

6 Limitations in Wire Chambers and Future Perspectives

We have seen that the sense wire spacing places a limit on the measurement accuracy of the MWPC, as well as on the two-track resolution being of the order of 1 mm. The maximum permissible length of the wire is also dominated by electrostatic and mechanical constraints such that the maximum stable length l_c is given by:

$$l_c = s \, (\sqrt{4\pi\varepsilon_0 T}) \, / \, CV_0$$

    s: wire spacing
    V_0:   cathode voltage
    T: wire tension

and is of the order of 10 cm for s = 1 mm.

The width of the cathode induced charge or the pad response function is limited by the geometry thus limiting the two-track resolution. The ion induced long tails of the signals also limit the two-track resolution in the coordinate given by drift time. The space charge accumulated around the wire due to the previous avalanches is not cleared up due to the slowly moving positive ions. This results in a variation (and reduction) of the local field, which yields a rate-dependent reduction of the gain. This limits the rate capability of the detector as shown in Fig. 50.

336

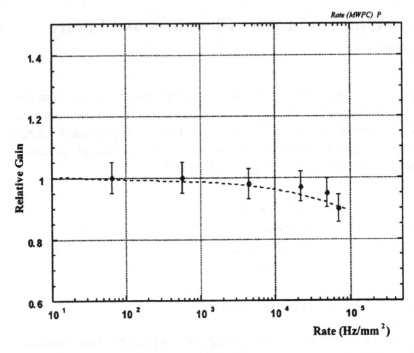

Fig. 50

Nevertheless, for large (several tens of m^2) and cheap tracking systems wire chambers are still the most robust and reliable detectors. In the last decade thanks to advances in photolithography and wet etching techniques, many of the aforementioned limitations have been overcome by the new generation of micro-pattern detectors which have been described in the review article [50].

References

[1]   Principles of operation of multiwire proportional and drift
        chambers, F. Sauli (CERN 77-09, 1977); Techniques and concepts
        of high-energy physics, ed. by Th. Ferbel (Plenum, New York
        1983); Radiation detection and measurements, G.F. Knoll (Wiley,
        New York 1989); Particle detection with drift chambers, W. Blum
        and L Rolandi (Springer-Verlag, Berlin 1993); Particle detectors,
        C. Grupen (Cambridge Monographs on Part. Phys. 1996); Particle
        detectors, C.W. Fabjan and H.G. Fischer, Rep. Progr. Phys. 43
        (1980) High-resolution electronic particle detectors, by g. Charpak
        and F. Sauli, Ann.Rev.Nucl.Part.Sci. 34 (1984) 285.

[2]   ARGUS: a universal detector at DORIS II, by H.Albrecht et al,
        Nucl. Instr. Methods A275 (1989) 1;
        The OPAL Detector at LEP, K. Ahmat et al, Nucl. Instr. Methods
        A305 (1991) 275;
        The DELPHI Detector at LEP, by DELPHI collaboration, Nucl.
        Instr. Methods A303 (1991) 233.

[3]   G. Charpak, R. Bouclier, T. Bressani, J. Favier and C. Zupancic
        Nucl. Instr. Methods 62 (1968) 262.

[4]   E. Babichev et al, Nucl. Instr. And Meth A323(1992)49

[5]   A. Pansky et al, Nucl. Instr. And Meth. A 392(1997)465
        A. Breskin, To be published in NIM Proceedings of SAMBA,
1999
        Preprint WIS-00/2/Feb-DPP.

[6]   GARFIELD A Drift Chamber Simulation Program by R. Veenhof.

[7]   HEED Particle Interactions in Gases, interfaced with Garfield, v
        1.01
        I. Smirnov (1996).

[8]   H. Breuker et al CERN EP/87-97, Nucl Instr. And Meth. (1987)
        72.

[9]   F. Sauli Academic Training CERN AS/CP-MP/1999.

[10]   A. Sharma and F. Sauli Nucl. Instr. And Meth. A 323(1993) 280.

[11]   Properties of some gas mixtures used in tracking detectors,
        A. Sharma SLAC-JOURNAL-ICFA-16-3, IIB Summer
        (www.stanford.slac/icfa/pubs).

[12]   G. Battistoni et al, Preprint LNF - 87/89;
        R. Albrecht et al, NIM sub, GSI-88-56;

[13]   H. Okuno, J. Fisher, V. Radeka and H. Walenta IEEE Trans. Nucl.
        Sci NS 26(1979)160-168;

[14]   J. S. Gordon and E. Matheison NIM 227(1984)267-276;
        J. S. Gordon and E. Matheison NIM 227(1984)282;

338

J. R. Thompson, J.S. Gordon and E. Matheison NIM A 234(1985)505 – 511;
E. Matheison NIM A270(988) 602-603.

[15]   G. Charpak, J. Favier and L. Massonet NIM 24(1963) 501-502.

[16]   K. Ahmat et al, Nucl. Instr. Methods A305 (1991) 275;
DELPHI collaboration, Nucl. Instr. Methods A303 (1991) 233.

[17]   C. Martin, P. Jelinsky, M. lampton, R. F. Malina and H. Anger
Rev. Sci Instr. 52(1981) 1067-1074.

[18]   T. Lohse and Witzeling, Internal Note CERN-ALEPH 91-156 and references

[19]   G. Charpak and F. Sauli Phys. Lett 78B(1978) 523.

[20]   D. F. Anderson IEEE Trans. Nucl. Sci. (1984).

[21]   G. Mikenberg, NIM A 265(1998)223, Y. Arai et al NIM A 367 (1995) 398

[22]   Honey Comb

[23]   Measured Drift Velocities Palladino and Sadoulet, Nucl. Instrum. Methods 128 (1975) 323, Computations A. Sharma 1998
Wire chamber gases, by J.Va'vra, Nucl. Instr. Methods A323 (1992) 34
B. Schmidt, Nucl. Instrum. Methods A252 (1986) 579, B. Schmidt
1998

[24]   Biagi, Nucl. Instrum. Methods A283 (1989) 716

[25]   H. W. Ellis, R. Y. Pal and E. W. McDaniel, At. Data and Nucl.
Tables 17 (1976) 177-210. Parts I and II
H. W. Ellis et al., Transport properties of gaseous ions over a wide energy range, Part III, At. Data and Nucl. Data Tables 31 (1984) 113-151.

[26]   A. H. Walenta t al NIM 92(1971)373, G. Charpak et al NIM 108(1973)413
M. Agnello . NIM A 385 (1997) 58-68

[27]   OPAL Jet Chamber, A. Wagner SLAC Workshop, STanford 1987

[28]   R. Veenhof Private Communication

[29]   The time projection chamber, ed. by J.A. Macdonald, AIP Conf. Proc. 108 (Am.Inst. of Phys. New York 1984)

[30]   MAXWELL 3D Finite Element Computation Package, by Ansoft SA,   Pittsburgh, USA

[31]   Time Projection Chamber Technical Design Report, CERN/LHCC 2000-001

[32]   H. J. Hilke Detector calibration with lasers: a review, Nucl. Instr. And Meth A252(1986)161

[33]   A. Lusiani The functioning and initial performance of the ALEPH
       TPC at LEP, INFN PI/AE 90/12
[34]   Heavy Ion Experiments at CERN: CERES (NA45), NA49, ALICE
[35]   S. Afanasiev et al. (NA49 Collaboration) Nucl. Instr. And Meth.
       A 430(1999)210
[36]   Study of fast gases, resolutions and contaminants in the D0 muon
system
       J. M. Butler et al FERMILAB-PUB-89/222-E
[37]   D. L. Fancher and A.C. Schaffer, Experimental Study of the
       signals from a segmented cathode Drift chamber IEEE Nucl. Sci.
       NS-26,(1979) 150
[38]   Simulation D. Decamp et al Nucl. Instr. And Meth
       A269(1990)121; Simulation A. Sharma 1995 (also see Thesis
       Univ. of Geneva 1996).
[39]   M. Hauschild – Opal web Pages1996
[40]   D. Friedich et al Positve Ions
[41]   Particle Detection with Drift Chambers, W. Blum and L. Rolandi
       eds F. Bonaudi and C. Fabjan Springer Verlag 1993
[42]   CERN R&D project, RD32 Decelopment of the Ring Cathode
       Chamber
[43]   R. Veenhof Private Communication 1999
[44]   J. Kent Internal Note LPC 84-17, College de France, Paris, 1984
       S. R. Amendolia et al, Nuc. Instr. And Meth. A234(1985)47-53
[45]   F. Sauli, C. del Papa Private Communication
[46]   Proc. workshop on radiation damage to wire chambers, Ed. J.
       Kadyk, LBL-21170 (1986).
[47]   Wire chamber aging, by J. Kadyk, Nucl. Instr. Methods A300
       (1991) 436, J. Va'vra SLAC PUB 5207, J. Kadyk NIM
       A300(1991) 436
[49]   R. Bouclier et al NIM A350(1994)464
[50]   Micro-pattern gaseous detectors, by F. Sauli and A. Sharma,
       CERN-EP/99-69 (1999), Ann. Rev. Nucl. Part. Sci. Vol
49(1999)341-388.

NUCLEI AT THE BORDERLINE OF THEIR EXISTENCE

Yu. Ts. Oganessian
Flerov Laboratory of Nuclear Reactions
Joint Institute for Nuclear Research
141980 Dubna, Moscow region
Russian Federation

Abstract   Throughout the 60 years that have passed since the discovery of the first man-made elements, neptunium and plutonium, investigations in the field of synthesis of new elements and of their properties have become one of the most important and rapidly developing fields in nuclear physics and chemistry. The transition from conventional methods of neutron capture for producing man-made elements to methods employing heavy-ion reactions has made it possible to synthesize many elements heavier than fermium ($Z = 100$). In the mid-1960s, a theoretical description of the masses and fission barriers for new nuclei led to the prediction of islands of stability of heavy and super-heavy nuclides near the closed proton and neutron shells. The results of the first experiments devoted to the synthesis of heaviest nuclides formed in nuclear reactions induced by ^{48}Ca ions are presented. For various individual nuclei, the observed decay chains consisting of successive events of alpha decay and ending in spontaneous fission, as well as the decay energies and lifetimes, are consistent with the predictions of theoretical models that describe the structure of heavy nuclei. These data furnish the first indication of the existence of superheavy elements being highly stable with respect to different decay modes.

The experiments in question employed the heavy-ion accelerator installed at the Flerov Laboratory for Nuclear Reactions at the Joint Institute for Nuclear Research (JINR, Dubna, Russia). They were performed in collaboration with physicists from the Lawrence Livermore National Laboratory (Livermore, USA); the Gesellschaft für Schwerionenforschung (GSl, Darmstadt, Germany); RIKEN (Saitama, Japan), the Institute of Physics and Department of Physics, Comenius University (Bratislava, Slovak Republic), and the Dipartimento di Fisica, Università di Messina (Messina, Italy).

H.B. Prosper and M. Danilov (eds.), Techniques and Concepts of High-Energy Physics XII, 341–374.
© 2003 *Kluwer Academic Publishers. Printed in the Netherlands.*

1   Introduction

According to QED, the well-known concept of the atom as a system consisting of a nucleus, which carries almost entirely the atomic mass, and electron orbits, occurring at a large distance from the charge center, is valid for very heavy atoms ($Z \leq 170$). However, the limit of existence of atoms (elements) is reached much earlier because of the instability of the nucleus itself.

Of approximately 2700 nuclear species known at the moment, only 287 nuclides have survived during the time interval between the completion of nucleosynthesis and the present day. It is well known that changes in the proton-to-neutron ratio in these nuclei generate beta decay. An excess of neutrons in a nucleus leads to the reduction of the neutron binding energy; the limit of existence of neutron-rich nuclei is reached at $E_n = 0$ (neutron drip line). Similarly, zero proton binding energy, $E_p = 0$ (proton drip line), determines the boundary of existence of proton-rich nuclei.

Another boundary is associated with the maximal possible number of nucleons in a nucleus. Formally, the limiting mass of a nucleus near the boundary of its stability – even at the most favorable value of the proton-to-neutron ratio (nuclei occurring on the beta-stability line) – is determined by spontaneous fission (SF). For the first time, this type of nuclear transformations of heavy nuclei, was observed for the ^{238}U isotope ($T_{SF} = 10^{16}$ y) by Flerov and Petrzhak in 1940 [1]. By that time, Hahn and Strassman had already discovered the induced fission of uranium. In order to describe this phenomenon, N. Bohr and J.A.Wheeler proposed the liquid-drop model of nuclear fission [2].

This beautiful theoretical model, which is essentially classical, is based on the assumption that the nucleus is a macroscopic structure less (amorphous) body similar to a drop of charged liquid. A deformation of the drop due to Coulomb forces – eventually, this deformation leads to the fission of this drop into two parts of approximately identical masses – arises upon overcoming the potential barrier opposing nuclear deformations. For the ^{238}U nucleus, the height of the fission barrier is $B_f \approx$ 6 MeV (Fig. 1a). With increasing Z, the height of the fission barrier decreases fast. As a result, the nucleus becomes absolutely unstable with respect to spontaneous fission ($T_{SF} \sim 10^{-19}$ s) at some critical value of the nuclear charge. According to the estimates of N. Bohr and J.A.Wheeler, this critical situation is realized as soon as the charge number reaches values of $Z = 104\text{-}106$.

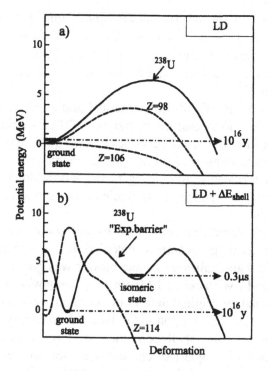

Fig. 1. Calculated potential energy as a function of nuclear deformation.
  a)   macroscopic model of a charged liquid drop
  b)   macro-microscopic model. The solid lines refer to the ^{238}U nucleus, the
       dashed lines – to the region of more heavy artificial elements.

It is interesting to note that, much later, when the first transuranium elements were synthesized at high-flux reactors, the radioactive properties of the new nuclides confirmed qualitatively the liquid-drop analogy of nuclear matter: the probability of spontaneous fission increased by more than 13 orders of magnitude when going from ^{238}U ($Z = 92$) to ^{257}Fm ($Z = 100$).

The discovery of isomers that can undergo spontaneous fission [3] was unexpected insofar as it was at odds with the liquid-drop model (Fig. 1b). Presently, it has been proven that shape isomerism in 33 nuclei known by that time (isotopes of nuclei occurring between U and Cm) arises due to the complicated shape of the nuclear potential-energy surface – in particular, due to the two-humped shape of their fission barrier. (The reader can find a comprehensive description of this phenomenon in the excellent review article of Bjornholm and Lynn [4].) Yet another contradiction with the liquid drop model was found in the significant variations of the partial lifetimes of spontaneous fission, best shown for the isotopes of Cf, Fm and No, which have been synthesized in heavy-ion induced reactions [5].

A more detailed analysis of the theoretical and experimental values of nuclear masses has shown that the deviations of the experimental nuclear binding energies from the theoretical ones behave quite regularly: they are maximal (highest binding energy) at specific magic numbers of protons and neutrons in a nucleus. According to the terminology adopted from atomic physics magic numbers in the nucleus correspond to closed proton and neutron shells.

As a rule, shell effects are described by correction terms in nuclear-mass formulas used in practical calculations. A phenomenological description of shell anomalies in nuclear masses was given in the works of Swiatecki [6a] and Swiatecki and Myers [6b]. Somewhat later (in 1967), Strutinsky proposed an original and, in my opinion, very physical method for calculating the shell correction to the liquid-drop energy of the nucleus [7]. In his approach, the shell correction is defined as the difference between the sum of single-particle energies for the actual quantum distribution of nucleons and the energy for some uniform distribution of levels in the mean nuclear potential characterizing the liquid drop. The total energy of the nucleus is represented as the sum $E_{tot} = E_{ld} + \Delta E_{shell}$, where E_{ld} is the macroscopic (liquid-drop) energy, while ΔE_{shell} is the microscopic correction taking into account shell effects and pair correlations of nucleons. Calculations performed within the macroscopic-microscopic model revealed regular shell phenomena in deformed nuclei. This made it possible to improve substantially the accuracy in determining their ground-state masses and shapes. In contrast to the widespread opinion that shell effects are smeared out with increasing nuclear deformation, it turned out that in highly deformed nuclei a substantial re-distribution of the nucleons takes place. As the deformation becomes more pronounced, shell effects change, rather than disappear, still considerably correcting the potential energy of the nucleus [8].

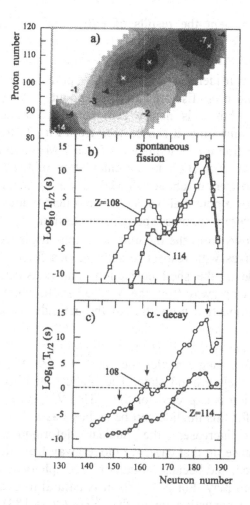

Fig. 2. a) Contour map of the amplitudes of shell corrections to the liquid-drop nuclear energy.

  b) and c) Half-lives with respect to spontaneous fission and α-decay calculated in the macro-microscopic nuclear model. The closed deformed nuclear shells at $N = 152, 162$ and the spherical shell $N = 184$ are shown by arrows.

2 Nuclear Shells and the Stability of Heavy Elements

Like any theory, the macro-microscopic model possessed some predictive power – for example, what concerned very heavy, hitherto unknown nuclei. Predictions on its basis were made in a number works.

Here, we shall present the results of Patyk and Sobiczewski [9] and Smolanczuk [10], who calculated the masses and fission barriers of even-even nuclei with $Z = 104\text{-}120$ and $N = 140\text{-}190$.

We shall first consider the probability for spontaneous fission of heavy nuclei. The liquid-drop fission barrier is about 1 MeV for the ^{254}No ($Z=102$) nucleus, but it is nearly zero for the heavier nucleus ^{270}Hs ($Z=108$). At the same time, it can be seen from Fig. 2a that the amplitude of the shell correction for these nuclei is 5 and 7 MeV, respectively. When including the shell correction in the calculation of the nuclear potential energy, fission barriers of about 6-8 MeV show up for the above nuclei. The appearance of a potential barrier, when a heavy nucleus is deformed, is expected to severely suppress spontaneous fission.

Indeed, it follows from the theoretical results displayed in Fig. 2b that the partial half-lives with respect to spontaneous fission depend strongly on the amplitude of the shell correction. The considerable increase in $T_{SF}(N)$ when moving away from the $N = 152$ shell, which manifests itself clearly in the radioactive properties of the actinide nuclei, is due to the effect of another neutron shell, that at $N = 162$. It should be noted that either shell is associated with deformed nuclei contrary to the case of the well-known doubly magic nuclei such as ^{208}Pb ($Z= 82$, $N= 126$), which are spherical in their ground state. The highest stability with respect to spontaneous fission is expected for the ^{270}Hs ($N = 162$) nucleus – the T_{SF} value predicted for this nuclide can be as large as a few hours. When the neutron number increases, the nuclear deformation becomes less pronounced because, in that case, we move away from the $N = 162$ shell, which is deformed, and because another closed spherical shell, that with $N = 184$, comes into play. For $N > 170$, it is natural to expect a significant growth of $T_{SF}(N)$ persisting up to the ^{292}Hs ($N = 184$) nucleus, whose partial half-life with respect to spontaneous fission is $T_{SF} \sim 3 \times 10^4$ yr, an enormous value indeed.

Here we come across a very interesting situation. If superheavy nuclei possess high stability with respect to spontaneous fission, they will undergo other modes of decay, such as alpha-decay and, possibly, beta decay. The probability of these decay modes and, hence, the corresponding lifetimes will be determined by the ground-state nuclear masses, which can be computed within theoretical models based on various assumptions about the fundamental properties of nuclear matter (about the nature of nuclear forces). Under such conditions, experimental results are of paramount importance for testing theoretical models. According to the calculations of Smolanczuk and Sobiczewski, performed within the macroscopic-microscopic model, the deformed nucleus ^{268}Sg ($Z=106$, $N= 162$) should undergo α-decay with a half-life of $T_a \approx 2$ h (according to the calculations of Möller and Nix [11], this half-life amounts to a few days). For the heavier spherical nucleus $^{294}110$ ($N = 184$), T_a becomes as large as a few hundred – or, maybe, a few thousand – years (Fig. 2c). It is worthwhile noting that, in the absence of nuclear structure (that is, in the liquid-drop model), this nucleus would fission spontaneously with $T_{SF} < 10^{-19}$ s. The above two values differ, as we can see, by more than 30 orders of magnitude!

Other calculations of the energy of the nucleus treated as a many-body system, carried out in the Hartry-Fock-Bogoliubov model using various forces, as well as calculations in the relativistic mean field model, also indicate a significant increase in the binding energy of the nucleus when approaching the closed neutron shell N = 184.

Among theoreticians there is presently no consensus on what the magic proton number is when the closed neutron shell is N = 184, at which the binding energy of a spherical doubly magic nucleus is expected to have the maximum value. For instance, in the macro-microscopic model, irrespective of the variations in the parameters used in the calculations, the amplitude of the shell correction is maximum for the $^{298}114$ (N = 184) nucleus [12, 13]. On the contrary, calculations in the HFB model indicate that in addition to Z = 114 there are other possible candidates: Z = 120, 122, 126 and even 138, depending on the chosen set of model parameters [14]. This, however, does not change the main conclusion that in the region of very heavy nuclei there may be "islands of enhanced stability", which extend significantly the limits within which elements may exist.

3   Reactions of Synthesis

It is well known that the first man-made elements heavier than uranium were synthesized in reactions of successive neutron capture during long-term exposures at high-flux nuclear reactors. The long lifetimes of the new nuclides made it possible to separate and identify them by radiochemical methods followed by the measurement of their radioactive-decay properties. These pioneering studies, performed by Professor G.T. Seaborg and his colleagues between 1940 and 1953 at the Lawrence Berkeley National Laboratory (for an overview, see, for example, [15]), resulted in the discovery of eight man-made elements with $Z = 93$-100. The heaviest nucleus obtained was ^{257}Fm ($T_{1/2} \sim 100$ d). The further advance toward the region of heavier nuclei was hindered by the extremely short lifetime of ^{258}Fm ($T_{SF} = 0.3$ ms). Attempts at overcoming this barrier in pulsed high-intensity neutron fluxes from underground nuclear explosions did not produce anything beyond ^{257}Fm.

Transfermium elements with masses $A > 257$ were produced in heavy-ion induced reactions. In this method, unlike the method of sequential neutron capture, heavy-ion reactions are advantageous in that in the fusion process the entire projectile mass is imported in the target nucleus.

The excitation energy of the compound nucleus is given by

$$E_x = E_P - [M_{CN} - (M_T + M_P)] = E_P - Q ,$$

where E_P is the projectile energy, while M_{CN}, M_T, and M_P are the masses of the compound nucleus, the target nucleus, and the projectile ion, respectively.

The minimal excitation energy is realized at the fusion reaction threshold energy, corresponding, in first approximation, to the Coulomb barrier: $E_x^{min} = B_c - Q.$ For heavy target nuclei, we have $B_c \sim 5$ MeV/nucleon.

In contrast to (n, γ) reactions, where the excitation energy of the nucleus amounts to about 6÷8 MeV, in fusion reactions induced even by an ion as light as ^{4}He, $E_x^{min} \approx 20$ MeV. With increasing the projectile mass, the excitation energy of the compound nucleus will increase due to the growth of the Coulomb barrier. The de-excitation of a hot nucleus - the transition to the ground state ($E_x = 0$) - will proceed predominantly through the emission of neutrons and gamma rays.

The production cross section of the evaporation residues (EVRs) can be written as

$$\sigma_{xn}(E_x) = \sigma_{CN}(E_x) P_{xn} \prod_{i=1}^{x} \left[\frac{\Gamma_n}{\Gamma_{tot}} (E_x) \right]_i,$$

where $\sigma_{CN}(E_x)$ is the cross section for the production of a compound nucleus with energy E_x, P_{xn} - the probability of its cooling via neutron emission, and Γ_n/Γ_{tot} - the ratio of the width (probability) with respect to neutron emission to the total width of decay along the cascade of sequential emission of x-neutrons at E_x.

The ratio Γ_n/Γ_{tot} can be calculated within the statistical theory under certain assumptions on the thermodynamic properties of a hot nucleus. The $\sigma_{xn}(E_x)$, which characterizes the survival probability of evaporation products, quickly decreases with increasing E_x (this is equivalent to an increase of the number x of the neutron evaporation steps). The situation is aggravated by the fact that the amplitude of the shell correction, which suppresses the fission of a nucleus in the ground state, decreases fast with increasing the compound-nucleus excitation energy. Both of these factors lead to an extremely small survival probability for heavy compound nuclei.

Definite advantages have fusion reactions of magic nuclei. When using ^{208}Pb and projectiles of mass A $\geq$ 50, E_x^{min} becomes as low as 10-20 MeV, which indeed increases the survival probability of the compound nucleus [16]. Unfortunately "cold-fusion" reactions cannot be used in our case. The reason is that in this case the compound nucleus with Z $\geq$ 114 would be neutron-deficient and the evaporation residues would be far from the boundary of the predicted "island of stability" of superheavy elements.

Nuclides with higher neutron number can in principle be obtained by using as target material an isotope of heavy actinides with Z = 94-98 and of the rare isotope ^{48}Ca as projectiles.

The compromise here consists in that, while abandoning the idea of using magic target nuclei upon going over from ^{208}Pb to neutron-rich isotopes of actinide elements, we regain magic proton and neutron numbers in the projectile ion [17].

Because of the considerable mass defect in the doubly magic nucleus ^{48}Ca, the compound nucleus excitation energy at the Coulomb barrier is about 30 MeV. The cooling of the nucleus occurs via the emission of three neutrons and gamma rays. It can be expected that, at this excitation energy shell effects are still noticeable in a heated nucleus and the probability of survival of evaporation products is larger than in the case of typical hot-fusion reactions ($E_x \geq$ 50 MeV). At the same time, the mass asymmetry of the nuclei in the entrance channel ($Z_1 \times Z_2 \leq$ 2000) should lead to the decrease of the dynamical limitation of nuclear fusion and, therefore, to the increase of the cross section for producing the compound nucleus as compared to the case of cold fusion reactions.

Despite these obvious advantages, all the preceding attempts made between 1977 and 1985 at various laboratories [18-20] to synthesize new elements with ^{48}Ca-projectiles yielded only upper limits on the cross sections of their formation. At the same time, the progress in experimental techniques over recent years and the possibility of obtaining intense beams of ^{48}Ca ions at new-generation heavy-ion accelerators make it possible to improve the sensitivity of relevant experiments by two and even three orders of magnitude. We have therefore chosen this way to advance toward the region of stability of superheavy elements.

4   Strategies of Experiments and Experimental Equipment

The strategy of experiments aimed at the synthesis of superheavy elements is determined to a considerable extent by their radioactive properties and, above all, by the lifetimes of the nuclides to be synthesized. As was indicated above, these lifetimes can vary within broad limits, their specific values depending on whether the theoretical predictions concerning the effect of nuclear shells on the stability of heavy nuclides with different values of Z and N are in fact true. Because of this, the operation of the experimental facility to be used must be extremely fast. At the same time, evaporation residues, whose yield is very small, must be separated within a short period of time from a huge background of reaction by-products whose formation probability is higher by eight to ten orders of magnitude. These conditions can be satisfied in the case of "in-flight" product separation (within a time interval 10^{-6}–10^{-5} s), taking into account the kinematical characteristics of the different reaction channels.

It should be noted that, in a fusion reaction leading to compound-nucleus formation, the projectile momentum is fully transferred to the compound system; as a result, the energy and momentum of recoil nuclei are well defined. Therefore, the problem amounts to separating the recoil

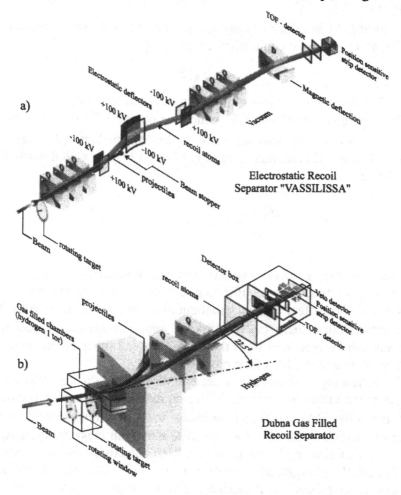

Fig. 3. Recoil separators used in FLNR (JINR) in experiments aimed at the synthesis and study of the radioactive decay of heavy nuclei.
  a) electrostatic separator – energy selector
  b) gas-filled separator.

atoms, emitted in the narrow angular interval $\theta_L = 0° \pm 2.5°$ with respect to the beam direction, according to their velocities (or energies). This function can be performed by Wien velocity filters (the separator SHIP at GSI) [21] or by an energy selector (the separator VASSILISSA at JINR) [22], where reaction products are separated according to the electric rigidities in transverse electric fields (see Fig. 3a). As a matter of fact, such functions can also be fulfilled by facilities of a different type – gas-filled separators, where the recoil nuclei are separated by magnetic rigidities in a hydrogen or a helium gas medium at a pressure of about 1 torr [23] (see Fig. 3b).

According to N.Bohr's theory as a result of multiple collisions the moving atom attains an equilibrium charge [24]

$$\bar{q} \sim (v/v_B)Z^{1/3},$$

where Z is the atomic number of the atom being considered.

The experimental values of $\bar{q}$, which we obtained for the recoil atoms with $Z = 89 - 116$ moving in a hydrogen medium [25] and for $Z = 69 - 110$ in a He medium (data of K.Morita [26]) are well described by the relation:

$$\bar{q} = 3.26(v/v_B) - 1.39 \qquad \text{for } H_2 \text{ and}$$
$$\bar{q} = 0.62(v/v_B)Z^{1/3} \qquad \text{for } He.$$

Since the heavy recoil nucleus and the bombarding ion move at different velocities, their equilibrium charges differ significantly from each other. This effect is most pronounced at low recoil-atom velocities close to the Bohr velocity ($v_B = 2.19 \times 10^8$ cm/s). In this case, evaporation products can be separated from the beam ions and other nuclei owing to the high magnetic rigidity of the recoil atoms.

The efficiency of kinematical separators depends on the ratio of the masses of the interacting nuclei. For fusion reactions induced by relatively light projectiles ($A_p \leq 20$) it amounts to only a few percent, but it becomes as high as 30-50% for projectile ions with mass numbers $A_p \geq$ 40. The facilities used also possess a high selectivity: in the separator focal plane, the background from the primary ion beam is removed almost completely, and the yield of products from incomplete-fusion reactions is suppressed by four to seven orders of magnitude depending on the kinematical characteristics of the different channels leading to the formation of these products. This is, however, insufficient for identifying extremely rare events of the production of atoms of new elements. For this reason, a further selection of the sought nuclei is accomplished with the aid of a sophisticated detecting device that is shown schematically in Fig. 3b.

Recoil atoms that have reached the focal plane are implanted into a multi-strip semiconductor silicon detector, which has an active area of about 50 cm^2. The length and the width of the strips, as well as their number, are determined by the size of the image in the focal plane of the separator. The facilities displayed in Fig. 3b and 3a employ 12 and 16 strips of lengths 40 and 60 mm, respectively. Each strip possesses longitudinal position sensitivity. The position resolution of each strip is determined experimentally. This is achieved by choosing reactions where known recoil atoms undergo successive α-decays or spontaneous fission. The position resolution depends on the type of the recorded particle (recoil nuclei, alpha particles, or spontaneous-fission fragments). However, more than 95% of all charged particles accompanying the decay of the implanted atom are confined, as a rule, within the range $\Delta_x \sim 1.5$ mm.

Thus, the entire area of the front detector is effectively partitioned into approximately 500 individual cells, each carrying information about the time of arrival and the energy of the recoil nuclei, as well as about the time of subsequent decays. The front detector is surrounded by side detectors. In such a way the entire array represents a box with an open front wall. This increases the efficiency of detection of α-particles from the decay of an implanted nucleus up to 85-87% of 4π. In order to separate signals of the recoil nucleus from those that are associated with particles from its decay, a time-of-flight (TOF) detector is installed before the front detector. The background conditions are substantially improved by selecting the events according to their generic decay. A parent nucleus implanted in the detector can be reliably identified if the successive α- and β-decays leads to nuclides with known properties. This method was used successfully in experiments devoted to the synthesis of elements with $Z = 107$-112 with $(N - Z \leq 53)$ in cold fusion reactions. As we advance further into the region of nuclei with greater neutron excess, the above advantage is lost. Here, the decay of a parent nucleus results in the formation of hitherto unknown isotopes, whose properties can be predicted only within the accuracy possible in theoretical calculations.

At the same time we can state that, if the basic theoretical prediction on the existence of the "island of stability" of superheavy elements is justified, the stability of nuclei approaching the $N = 184$ shell will be attaining higher stability with respect to spontaneous fission. Since the proton-to-neutron ratio in these nuclides is close to the β-stability line, they should be more stable with respect to β-decay as well. In this case α-decay becomes the main decay mode (Fig. 4). The daughter nucleus ($A-4$ and $Z-2$) will also be an α-emitter, after the emission of an α-particle. Consequent α-decays will be taking place until it happens that $T_\alpha < T_{SF}$. Eventually, the decay chains will end in the formation of spontaneously fissioning nuclei.

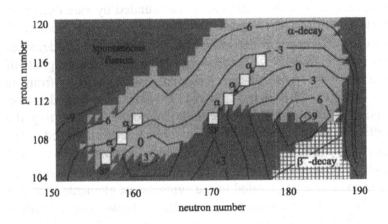

Fig. 4. Radioactive decay modes of heavy elements in different regions of the chart of nuclides. The contour lines are drawn through the calculated values of Log($T_{1/2}$). The expected decay sequences for $Z = 110$, $N = 160$ and $Z = 116$, $N = 176$ are shown as open squares.

In principle, such a decay pattern is convincing evidence for the formation of a superheavy nucleus. From a technical point of view, such an event differs substantially from those of other possible correlated decays. After observing a signal indicating the implantation of a nucleus at a certain place (strip and position) of the front detector, with the time of its flight being measured in the TOF detector, in the same position, after certain time intervals, signals from the emission of α-particles with amplitudes corresponding to their energies of about 8.5-10 MeV will arise. After that, a large-amplitude signal from spontaneous-fission fragments with total kinetic energy TKE of about 200 MeV will be recorded.

On the contrary, if the theoretical predictions are wrong in consistent, the nuclei will undergo spontaneous fission with a short half-life. When T_{SF} for these nuclei turns out to be longer than the flight time from the target to the detector, they can be registered by our setup as two fission fragments with high energy release in the detector array.

5   Experiments Devoted to the Synthesis of Superheavy Nuclei ^{48}Ca + ^{244}Pu Reaction

In the ^{48}Ca + ^{244}Pu fusion reaction, isotopes of element 114 can be synthesized. They will be closest to the top of the island of stability and will possess maximum neutron excess. In this reaction, at a bombarding energy close to the Coulomb barrier, we expect a maximum yield for the 3n- and 4n-evaporation channels. According to calculations performed within various theoretical models, the ground-state shapes of the $Z = 114$ nuclei with $N = 174$ and $N = 175$ are expected to be spherical; in these nuclei, the stabilizing effect of the strong neutron shell $N = 184$ must manifest itself.

Experiments aimed at synthesizing these isotopes were performed by using a gas-filled separator of recoil nuclei [23].

A thin layer (about 0.35 mg/cm^2 thick) of a plutonium target material enriched in ^{244}Pu to 98.6% in the form of oxide was deposited onto a 1.5 μm Ti-backing foil. Each of the 9 manufactured targets had the shape of a segment with an angular extension of 40° and an area of 3.5 cm^2. They were mounted on a disk of radius $R = 60$ mm, rotating with a speed of 2000 rpm about an axis orthogonal to the beam direction.

The projectile energy in the middle of the target was chosen equal to 236 MeV. With allowance for energy losses in the depth of the target layer and for slight changes in the beam energy during the long-term exposures, the excitation energy of the ^{292}114 compound nuclei was estimated to lie between 31.5 and 39 MeV. Actually, for each recoil nucleus recorded by the detector array, it was possible to determine the exact (instantaneous) beam-energy value and the target-segment in which the event originated.

Under these conditions, two virtually identical experiments were performed.

In the first experiment, which went on for two-months in the end of 1998, at a total beam dose of 5.2×10^{18} ions, a sequence of α-decays terminated by spontaneous fission was observed. The total decay time along the chain amounts to 35 min [27]. We will come back to the analysis of these results later in section 7, where we shall discuss the properties of even-odd nuclei.

In the second experiment, carried out in the period between June and October 1999, the total beam dose amounted to 1.0×10^{19} ions. Yet two identical chains of successive alpha decays, each ending in a spontaneous-fission event (see Fig. 5a), were observed here [28, 29]. Below, we consider these chains individually.

Following the implantation of a nucleus with $E_r = 11.1$ MeV into strip no. 2, after about 0.8 s, an α-particle with an energy $E_{\alpha 1} = 9.87$ MeV was registered. A second alpha particle, with energy $E_{\alpha 2} = 9.21$ MeV, was emitted 10.3 s later. Finally, 14.3 s later, spontaneous fission into two fragments depositing energies $E_{f1} = 156$ MeV and $E_{f2} = 65$ MeV ($E_{tot} = 221$ MeV) in the front and the side detector, respectively, was observed. All four signals (ER, α_1, α_2, SF) fall within a position interval of 0.5 mm, a fact that suggests a strict correlation between the observed decays.

In the second chain, an α-particle with $E_{\alpha 1} = 9.80$ MeV was recorded 4.6 s after the implantation of a recoil nucleus with an energy of 7.8 MeV into strip no. 8 (as in the first case, the energy of the recoil nucleus and its time of flight in the TOF detector are close to the values expected for evaporation products with $Z = 114$). A second α-particle, with energy $E_{\alpha 2} = 9.13$ MeV, was emitted 18 s later. Finally, spontaneous fission into two fragments, which deposited the energies of $E_{f1} = 171$ MeV and $E_{f2} = 42$ MeV ($E_{tot} = 213$ MeV) in the front and the side detector, respectively, was observed after the next 7.4 s. The fact that all four signals (ER, α_1, α_2, SF) were recorded within a position interval of 0.4 mm again indicates that the observed decays are strictly correlated. Within the detector resolution of $\Delta E \sim 0.06$ MeV and the statistical uncertainty in the decay times, the two events are consistent in all 9 parameters measured experimentally. The probability of random signal coincidences that could imitate recoil nuclei and their correlated decays is estimated at less than 5×10^{-13}. The probability of losing an α-particle in the observed decays is less than 3%.

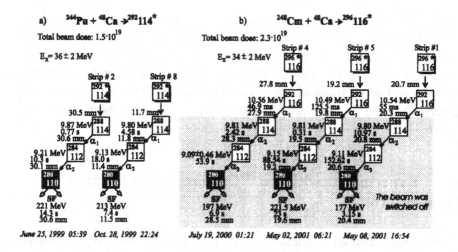

Fig. 5. Decay chains observed in the reactions: a) $^{48}Ca+^{244}Pu$ and b) $^{48}Ca+$ ^{248}Cm. The experimental conditions - beam dose and calculated excitation energies of the compound nuclei at which the shown decays take place - are shown in the figure.

It should be noted that only two spontaneous-fission events were recorded in this long-term experiment. They are characterized by large energy depositions from the fission fragments and are preceded, in either case, by identical chains of successive α-decays. The projectile energy measured at the instant when the given events were detected corresponds to exciting the $^{292}114$ compound nucleus to energies $E_x = 36$ to 37 MeV. The channel involving the evaporation of four neutrons and leading to the formation of the $^{288}114$ isotope in the ground state is the most probable one at this excitation energy.

In order to check this conclusion we shall try to observe the same decay chain in a independent experiment. From all possible ways the most direct one (but, obviously, the most difficult one) seems to be the experiment on the synthesis of element 116 in the $^{48}Ca+$ ^{248}Cm reaction.

Changing the ^{244}Pu target to the ^{248}Cm one, all other conditions of the experiment being kept, should lead to the formation of the new heavy nucleus with $Z = 116$ and mass 292 in the 4n-evaporation channel. As a result of the α-decay of this nucleus, expected with a high probability, we should obtain the daughter nucleus, the isotope $^{288}114$, produced earlier in the reaction $^{48}Ca + ^{244}Pu$. For this reason, after the decay of the $^{292}116$ nucleus the whole chain of the daughter nucleus decay, i.e. $^{288}114 \rightarrow ^{284}112 \rightarrow ^{280}110(SF)$, should be observed in the experiment.

In other words, if the even-even nuclide $^{288}114$, undergoing sequential decay such as α-α-SF, is formed in the $^{48}Ca+$ ^{244}Pu –reaction at $E_x \approx 36$

MeV in the 4n-evaporation channel, then the reaction ^{48}Ca + ^{248}Cm at $E_x \sim$ 34 MeV will similarly lead to the formation of the nucleus ^{292}116. This nucleus should undergo three consequent α-decays and SF. The energies and decay times of the three steps in the chain should correlate with the energies and decay times of the ^{288}114 nucleus. At the same time the first α-particle could be attributed to the decay of the new, heavier nuclide ^{292}116.

6 Experiments with the ^{248}Cm-target. Synthesis of Element 116

The target material - enriched isotope of ^{248}Cm ($Z = 96$) - was produced at the high-flux reactor in Dimitrovgrad (Russia) in the quantity of 10 mg. The other target material of enriched isotope ^{248}Cm was provided by the Lawrence Livermore National Laboratory (USA).

Usually the accelerator operates with a continuous ^{48}Ca beam. In the experiment on the synthesis of element 116 this regime was changed.

After the implantation into the focal plane detector of the heavy nucleus with the expected parameters (energy and velocity) and its decay with emission of an α-particle with $E_\alpha \geq 10$ MeV (the two signals are strictly position-correlated) the beam was switched off.

Measurements made immediately after switching off the beam showed that the rate of the α-particles ($E_\alpha \geq 9$ MeV) and spontaneous fission fragments in any strip within $\Delta x = 0.8$ mm, which is determined by the position resolution of the detector, is 0.45/year and 0.1/year, respectively. Incidental coincidences of the signals simulating the 3-step 1.5 min.-chain of the decay of the nucleus ^{288}114 ($\alpha - \alpha - $ SF) are practically excluded even for a single event [30].

In these conditions, at a beam dose of $2.25 \cdot 10^{19}$ ions, three decay chains of element 116 were registered (Fig. 5b) [31].

After the emission of the first α-particle ($E_\alpha = 10.53 \pm 0.06$ MeV) the following decay occurred in the absence of the beam (see the grey area). As is seen from Fig. 6a, two decay chains of the ^{288}114 nucleus produced in the reaction ^{48}Ca + ^{244}Pu and three new decay chains observed in the reaction ^{48}Ca + ^{248}Cm are strictly correlated: the five signals arising in the front detector, i.e., the recoil nucleus, three α-particles and fission fragments (in the case of the decay of the ^{288}114 nucleus: $R - \alpha - \alpha - $ SF), differ in their position by no more than 0.6 mm.

For each decay chain of a nucleus, the energy and decay time, as well as the position (coordinate) on the front detector are measured. Let us first compare the three decay sequences, obtained in the ^{48}Ca + ^{248}Cm reaction. Further, we can perform analysis of the data from the two reactions on the ^{244}Pu and ^{248}Cm targets (Fig. 6).

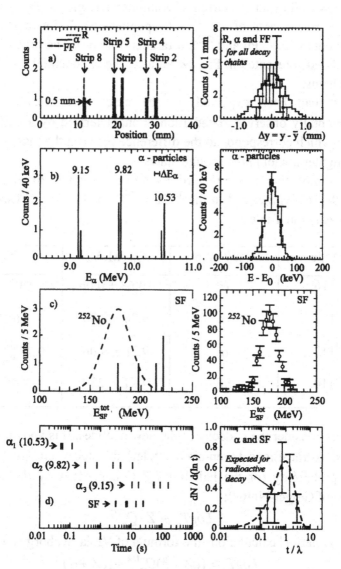

Fig. 6. Position, energy and time correlations in the decay chains (Fig. 5.)
a) position of the R-α-α-α-SF signals in the front detector
b) energy spectra of α-particles from 5 decay chains
c) spectrum of the total kinetic energy of spontaneous fission fragments
d) Time distribution of the signals from α-particles and SF.
Calibration spectra, obtained in test reactions, are shown in the right panels. The
symbols with error bars are the experimental data from the left panels.

The energies of the α-particles and the half-lives, obtained in the reaction ^{48}Ca + ^{248}Cm, coincide with each other within the limits of the detector energy resolution ($\Delta E_\alpha \sim 60$ keV) and the statistical fluctuations of the time measured in the nuclear decay chains. Fig. 6b shows the α-particle energy spectra for the three events corresponding to the ^{292}116 α-decay, the five events corresponding to ^{288}114 and five events to ^{284}112 decays, as well as the spectrum of combined energies of fission fragments from the five events of the ^{280}110 spontaneous fission obtained in the experiments with ^{244}Pu and ^{248}Cm targets (Fig. 6c). The time distributions of the signals presented in Fig. 6d indicate the exponential decay of nuclei in the chains.

As it had been expected, in the case of even-even nuclei the experimentally observed α-radiation is characterized by a definite decay energy, which corresponds to the difference between the masses of mother and daughter nuclei in the ground states.

The radioactive properties of even-even isotopes of elements with $Z = 110$-116, synthesized in the ^{48}Ca + ^{244}Pu and ^{48}Ca + ^{248}Cm reactions, are presented in Table 1.

Isotope	Decay mode	Q_α (MeV)	T_α
^{292}116	α	10.68 ± 0.06	53^{+62}_{-19} ms
^{288}114	α	9.96 ± 0.06	$2.6^{+2.0}_{-0.8}$ s
^{284}112	α	9.28 ± 0.06	$0.75^{+0.57}_{-0.23}$ min
^{280}110	SF	TKE ~ 230	$7.6^{+5.8}_{-2.3}$ s

Let us consider the α-decay of the new nuclides. It is known that for the allowed α-transitions (even-even nuclei) the decay energy Q_α and the decay probability (or the half-life T_α) is connected by the well-known relation of Geiger-Nuttall:

$$LogT_\alpha \sim Z \cdot Q_\alpha^{-1/2} .$$

For numerical calculations the formula of Viola-Seaborg is used:

$$LogT_\alpha = (aZ + b)Q_\alpha^{-1/2} + (cZ + d) .$$

This relation is strictly fulfilled for all known by now 60 even-even nuclei heavier than Pb, for which Q_α and T_α have already been measured. Fig. 7

shows experimental data and calculations for the region of heavy nuclei with $Z \geq 100$.

The results obtained for the new even-even isotopes with $Z = 112, 114$ and 116 are in good agreement with the theory of α-decay.

Since in the experiments the E_α and T_α values were measured simultaneously for every nucleus in the decay chains, from the relation $\mathrm{Log}T_\alpha$ (Q_α) one can determine the atomic number of the α-emitters. Taking into account that all the decays observed in the experiment are correlated, it follows that this radioactive family belongs to the sequential decay chain of nuclei with $Z = 116 - 114 - 112 - 110$ with a probability higher than 96%.

Finally, in the spontaneous fission of the $^{280}110$ nuclei the total energy release from the fission fragments in the detectors is about $E = 206$ MeV (Fig. 6c). This value with a correction for the energy loss in the dead layers of the front and side detectors as well as in the pentane gas corresponds to the mean fission-fragments total kinetic energy TKE ~ 230 MeV. Such a high value of TKE points to the spontaneous fission of a rather heavy nucleus. It is worthwhile mentioning that in the fission of ^{235}U induced by slow neutrons TKE $= 168$ MeV.

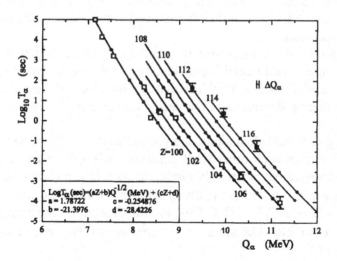

Fig. 7. $\mathrm{Log}T_\alpha$ as a function of Q_α for even-even nuclei with $Z \geq 100$. The solid lines are calculations done using the Seaborg-Viola formula with the parameters shown in the insert. The black squares correspond to the data given in the Table.

7 Even-odd Isotopes. Experiments with ^{238}U and ^{242}Pu

Let us remind ourselves that in the first experiment on the synthesis of element 114 in the ^{48}Ca + ^{244}Pu reaction a long decay sequence α_1 - α_2 − α_3 - SF with total decay period of about 34 min was observed. It is presented in Fig. 8a. On the basis of the above-performed analysis for the even-even nucleus ^{288}114 it can be attributed to the decay of the heavier, even-odd nucleus ^{289}114, produced in the 3n-evaporation reaction channel. It is not a surprise that the heavier nucleus has a longer lifetime; in addition, the odd neutron number, as a rule, introduces a hindrance both on α-decay and on spontaneous fission. The problem consists in that according to calculations the expected cross section of the 3n-channel is higher than $\sigma(4n)$, while in the ^{48}Ca + ^{244}Pu, ^{248}Cm experiments rather the opposite situation takes place. This may be the result of the unusual properties of this nucleus.

It is noteworthy that, unlike the even-even nuclei, for which as mentioned earlier, α-decay takes place between the nuclear ground states with spin $J = 0$ and positive parity, in the case of odd nuclei it is necessary to take into account the different quantum numbers of the ground states of the parent and daughter nuclei. Thus, here the characteristics of the low-lying states, which are most probably involved in the decay of the nucleus, can play a significant role.

Unfortunately, in the macro-microscopic theory the calculation of the nuclear structure of odd nuclei is problematic. At the same time, in purely microscopic models such as HFB calculations have been done for a series of nuclei, including the nuclei in the decay sequence of ^{289}114 and ^{293}116 [32].

According to these calculations, the ground states of ^{289}114 and ^{285}112 strongly differ from each other what concerns their spin. As a result the transition between them is strongly hindered. As it follows from the decay schemes shown in Fig. 8a, the α-decay takes place from the first excited state at $E = 0.52$ MeV ($J = 1/2^+$). This, in turn, determines the high energy of the first α-particle. The situation is repeated for the second α-decay ^{285}112 $\rightarrow$ ^{281}110.

We should note that the calculated values Q_α for all three α-transitions are in good agreement with the experiment. A similar situation arises in the decay ^{292}116 $\rightarrow$ ^{289}114, where the transition between the ground states is blocked due to the same reasons.

In such conditions the observation of the decay of the ^{289}114 nucleus depends on the probability of population of its first excited state with $E = 0.52$ MeV and $J = 1/2^+$ in the process of γ-decay following the emission of the last neutron, Fig. 8b. On the other hand, this probability depends strongly on the characteristics of the high-lying states and can be strongly suppressed by the competition from populating the ground state with high spin, the decay time of which may be tens of hours. The registration of such long decay sequences is outside the capability of our setup due to the background, which causes random coincidences.

All that was said above refers to the individual structure of the ^{289}114 nucleus. For the other odd isotopes, in particular for the more neutron-deficient nucleus ^{287}114, formed in the ^{48}Ca + ^{242}Pu reaction, this scenario may change.

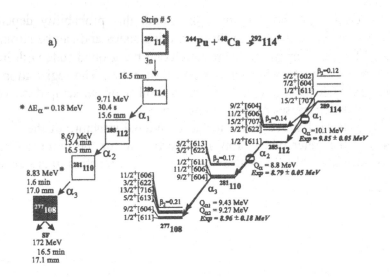

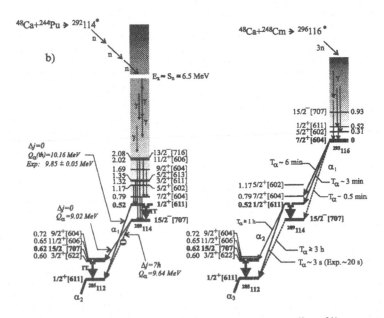

Fig. 8. a) The decay chain R-α-α-α-SF observed in the ^{48}Ca+^{244}Pu reaction. The calculated level scheme from ref. [32] is shown on the right.

b) Population of states in the ^{289}114 and ^{293}116 nuclei after emission of the last neutron. For α-transitions the values of Δj and Q_α are shown.

Experiments pursuing this goal were conducted in March 1998 with the VASSILISSA separator [33]. A rotating target from enriched ^{238}U in the form of a layer 0.3mg/cm^2 thick deposited on an aluminum substrate 1.6

mg/cm^2 thick was used to receive an intense ion beam. Two long-term exposures were performed at the beam energies of E_p = 231 and 238 MeV, which corresponded to the compound-nucleus excitation energies of E_x = 33 and 39 MeV. Under these conditions, we may expect a maximum yield of isotopes originating from reactions involving the evaporation of three and four neutrons from the ^{286}112 compound nucleus.

In the first experiment (E_x = 33 MeV), two events of spontaneous fission in the form of two coincident fragments with energies of (162 + 28) MeV and (191 + 21) MeV in the front detector and in the side detectors, respectively, were recorded at a total beam dose of 3.5 × 10^{18} ions (Fig 9a).

In each case, all signals were analyzed in the corresponding position windows over a broad time interval from 5 μs to 10000 s in order to find the recoil nucleus and to perform a search for α-particles preceding spontaneous fission. Signals from the implanted ions that were expected for heavy nuclei, were recorded 52 and 182 s prior to the emergence of the fragments. Within these time intervals, there were no signals from α-particles preceding spontaneous fission. Other successive α-decay chains that did not end in spontaneous fission and which could be associated with the decay of heavy nuclei (in the energy range E_α = 8-13 MeV with a half-life of $T_{1/2} \leq$ 1000 s) were not discovered in the experiment, too.

Neither spontaneous fission nor α-particle chains that could be attributed to the decay of a heavy nucleus were observed in the second experiment (E_x = 39 MeV), where the beam dose was 2.2 × 10^{18} ions.

The half-life determined for the new spontaneously fissioning nuclide on the basis of the two aforementioned events is $T_{1/2}$ = 81$^{+200}_{-35}$ s. At the excitation energy of E_x = 33 MeV, it is formed with a cross section σ = 5$^{+6}_{-3}$ pb; at E_x = 39 MeV, the cross section for its formation was deduced as σ ≤ 7.5 pb.

It should be noted that the properties of nuclei around ^{238}U are well known. Only very heavy isotopes with $Z \geq$ 96, which are offset from ^{238}U by 12-16 nucleons, undergo spontaneous fission. The formation of these or still heavier isotopes in ^{48}Ca + ^{238}U interactions is energetically forbidden.

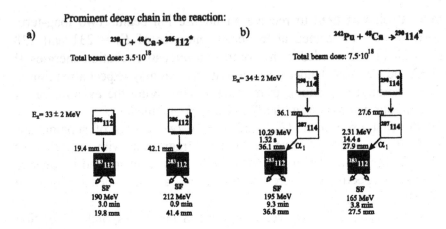

Fig. 9. Decay chains observed in the reactions given in the figure.

Most probably, the observed spontaneously fissioning nuclei are formed in the fusion channel as the result of the de-excitation of a compound system with energy $E_x = 33$ MeV. Here, there are also only very few possibilities. The evaporation of charged particles from a slightly heated heavy nucleus is strongly suppressed by the Coulomb barrier; fusion-reaction products formed upon the emission of α-particles and heavier nuclei are additionally suppressed by the separator more than two orders of magnitude. Since the effect was observed at $E_x = 33$ MeV and since it did not become more pronounced at a higher energy, it is most probable that the observed spontaneous-fission events are associated with the decay of the $^{283}112$ ($N = 171$) even-odd nucleus.

If the identification of the nuclei obtained in the previous experiments is valid, we can easily predict the properties of another isotope, viz. $^{287}114$ ($N = 173$). It must undergo predominantly α-decay into the daughter nucleus $^{283}112$, which was previously observed among the products of the ^{48}Ca + ^{238}U reaction. In the case being considered, we could expect a short decay chain (α-SF) involving α-decay, characterized by a half-life of a few seconds and followed by spontaneous fission, whose half-life is much longer in the present case (a minute or a few minutes). This isotope of the $Z = 114$ element can be synthesized in the channel of ^{48}Ca + ^{242}Pu reaction that involves the evaporation of three neutrons.

The implementation of the experiment conducted in March and April 1999 [34] was nearly identical to that described above when discussing the synthesis of the $^{283}112$ isotope in the ^{48}Ca + ^{238}U reaction.

A rotating target that was made of ^{242}Pu and which had a thickness of about 0.2 mg/cm^2 was exposed to a beam of 235 MeV ^{48}Ca ions; the total beam dose was 7.5×10^{18} ions. The most probable channel of de-excitation of the compound nucleus ^{290}114 ($E_x \approx 33.5$ MeV) involves the emission of three neutrons and must lead to the formation of the even-odd isotope ^{287}114 ($N = 173$).

Coincident signals from the two fragments were observed for the two events with energy release of 195 MeV and 165 MeV and $T_{SF} > 0.1$ ms.

Searches for α-decays preceding spontaneous fission resulted in the discovery of the two chains shown in Fig. 9b.

In one of the chains, only one α-particle ($E_\alpha = 10.29$ MeV) was recorded by the front detector 1.32 s after the implantation of a heavy recoil atom with energy $E_r = 10$ MeV. The spontaneous-fission event was observed 559.6 s later. All three signals (ER,α, SF) fall within the position interval of 0.82 mm, from where we conclude that the observed decays are correlated.

In the second case, spontaneous fission was observed 243 s after the detection of the $E_r = 13.5$ MeV recoil nucleus. Within this interval, 14.4 s after the implantation of the recoil nucleus, the front detector recorded only part of the energy ($E_{\alpha1} = 2.31$ MeV) of the α-particle emitted into the backward hemisphere (open window). That all three signals (ER,α, SF) fall within the position interval of 1.0 mm again indicates a correlation of the observed decays. The probability that both events are due to a random coincidence of signals that imitate the decay chains (ER, α, SF) in the above position intervals is less than 10^{-9}.

Since the α-particle energy is not defined in the second case and since the daughter nuclei possess identical properties, we have to assume that, in the two cases, α-decay proceeds from the same state of the parent nucleus. Its half-life as determined on the basis of the two cases in question is $T_\alpha = 5.5^{+10}_{-2}$ s. The daughter nuclei undergo spontaneous fission. Their decay properties are close to those observed previously in the ^{48}Ca + ^{238}U reaction. All four spontaneous-fission events recorded in the two experiments possess, within the statistical uncertainties, the same half-life of $T_{SF} = 3.0^{+2.8}_{-1.0}$ min and are associated with the decay of the same nucleus. In the ^{48}Ca + ^{238}U reaction, this nuclide is formed as the evaporation product in the 3n-evaporation channel, whereas, in the ^{48}Ca + ^{242}Pu reaction, it appears as the daughter product of the α-decay of the parent nucleus ^{287}114 ($E_\alpha = 10.29$ MeV).

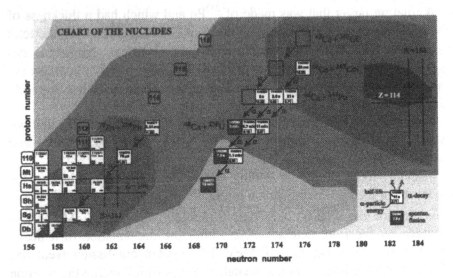

Fig. 10. Nuclide chart for the region $Z \geq 105$. The different shading are taken from Fig. 2a to guide the eye. The dashed lines denote nuclei with closed proton and neutron shells. The decay chains $Z \geq 108$ and $N \geq 169$ are obtained in reactions induced by ^{48}Ca.

The cross section for the production of the new isotope of element 114 is about 2 pb. Its half-life is less than that of the heavier isotope ^{289}114 formed in the ^{48}Ca + ^{244}Pu reaction, and the decay sequence that follows is shorter for the former than that for the latter (Fig. 8*a*). According to theoretical predictions, such a trend is expected as the number of neutrons is decreased, i.e., as the nuclear species being studied move away from the $N = 184$ closed shell.

All decay sequences, obtained in the described experiments with ^{48}Ca-beams, are shown on the nuclide chart, including the region $Z \geq 105$ (Fig. 10). For easy observation, they are drawn on the chart of the amplitude of the shell correction to the liquid-drop nuclear energies. With respect to known isotopes of the heaviest elements with $Z \leq 112$, the new nuclides possess a neutron excess of $\Delta N \geq 8$ atomic mass units. This makes it possible to check the basic theoretical conclusion on the increase of the stability of heavy nuclei when approaching the closed neutron shell $N = 184$.

8   Comparison with Theoretical Predictions

Let us try to compare the obtained experimental results on the properties of the synthesized nuclei with the theoretical predictions from different models. In the first place, it seems reasonable to make a comparison with the macro-microscopic model in the framework of which calculations have been carried out both for the partial half-lives and α-decay energies and for the probability of spontaneous fission.

First of all, we note that the heaviest isotopes of $Z = 110$, 112, and 114 elements from reactions induced by ^{48}Ca projectiles undergo α-decay. In this region of nuclei, spontaneous fission is observed only for $(N - Z) \leq 61$ nuclei. For the even-odd nuclei $^{277}108$ ($N = 169$) and $^{283}112$ ($N = 171$), the half-lives with respect to spontaneous fission proved to be, respectively, five and three orders of magnitude larger than the values predicted for the neighboring even-even isotopes. This difference may be due to the presence of an odd neutron, which reduces significantly the probability of the spontaneous fission of a heavy nucleus. For the even-even nucleus $^{280}110$ ($N = 170$), the experimental value ($T_{SF} \sim 7$ s) is also approximately three orders of magnitude larger than the value calculated in [35]. Although the calculations of the probability of spontaneous fission, which is associated with tunneling through a potential barrier, are connected with considerable uncertainties, the mentioned difference may suggest a greater contribution of the shell structure to the deformation energy of the nucleus.

Some conclusions can be drawn from an analysis of the ground-state properties of superheavy nuclei. Let us first consider the even-even nuclides for which the energy and decay probability is directly connected to the nuclear ground-state masses.

Fig. 11 presents the half-lives T_α and T_{SF} for nuclei with $Z = 110$-114 and $N = 160$-178, calculated in the macro-microscopic model, and the experimental values, obtained in our experiments. As it can be seen from this figure, the increase of the neutron number leads to a considerable increase of the half-life. For $Z = 110$, when ten neutrons are added to the $^{270}110$ nucleus ($N = 160$), T_α increases more than by a factor of 10^5. A similar situation takes place for $Z = 112$. The transition from the even-odd nucleus $^{277}112$ ($N = 165$) to the even-even isotope $^{284}112$ ($N = 172$) involves an increase in T_α by a factor of $2 \cdot 10^5$.

It should be noted that for all even-even nuclides with $Z = 110\text{-}116$ the experimental values $T_\alpha(\exp)$ exceed the theoretical ones by about one order of magnitude. This discrepancy, although insignificant, is due to the fact that the experimental values Q_α are smaller than the calculated ones.

The comparison of the experimental data with calculation done within

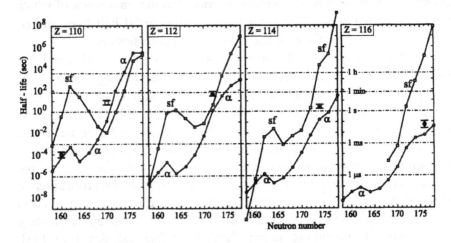

Fig. 11. Partial half-lives T_{SF} (connected by solid lines) and T_α (dashed lines) as a function of neutron number, calculated in the macro-microscopic model for even-even nuclei with $Z = 110$, 112, 114 and 116. The experimental points denote: open square - spontaneous fission of $^{280}110$, black points - α-decay.

different theoretical models is shown in Fig. 12 in terms of the difference $Q_\alpha(\text{th}) - Q_\alpha(\exp)$ as a function of the neutron number in the range $154 \leq N \leq 176$. As it is seen from Fig. 12, the experimental values Q_α for the even-even nuclei differ from the ones, calculated within different approaches, as much as $\Delta Q_\alpha \approx \pm 0.5$ MeV ($\Delta Q_\alpha / Q_\alpha \approx 5\%$).

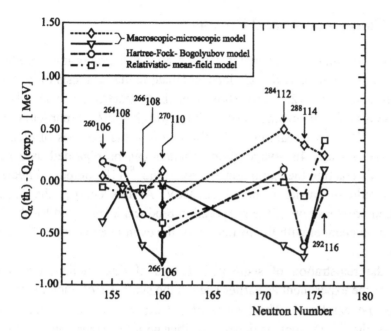

Fig. 12. Comparison of decay energies of even-even nuclei $Q_\alpha(\exp)$ with calculated values $Q_\alpha(\text{th})$, obtained in different models (indicated in the insert).

The cross section of the evaporation residues with $Z = 114,116$ formed in the fusion reactions ^{48}Ca + ^{244}Pu, ^{248}Cm in the 4n-evaporation channel is about 0.5 pb. On the basis of the experimental cross section values for the excited nuclei with $Z = 114$ and 116, one can determine the lower limits of the fission barrier heights. Using the theoretical model for calculations described in [36], for the nuclei $^{288-292}$114 and $^{292-296}$116 we obtained the values $B_f(E_x = 0) \geq 6.7$ and 6.4 MeV, respectively.

Without making so far a more detailed analysis, one can conclude that the theoretical predictions about the decisive influence of the nuclear structure on the stability of superheavy nuclides have been confirmed not only qualitatively but also to some extent quantitatively.

The increase in the half-lives of the heaviest nuclei up to tens of seconds and minutes substantially extends the area for the investigation of heaviest elements, including the study of their chemical properties, the measuring of atomic masses, etc. On the other hand, development of accelerator and experimental techniques will allow us to make an advent into the region of even heavier nuclei in the future.

9   Consequences and Prospects

1. Experimental and calculated cross sections of evaporation products of the fusion reactions with ^{48}Ca ions, obtained in the synthesis of heaviest nuclides with $Z = 112 - 116$, show that these reactions can also be in principle used for the synthesis of nuclides with $Z > 116$. In particular, the reaction ^{249}Cf + ^{48}Ca leading to the formation of the isotopes of element 118 in the 3n- and 4n- evaporation channels can be regarded as a next step. In this reaction the excitation energy of the ^{297}118 compound nucleus at the Coulomb barrier is about 26 MeV. At this energy, the largest formation cross section of the evaporation products is expected for the 3n-evaporation channel with the formation of the even-even isotope ^{294}118 ($N = 176$).

For the registration of sequential decays of this nuclide, the same setting of the experiment as earlier used in the synthesis of nuclei with $Z = 116$ can be employed. The energy of the allowed α-transitions in the chains of the ^{294}118 nucleus decay, together with the earlier obtained data for $Z = 114$ and 116 nuclei, can yield more information on the location and strength of the closed proton shell in the region of superheavy nuclei.

According to calculations made with the use of the macro-microscopic model [10] the ^{294}118 nucleus will be undergoing α-decay with $Q_\alpha = 12.2$ MeV ($T_\alpha \sim 38$ μs), whereas predictions of microscopic models - relativistic mean field [14] and Hartree-Fock-Bogoliubov [13] - models yield much smaller values: $Q_\alpha = 11.03$ MeV ($T_\alpha = 25$ ms) and $Q_\alpha = 11.31$ MeV ($T_\alpha = 4.5$ ms), respectively. In the event of synthesizing the even-even nucleus with $Z = 118$, $N = 176$ in the reaction ^{249}Cf(^{48}Ca,3n)^{294}118, such a high difference in the Q_α and T_α values can be reliably determined in the experiment.

We started these experiments in March and now for four months a ^{249}Cf target is being irradiated by a ^{48}Ca beam. The results show that, in spite of the very low cross section, this reaction may lead to the synthesis of the nuclide ^{294}118.

2. Relatively long half-lives of the new nuclides amounting to several seconds in the case of $Z = 114$ and nearly one minute in the case of $Z = 112$ open up possibilities for the investigation of the chemical properties of new elements. In our opinion, the most suitable candidates are the isotopes of element 112 (EkaHg) due to their relatively long half-lives and the expected unique chemical properties of Hg. One possible approach to this task is to determine the probability of formation of intermetallic bonds

between EkaHg and Pd, Au or Pt, which are well-known for Hg. On-line experiments of this type have been already started at the ^{48}Ca beam.

3. The enhanced stability of heaviest nuclei changes the entire approach to their synthesis in fusion reactions with heavy ions. The necessity of separating in-flight the reaction products for a speedy delivery of recoil atoms to the detector array is not a decisive factor in the synthesis of nuclei with $Z=112-114$ due to their long half-lives. The use of classical on-line separators is more preferable since the latter alongside their high efficiency, connected with the possibility of using thicker targets and higher selectivity, allow measuring the nucleus mass. This idea will be implemented in the design of the set-up MASHA (Mass Analyzer of Superheavy Atoms), which is now being developed at the Flerov Laboratory.

4. The use of fusion reactions with ^{48}Ca ions, presently employed for the synthesis of superheavy elements, is restricted by the availability of the target material. The isotope ^{249}Cf, which is to be used as a target in the experiments on the synthesis of nuclei with $Z = 118$, is apparently the last one in the row of actinides. At the same time, according to predictions of microscopic models of the type of Hartree-Fock-Bogoliubov or Relativistic Mean Field, a change in the nuclear density (the so call "semi-bubble nuclei") also leading to the stability of hyperheavy nuclei is expected in the case of nuclei with $Z \geq 120$ [37].

For the synthesis of such nuclei it will be necessary to use fusion reactions with ions heavier than ^{48}Ca, i.e. ^{58}Fe, ^{64}Ni, etc. Two fusion reactions, ^{248}Cm + ^{54}Cr and ^{238}U + ^{64}Ni, could lead to the formation of the weakly excited compound nucleus ^{302}120 with $N = 182$. It is planned to study the fusion and fission processes involving such massive nuclei.

Acknowledgements

In a two-hour lecture it is difficult to encompass all questions concerning the atomic- nucleus physics issue, which has a 60-year history. I hope that the reader will find in it the basic sides of the problem and the newest results on the extremely heavy nuclei, which can be considered as an object connected not only with nuclear physics, but also with many other interdisciplinary scientific fields: atomic physics, astrophysics, nuclear chemistry, etc.

It is my great pleasure to thank the Organizers of the Advanced Study Institute for the excellent organization of the school in St. Croix and the very interesting scientific program.

References

1. G. N. Flerov and K. A. Petrzhak, Phys. Rev. **58**, 89 (1940); J. Phys. (Moscow) **3**, 275 (1940).
2. N. Bohr and J. Wheeler, Phys. Rev. **56**, 426 (1939).
3. S. M. Polikanov *et al.*, Zh. Eksp. Teor. Fiz. **42**, 1464 (1962) [Sov. Phys. JETP **15**, 1016 (1962)].
4. S. Bjornholm and J. Linn, Rev. Mod. Phys. **52**, 725 (1980).
5. Yu. Ys. Oganessian *et al.*, Nucl. Phys. **A239**, 157 (1975).
6. (a) W. J. Swiatecki, in *Proceedings of the 2nd International Conference on Nuclear Masses, 1964* (Springer, Vienna, 1964), p.52 (b) W. D. Myers and W. J. Swiatecki, Nucl. Phys. **81**, 1 (1966).
7. V. M. Strutinsky, Nucl. Phys. **A95**, 420 (1967); 122, 1 (1968).
8. M. Brack *et al.*, Rev. Mod. Phys. **44**, 320 (1972).
9. Z. Patyk and A. Sobiczewski, Nucl. Phys. A **533**, 132 (1991).
10. R. Smolanczuk, Phys. Rev. C**56**, 812 (1997).
11. P. Möller and J. R. Nix, J. Phys. G**20**, 1681 (1994).
12. A. Sobiczewski, F. A. Gareev and B. N. Kalinkin, Phys. Lett. **22**, 500 (1966).
13. H. Meldner, Ark, Fys. **36**, 593 (1967).
14. *Heavy Elements and Related New Phenomena*, Ed. by W. Greiner and R. K. Gupta (World Sci., Singapore, 1999), Vol. 1.
15. G. T. Seaborg, *Man-Made Transuranium Elements* (Prentice Hall, New York, 1963).
16. Yu. Ts. Oganessian, Lect. Notes Phys. **33**, 221 (1974).
17. Yu. Ts. Oganessian, Phys. of Atomic Nuclei **63**, 1315 (2000).
18. E. K. Hulet *et al.*, Phys. Rev. Lett. **39**, 385 (1977).
19. Yu. Ts. Oganessian *et al.*, Nucl. Phys. **A294**, 213 (1978).
20. P. Armbruster *et al.*, Phys. Rev. Lett. **54**, 406 (1985).
21. G. Münzenberg *et al.*, Nucl. Instrum. Methods **161**, 65 (1979).
22. A. V. Yeremin *et al.*, Nucl. Instrum. Methods Phys, Res, A**274**, 528 (1989).
23. Yu. A. Lazarev *et al.*, in *Proceedings of the International School-Seminar on Heavy Ion Physics, Dubna, Russia, 1993*, JINR Report E7-93-274 (Dubna, 1993), Vol. II. p. 497.

24. N. Bohr, Phys. Rev. **58**, 654 (1940); **59**, 270 (1941).
25. Yu. Ts. Oganessian *et al.*, Phys. Rev. C**64**, 064309 (2001).
26. K. Morita WE-Heraeus-Seminar 4-8 Sept. 2002 (Germany).
27. Yu. Ts. Oganessian *et al.*, Phys. Rev. Lett. **83**, 3154 (1999).
28. Yu. Ts. Oganessian *et al.*, Phys. Rev. C**62**, 041604 (R) (2000).
29. Yu. Ts. Oganessian *et al.*, Phys. of Atomic Nuclei **63**, 1679 (2000).
30. Yu. Ts. Oganessian *et al.*, Phys. Rev. C**63**, 011301 (R) (2000).
31. Yu. Ts. Oganessian, V. K. Utyonkov and K. Moody, Phys. of Atomic Nucl. **64**, 1349 (2001).
32. S. Čwiok, W. Nazarewicz and P. H. Heenen, Phys. Rev. Lett. **83**, 1108 (1999).
33. Yu. Ts. Oganessian *et al.*, Eur. Phys. J. A**5**, 63 (1999).
34. Yu. Ts. Oganessian *et al.*, Nature **400**, 242 (1999).
35. R. Smolanczuk, J. Skalski and A. Sobiczewski, in *Proc. of the Workshop on Cross Proper of Nuclei and Nuclear Excitations "Extremes of Nuclear Structure" Hirshegg, 1996* (GSI, Darmstadt 1996).
36. M. G. Itkis, Yu. Ts. Oganessian and V. I. Zagrebaev, Phys. Rev. C**65**, 044602 (2002).
37. J. F. Berger *et al.*, Nucl. Phys. A**685**, 1c (2001).

Index

LECTURERS

J. Ellis	CERN, Geneva, Switzerland
S. S. Willenbrock	University of Illinois, Urbana-Champaign, USA
W. H. Kinney	Columbia University, New York, USA
M. C. Gonzalez-Garcia	Universitat de Valencia, Valencia, Spain
J. J. Engelen	NIKHEF, Amsterdam, The Netherlands
J. Rich	DAPNIA, Paris, France
A. Honma	CERN, Geneva, Switzerland
A. Sharma	CERN, Geneva, Switzerland
G. Hanson	University of California, Riverside, USA
J. T. Rogers	Cornell University, Ithaca, USA
Y. T. Oganessian	JINR, Dubna, Russia

ADVISORY COMMITTEE

R. Cashmore	University of Oxford, Oxford, UK
C. Jarlskog	Lund University, Lund, Sweden
C. Quigg	Fermilab, Batavia, USA
F. Sauli	CERN, Geneva, Switzerland
A. Skrinsky	BINP-Novosibirsk, Russia
M. Spiro	Saclay, Paris, France

CO-DIRECTORS

M. Danilov	Institute for Theoretical and Experimental Physics, Moscow, Russia
H. B. Prosper	Florida State University, Tallahassee, USA

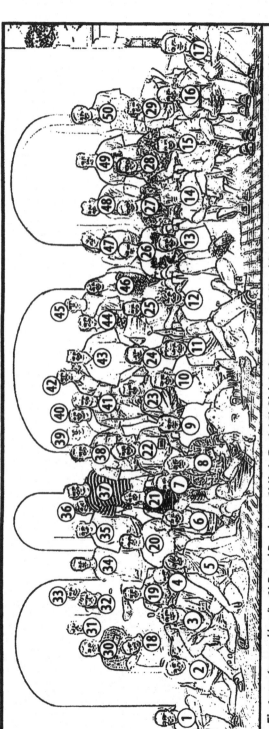

First row *(ground level)*: **1)** Pawel Bruckman **2)** Katia Fratini **3)** Oleksiy Atramentov **4)** Karel Smolek **5)** Antonia Lavorato **6)** Olga Lobban **7)** Roman Otec **8)** Kumar Prolay Mal **9)** Mircea Coca **10)** Martin Hennecke **11)** Gregory Schott **12)** Maxim Titov **13)** Jasmine Hasi **14)** Harrison Prosper **15)** Reinhard Schwienhorst **16)** Marc Buehler **17)** Yildirim Mutaf

Second row *(seated in chairs)*: **18)** Robertsen Riehle **19)** Muge Karagoz Unel **20)** Christian Schuetz **21)** Miriam Giorgini **22)** Joe Rogers **23)** Wolfgang Menges **24)** Paul Laycock **25)** Thomas Kluge **26)** Andrew Lyon **27)** Robert Bainbridge **28)** Matthew Ryan **29)** Gary Taylor

Third row *(everyone standing)*: **30)** Yuri Oganessian **31)** Ian Blackler **32)** Petr Travnicek **33)** Michele Petteni **34)** Simon Waschke **35)** Marcin Kucharczuk **36)** Gail Hanson **37)** Michael Eads **38)** Jos Engelen **39)** Jim Rich **40)** Andrew Ivanov **41)** Henning Flaecher **42)** Scott Willenbrock **43)** Gavin Nesom **44)** Alexei Sedov **45)** Andrea Bocci **46)** Concha Garcia **47)** Freya Blekman **48)** Brian Davies **49)** Alessio Sarti **50)** Alan Honma

Attendees who were not in the group photograph:
John Ellis   William Kinney   Pawel de Renstrom   Gavin Ward
Fedor Ignatov   Andrei Loginov   Archana Sharma